Progress in Physics
Vol. 9

Edited by
A. Jaffe, G. Parisi,
and D. Ruelle

Springer Science+Business Media, LLC

Fourth Workshop on Grand Unification

University of Pennsylvania, Philadelphia
April 21-23, 1983

H.A. Weldon,
P. Langacker, and
P.J. Steinhardt, editors

Springer Science+Business Media, LLC 1983

Editors:

H.A. Weldon
P. Langacker
P.J. Steinhardt
Department of Physics
University of Pennsylvania
Philadelphia, PA 19104

Library of Congress Cataloging in Publication Data

Workshop on Grand Unification (4th : 1983 : University of Pennsylvania)
Fourth Workshop on Grand Unification, University of Pennsylvania, Philadelphia, April 21-23, 1983.

(Progress in physics ; v. 9)
1. Grand unified theories (Nuclear physics) — Congresses. 2. Quantum gravity — Congresses. 3. Protons — Decay — Congresses. 4. Magnetic monopoles — Congresses. 5. Cosmology — Congresses. I. Langacker, P. II. Steinhardt, P. J. III. Weldon, H. A. IV. Title. V. Series: Progress in physics (Birkhäuser) ; v. 9)
QC794.6.G7W67 1983 530.1 83-22512
ISBN 978-1-4757-1814-0 ISBN 978-1-4757-1812-6 (eBook)
DOI 10.1007/978-1-4757-1812-6

CIP-Kurztitelaufnahme der Deutschen Bibliothek

Workshop on Grand Unification:
... Workshop on Grand Unification.

4. University of Pennsylvania, Philadelphia, April 21 - 23, 1983. - 1983.
(Progress in physics ; Vol. 9)
ISBN 978-1-4757-1814-0

NE: University of Pennsylvania (Philadelphia, PA.); GT

Originally published by Birkhäuser Boston, Inc., in 1983
Softcover reprint of the hardcover 1st edition 1983

ISBN 978-1-4757-1814-0

9 8 7 6 5 4 3 2 1

TABLE OF CONTENTS

OTHER EXOTIC PHENOMENA

COSMOLOGY

SUPERSYMMETRY, SUPERGRAVITY, AND KALUZA-KLEIN THEORIES

FOREWORD

It has been sixteen years since the unification of electromagnetism with the weak interactions was developed by Glashow, Salam, and Weinberg. Well before that proposal was fully confirmed by experiment, work began on unifying strong interactions with the electroweak. Now there is a growing effort to incorporate some theory of quantum gravity into the scheme. This enormous complex of theoretical and experimental efforts was the subject of the Fourth Workshop on Grand Unification held in Philadelphia and attended by over two hundred physicists.

During the workshop, experimental and theoretical talks alternated as shown by the program summary on page 409 . However, to display the logical scope of the workshop the proceedings are organized into five subject areas.

Howard Georgi opened the workshop with a keynote address in which he reminds us of some of the simple properties of the particle spectrum that have not yet been understood.

The first subject area, and also the largest, is proton decay and underground physics. This is introduced by William Marciano's review of the SU(5) predictions with particular attention paid to the theoretical uncertainties. Spokesmen for the major underground experiments present current results on proton decay, $n\bar{n}$ oscillations, and magnetic monopole flux: B.V. Sreekantan for the Kolar gold field experiment after 1.9 years of operation, Earl Peterson for the Soudan 1 detector after 0.55 years, and Bruce Cortez for the IMB detector after 0.22 years. A second generation of detectors is already underway and Alex Grant reviews the status of the Kamioka, Frejus, Soudan 2, Gran Sasso, and future designs. In all such detectors it is crucial to understand the neutrino background. Tom Gaisser and Arnon Dar report on new calculations of neutrino fluxes and on what new physics might be extracted from more precise measurements.

A second prediction of grand unification is the existence of very heavy magnetic monopoles. Alfred Goldhaber explains the current ideas on how monopoles can catalyze nucleon decay. The detection of monopoles by a superconducting current loop has been improved and Claudia Tesche reports on planar detectors with many loops. Peter Bosetti provides an overview of all the present and future monopole experiments.

There are several other exotic processes associated with grand unification. Felix Boehm reviews the evidence for neutrino masses from Tritium decay and from various neutrino oscillation experiments. It is also possible that the neutrino mass term could violate lepton number and Frank Avignone reports on neutrinoless double beta decay. If grand unification allows baryon number violation, it may allow oscillations of neutrons into antineutrons at some level. Giuseppe Fidecaro reports on these experiments. An outstanding puzzle in all these theories is how to eliminate a term that would violate CP in the strong interactions. Removing the CP violation generally requires introducing a new pseudo-scalar meson with a very small mass, referred to as the axion. Pierre Sikivie explains the properties of this particle and the profound astrophysical consequences it may have.

The fourth major section of the conference is cosmology. Michael Turner reviews the successes of the hot big bang model incorporating a period of exponential inflation and the possibilities of obtaining the proper proto-galactic density fluctuations. Joel Primack provides extensive evidence that the present density fluctuations comprising clusters and superclusters of galaxies are predominantly dark, nonradiating matter that is probably neither bayonic nor leptonic. Alex Szalay explains how the initial density fluctuations might evolve into galactic clusters with voids between.

The last group of papers report on supersymmetry, supergravity, and theories in more than four dimensions. Gordon Kane first outlines the experimental tests for supersymmetry. Making supersymmetry local yields a supersymmetric theory of gravity and Joseph Polchinski describes the simplest class of supergravity models. Michael Duff proposes that all ordinary physics is embedded in eleven dimensional supergravity and the ordinary internal symmetries are geometrical in the larger space. It is even possible to compute the fine structure constants of the four dimensional theory and Steven Weinberg relates these to the geometry of the compact manifold. Finally, Edward Witten explains the advantages of the superstring formulation of supergravity.

The papers in this volume cover an enormous range of physical phenomena. We feel that they both represent the current status of and outline the future directions of the field of grand unification.

Support for this conference was provided by the National Science Foundation, the Department of Energy, and the University of Pennsylvania Physics Department.

July 1983

H.A. Weldon
P. Langacker
P.J. Steinhardt

WELCOME

Thomas Ehrlich
Provost of the University of Pennsylvania

Welcome to the University of Pennsylvania. We are pleased and proud to have you with us and to have this conference here at Penn. Every university administrator, as you might well imagine, spends a fair share of time looking for a grand unified theory--some set of explanations, however larded with sophistry, that explain how it is possible that an atomistic institution such as a university, an institution in which each proton or faculty member - though provided with the absolute stability of tenure - nonetheless seems to be steadily decaying. On that basis, all universities ought to disintegrate and perhaps they will.

In the interim, however, today we have a special chance to hear from an extraordinary collectivity of physicists about work on grand unified theories on the cosmic scale. Last fall I read that remarkable novel "Night Thoughts of a Classical Physicist". I suspect that many, as they hear the work described today, must feel as Victor Jacobs in that novel felt as he saw the old world passing. But, at least to a layman such as myself, the theories at issue today are breathtaking.

Even the possibility of grand unification suggests a coherence in the scheme of things that is exhilarating. That cohesion may not exist in the university, but even the possibility that it exists in the universe has enormously exciting potential. You may have not yet found all the answers, or at least proven them, but a university is, after all, an institution primarily of questions. In every sense, no more difficult questions - or more important or exciting ones - exist than those at issue in this conference. We look forward to the proceedings with great expectations. Thank you very much.

OPENING REMARKS*

or

FLAVOR DEMOCRACY AND OTHER SPECULATIONS ABOUT THE STATE OF PARTICLE PHYSICS

or

BETWEEN CHEMISTRY AND MATHEMATICS

Howard Georgi

Lyman Laboratory of Physics, Harvard University
Cambridge, MA 02138

One of the disadvantages I have found in collaborating with Nobel prizewinners is that I seldom get to give talks like this. I am always second or third choice. At any rate, I am glad to have this opportunity to talk to you, even if I am only here because Shelly Glashow is too depressed to come.

Admittedly, there is much going on in the world of physics which might cause depression. We particle theorists are spinning our wheels, stuck in the rut produced by the success of the standard SU(3) x SU(2) x U(1) model. Our experimental friends are trying to help us out, but their results either confirm our faith in SU(3) x SU(2) x U(1), (for example, the discovery of the W) or else are negative or inconclusive (for example, proton decay).

I get worried about the state of theoretical physics when I see things like this:

*This research is supported in part by the National Science Foundation under Grant No. PHY-82-15249.

HELP WANTED
Young Particle Theorist
to work on
Lattice Gauge Theories
Supergravity
and
Kaluza-Klein Theories

I have nothing against any of these subjects. They represent two extremes in physics: Chemistry on the one end and metaphysics and mathematics on the other. The trouble is that at the moment there is not all that much physics going on in between.

But I'm an optimist. I will try to be upbeat and to suggest how one can do some physics between the two extremes.

Let me start by discussing the logical structure of modern particle theory. Theorists nowadays see physics as a tower of effective theories, each of which is appropriate to describe physics in a particular range of energies. Sometimes we can understand apparently arbitrary or nonsensical structure in a theory at one energy scale by discovering how it can be embedded in a more restrictive theory at a higher scale.

The SU(2) x U(1) model is an example of this process. Below the mass of the W and Z bosons, particle physics looks like an SU(3) x U(1) gauge theory of strong and electroweak interactions with some puzzling, nonrenormalizable weak interactions. Above M_W and M_Z, the puzzle is resolved by the appearance of extra gauge interactions associated with the larger SU(3) x SU(2) x U(1) gauge symmetry. Apparently, to judge from the discovery of the W, this is, in fact, the solution to the puzzle.

The SU(2) x U(1) theory itself introduces another puzzle. What is it that breaks the SU(2) x U(1) symmetry. Presumably, it is some physics associated with another energy scale not too different from M_W. The physics could be a fundamental Higgs field, with or without supersymmetric partners, or it could be technicolor and extended technicolor interactions which break the symmetry dynamically. Eventually we will find out which of these possibilities is actually realized in nature, when we have an accelerator which can study particle interactions in the multi-TeV region.

There are other puzzles which are evident in physics today that are not resolved by the partial unification of weak and electromagnetic interactions at M_W. Most of these are "Why?s". Why SU(3) x SU(2) x U(1)? Why three families (or however many there are) of quarks and leptons? Why do the fermion masses and mixing angles take the values they do? We expect the questions to be answered in the structure of the physics at larger scales.

It seems clear that at least one such large scale exists, the Planck scale, $M_P \simeq 10^{19}$ GeV associated in some way with the gravitational interaction.

Another scale which is presumed to exist is the grand unification scale M_G, at which the SU(3) x SU(2) x U(1) gauge interactions appear as different manifestations of the same unifying gauge symmetry, such as SU(5). The physics of grand unification can provide a partial answer to some of the questions left unanswered at lower scales by correlating the properties of quarks and leptons. But the fundamental problem still remains: What makes the gauge structure and fermion content of the world special and unique?

This puzzle, which I will call the "uniqueness" puzzle, simply cannot be answered in the context of conventional quantum field theory (QFT). Conventional QFT does not single out any particular gauge structure. Thus we might expect the uniqueness question to be answered at some large scale M_B where conventional quantum field theory breaks down.

It is reasonable to assume that M_B and M_P are related in some way, because conventional QFT is inadequate in both cases. Some of the wilder current theoretical ideas, such as extended supergravity, superstrings and Kaluza-Klein models, are attempts to implement this physics. My own feeling is that these ideas are not wild enough and that it may be very difficult to guess what is going on at these large scales.

Let me not give the impression that I am suggesting that theorists spend their time working on bizarre generalizations of QFT. Quite the contrary, I believe that the questions are too hard. Nature is much more imaginative than we and we simply do not have enough experimental information about these short distances to guide us. It is okay to spend an hour a week on such metaphysics. But it seems to me that a theorist who spends all his time speculating about the physics of the Planck scale is suffering from either an overdeveloped Einstein complex or an underdeveloped instinct for self-preservation.

In the simplest view of grand unification, the unification scale M_G is smaller than M_P and M_B, so that we can use conventional QFT. As you all know, the most straightforward realization of this idea in a minimal SU(5) model leads to the prediction of proton decay at a (possibly) observable rate. I expect that we will hear a great deal about the theoretical and experimental status of proton decay in the talks to come. But I do want to make a few comments first. The experimental situation is a little confusing, with the Kolar Gold Field (KGF) and Mont Blanc (NUSEX) experiments reporting candidates at the level corresponding to a proton lifetime of the order of 10^{31} years, but with the Irvine-Michigan-Brookhaven (IMB) experiment quoting a limit of the lifetime, $\tau_p/B_{e^+\pi^0} > 6.5 \times 10^{31}$ years. Of the four candidate events in the other detectors, only one looks as if it could be $e^+\pi^0$, so there is probably not an enormous discrepancy here. Presumably we will hear more about these issues in future talks.

We theorists, I believe, should beware of taking these early results too seriously, either positively or negatively. In particular, while it is certainly disappointing that the IMB experiment does not already have lots of candidate events, it is still very early in the career a new kind of experiment. And experimental physics is hard. I think that they should keep looking and that we should wait and see.

It is true that we theorists tend to underestimate the difficulties involved in experimental physics. One of my favorite examples of this involved my wife's first trip with me to the Coral Gables conference. This trip took place after the discovery of the J/ψ, but before the convincing discovery at SPEAR of naked charm. We usually remember these events as having happened at practically the same time. But in fact, it took almost two years after the J/ψ before the evidence for naked charm was conclusive. At the time of the Coral Gables conference, the preliminary indications from SPEAR were negative, and many theorists had given up on charm as an explanation of J/ψ.

Anyhow, my wife had been getting all her information from people at Harvard, and it had not occurred to her that there were unbelievers in the world. But we sat down to dinner the first night of the conference with a group of old friends and they immediately started snickering, suggesting that charm was dead. My wife said that it reminded her of the time she first learned that there were people who did not use Colgate toothpaste.

The good news is that the IMB machine seems to be working very well. Let us hope that they see something soon.

For proton decay, in addition to the formidable experimental difficulties, there are theoretical uncertainties which are difficult to estimate. The strong dependence of the proton lifetime on the value of the QCD scale parameter Λ is a commonplace. Here I want to comment briefly on some of the less well known sources of confusion.

One important parameter which we do not really know much about theoretically is ratio of decays into pseudoscalar versus vector mesons. Most theoretical calculations based on various quark models give sizable or dominant pseudoscalar branching fraction. But that doesn't mean that they are correct. We really do not understand the pseudoscalar mesons, which are both quark-antiquark bound states and almost Goldstone bosons. If, for some reason, the proton decays into pseudoscalar mesons were very suppressed, it would be harder for IMB to see the effect.

The other theoretical uncertainty I want to call to your attention is something which I call a matching correction. The overall scale of the proton lifetime is determined in two steps. First one uses the renormalization group to evolve the $\Delta B = -1$ Hamiltonian from the unification scale (where it is given by the standard SU(5) model) down to low energy, where it is given by an operator involving three quark and one lepton fields. Then one calculates the matrix element of the three quark part of the operator between a nucleon state and a meson state in some type of quark model. The trouble with this procedure is that the three quark operator in your favorite quark model (non-relativistic or bag or whatever) is not necessarily the same as the three quark operator involving short distance QCD quarks. All useful quark models incorporate somehow the effect of spontaneous chiral symmetry breaking, whether in the constituent quark mass of the non-relativistic quark model or the boundary conditions of the bag model. They are thus, at best, effective theories which are useful (perhaps) at small momentum scales. In general, the three-quark QCD operator will have to be matched onto some combination of three-quark quark model operators which give the same physics. And we have no idea how to do this. None! If anyone tells you that this is true, but it is unlikely to make a difference of more than a factor of two or so, ask him about the $\Delta I = 1/2$ rule. We just do not know.

These questions are really interesting chemistry! It may be that numerical lattice gauge theory calculations can someday give us useful information.

Well, I'm not going to have time to slander everyone, although there is much that deserves comment. Monopoles and cosmology are fascinating and beautiful subjects, but they continue to draw theorists towards metaphysics. The epidemic of low energy supersymmetry continues unabated, although if anything, the theories are getting farther from the original goal of an explanation of the large ratio M_G/M_W. But I want to proceed to a discussion of the flavor problem, based on some recent work I have done in collaboration with two extraordinary graduate students, Ann Nelson and Aneesh Manohar.

I have already said that I believe that the answer to the puzzle of the uniqueness of the flavor structure of our world lies beyond the realm of QFT at high energy. It is possible that the structure of the quark and lepton mass matrices is likewise determined at these large energies. But it is also possible that the key to the masses can be found at an intermediate energy scale, beyond the domain of validity of SU(3) x SU(2) x U(1), but below the scale at which QFT breaks down. In this case, we may hope to gain some insights by clever use of conventional ideas.

If this latter possibility is to be interesting, the Yukawa couplings which are ultimately responsible for the fermion masses must be related in a very simple way at the large scale, in order that the intermediate scale physics produces all the interesting structure in the mass matrix. Two simple relations come to mind immediately. One possibility is that only one fermion or family of fermions get mass from the large scale physics, while the physics at intermediate scales allow mass to cascade down from one family to the next, each time suppressed by a power of some small parameter (such as a gauge coupling constant). This idea has a long history, going back to attempts to relate the electron-muon mass ratio to the electromagnetic coupling α [1]. It still looks attractive because it could explain the apparently hierarchical nature of the flavor masses. Nevertheless, it has not led to any convincing results for the mass matrices and we believe that it is worth exploring the other possibility, <u>that all fermions are created equal</u>. Here the idea is that at the large scale, the families are not distinguished. Only the interactions at intermediate energies differentiate among the various fermions and produce the nontrivial mass matrix. Thus in such a picture, all mass ratios are quantum renormalization effects like m_b/m_τ in the simplest SU(5) model [2]. In this talk, I will investigate a simple version of this idea. We will find that it can naturally account for many features of the fermion mass matrix, but that it cannot produce large enough mixing between the families in a satisfactory way. We will conclude by discussing modifications of the idea which could produce a fully realistic model.

Our first task is to write down a model of the high energy physics. As discussed above, in the real world we do not expect physics at high energy ($\sim M_p$) to be described by conventional QFT. However, here I will model the high energy world with purely conventional ideas for two reasons. I don't know anything better and want to concentrate on the physics of intermediate energies that produces the nontrivial structure of the mass matrix. One of the simplest models that produces equal masses for all the fermions is a left-right symmetric model based on a gauge group

$$SU(2)_L \times SU(2)_R \times G, \tag{1}$$

under which the fermions are a left-handed (LH) multiplet ψ_L which transforms as

$$(2,1,R) \tag{2a}$$

and a right-handed (RH) multiplet ψ_R which transforms as

$$(1,2,R) \tag{2b}$$

where R is some complex irreducible representation of G. Here $SU(2)_L$ is the SU(2) of weak interactions. If the dimension of R is N, (2) describes N SU(2) doublets of LH fermions with the corresponding right-handed (RH) particles in singlets. Note that G must contain color SU(3) because some of these doublets must be quarks. For simplicity, we will take G = SU(N), with R as the defining representation. Now if the only scalar mesons in the theory which transform as $SU(2)_L$ doublets are in a real four-component representation

$$(2,2,1), \tag{3}$$

the symmetry G requires that all the Yukawa couplings between (2) and (3) are equal. If (3) got a vacuum expectation value (VEV) with no breaking of G symmetry, all the fermions would be degenerate.

What I have in mind is breaking (1) at the large scale, M_G, down to

$$SU(2)_L \times U(1)_H \times H \tag{4}$$

where H is a subgroup of G and the $U(1)_H$ is a combination of the neutral generator of $SU(2)_R$ and a U(1) subgroup of G. At a much smaller scale M_I, perhaps 100 TeV, (4) breaks down to SU(2) x U(1) x SU(3). It is in this intermediate region between M_G and M_I that the flavor masses can split apart.

A nontrivial mass matrix results if the different generations transform differently under H. For example, H might contain an SU(5) factor under which R transforms reducibly. Suppose that some of the components of R transform like SU(5) 10's (or $\overline{10}$'s), some transform like 5's, and some transform like singlets. Each of these components is renormalized differently by the SU(5) gauge interactions. Just as the SU(3) gauge interactions make the b quark heavier than the τ in the simplest SU(5) model, so in a model of this kind, the 10's are heavier than the 5's which are heavier than the singlets. To leading order the renormalization due to some factor h of the group H has the form:

$$\left(\frac{g(M_I)}{g(M_G)}\right)^{-3(T_{La}^2+T_{Ra}^2)/16\pi^2 B} \tag{5}$$

where T_{La} (T_{Ra}) are the generators of h on the LH (RH) and

$$g(\mu) = g/(1-2\ B\ g^2\ \ell n\ \mu/\mu_0)^{1/2} \tag{6}$$

is the gauge coupling at scale μ.

Can we generate mass renormalizations in this way which are large enough to account for the large ratio (at least $\sim 10^5$) between the heaviest and lightest Dirac fermions? In the familiar example of m_b/m_τ in SU(5), the renormalization factor is only about a factor of 3. But here the renormalization is due primarily to color SU(3) which is asymptotically free. Most of the effect comes from small momenta, near the b mass, where the SU(3) coupling constant is not very small. Likewise, for a U(1) gauge group, or any other gauge group whose coupling grows with increasing energy, the coupling must be very small at low energies, otherwise the system becomes nonperturbative at high energies. Then one gets a sizable renormalization only from the high energy region and the total effect cannot be very large. To get the maximum effect, we want at least one factor of H whose coupling is approximately asymptotically flat. Call the factor f. Then the renormalization due to f comes from the entire region from M_G to M_I In the limit $B \to 0$, g is constant and the renormalization factor (5) becomes

$$\left(\frac{M_G}{M_I}\right)^{3g^2(T_{La}^2+T_{Ra}^2)/16\pi^2} \tag{7}$$

Clearly, if M_G/M_I is enormous, the renormalization due to f can be sizable, even if the coupling is not large. On the other hand it is also clear that the mechanism will not work unless M_G/M_I is very large. This is exciting because it suggests that the onset of flavor physics cannot be too far away. An M_I of 100 TeV might be detectable in experiments to detect GIM violating flavor changing neutral current processes, such as $K_L \to \mu e$, $\mu \to e\gamma$ or 3e, etc.

In this simple model, the nonabelian components of H (including f) come exclusively from G. Thus $T_{La} = T_{Ra}$ is the same for both members of an $SU(2)_L$ (or $SU(2)_R$) doublet. These renormalizations do not split charge 2/3 quarks from charge -1/3 quarks or neutrinos from charged leptons. However, the $U(1)_H$ coupling does distinguish ups from downs. If the $U(1)_H$ charge of some LH doublet is q, the $U(1)_H$ charges of the corresponding RH singlets are $q \pm 1/2$. Thus the $U(1)_H$ renormalization splits the up and down components of each SU(2) doublet. Up is heavier than down if $q > 0$ and down is heavier than up for $q < 0$. Because the $U(1)_H$ coupling is not asymptotically flat, these splittings within doublets are not as large as the splittings between families which come from f.

This picture can account very well for the mass spectrum of the quarks and charged leptons. However, it is incomplete in three ways.

(1) The neutrinos are Dirac particles with masses similar to those of the charged leptons.

(2) There are extra fermions, not associated with the light families, but which so far have only ordinary SU(2) x U(1) breaking masses.

(3) There is no flavor mixing.

All of these difficulties can be solved by enlarging the Higgs structure [3]. But I will not discuss this enlargement in technical detail, because it fails to produce a satisfactory model. Either the enlargement of the Higgs structure destroys the original degeneracy of the light fermions, or else the mixing angles produced by renormalization is very small. Evidentally, we are missing some of the physics of flavor in the intermediate region. Something else is needed to produce mixing angles as large as the Cabibbo angle. The difficulty may be associated with the simple SU(2) x SU(2) x G gauge structure. We are studying other possible gauge structures.

The idea of flavor democracy is interesting and new. I hope that it will lead to interesting physics at intermediate mass scales, between chemistry and metaphysics.

REFERENCES

[1] T. Hagiwara and B.W. Lee, Phys. Rev. D7 (1973) 459;
S. Weinberg, Phys. Rev. Lett. 29 (1973) 388;
B.W. Lee, Proc. 16th Int. Conf. on High-energy physics, Chicago-Batavia, Illinois, 1972, eds. J.D. Jackson and A. Roberts, Vol. 4, p. 249, (NAL, Batavia, IL 1973);
H. Georgi and S.L. Glashow, Phys. Rev. D6 (1972) 2977; D7 (1973) 2457;
H. Georgi and A. Pais, Phys. Rev. D10 (1974) 539.
See S. Barr, Phys. Rev. D21 (1980) 1424 and references therein for more recent work.

[2] M. Chanowitz, J. Ellis, and M.K. Gaillard, Nucl. Phys. B128 (1977) 506;
A. Buras, J. Ellis, M.K. Gaillard, and D.V. Nanopoulos, Nucl. Phys. B135 (1978) 66.

[3] H. Georgi, A. Manohar and A. Nelson, "On the Proposition That All Fermions are Created Equal," Physics Letters B, to be published.

PROTON DECAY THEORY

William J. Marciano

Physics Department
Brookhaven National Laboratory
Upton, NY 11973

1.1 MINIMAL SU(5) PREDICTIONS

The SU(5) Georgi-Glashow [1] model provided much of the motivation for ongoing proton decay experiments as well as a theoretical framework for estimating expected rates and branching ratios. In the so-called "minimal" model, one assumes the existence of a "great desert", i.e. no new particles up to m_X, the unification mass scale. This simplistic assumption has an appealing consequence; it leads to rather definite predictions. If those predictions turn out to be wrong, it doesn't necessarily imply that the concept of grand unification or even that the SU(5) model is invalid. Instead, it would most likely suggest that new physics populates the desert and modifies the predictions.

The renormalization group is the principle tool used to study grand unified theories (GUTS) [2]. That formalism has been refined to the next-to-leading log level, so theoretical uncertainties are negligible [3]. Employing $\alpha = 1/137.035965$, the known fermion masses, and a specific value for $\Lambda_{\overline{MS}}$, the QCD mass scale, one predicts (assuming a great desert) m_X and $\sin^2\theta_W(m_W)$, the weak mixing angle defined by modified minimal subtraction at the $W^\pm$ mass [4,5]. In addition, the $SU(2)_L \times U(1)$ model formulas [4,6]

$$m_W = 38.5 \text{ GeV}/\sin\theta_W(m_W) \tag{1a}$$

$$m_Z = 77.1 \text{ GeV}/\sin 2\theta_W(m_W) \tag{1b}$$

provide precise predictions for m_W and m_Z. Results of such an analysis [3] are illustrated in Table I.

Notice that m_X exhibits a sensitive dependence on $\Lambda_{\overline{MS}}$ ($m_X \simeq 1.3 \times 10^{15}\Lambda_{\overline{MS}}$). What is the currently accepted value of $\Lambda_{\overline{MS}}$? A 1981 survey by A. Buras [7], which I believe is still valid, found

$$\Lambda_{\overline{MS}} = 0.16^{+0.10}_{-0.08} \text{ GeV}. \tag{2}$$

$\Lambda_{\overline{MS}}$ (GeV)	m_X (GeV)	$\sin^2\theta_W(m_W)$	m_W (GeV)	m_Z (GeV)
0.10	1.3×10^{14}	0.2164	82.8	93.6
0.16	2.1×10^{14}	0.2136	83.3	94.1
0.20	2.7×10^{14}	0.2124	83.5	94.3
0.40	5.5×10^{14}	0.2084	84.3	94.9
0.50	6.9×10^{14}	0.2070	84.6	95.1

TABLE I: Predictions of the minimal SU(5) model for a variety of $\Lambda_{\overline{MS}}$ values.

Accepting that range, one obtains the following predictions [3]:

$$m_X = \left(2.1 \begin{array}{l} +1.7 \\ -1.2 \end{array}\right) \times 10^{14}\ \text{GeV} \tag{3a}$$

$$\sin^2\theta_W(m_W) = 0.214 \begin{array}{l} +0.004 \\ -0.003 \end{array} \tag{3b}$$

$$m_W = 83.3 \pm 0.7\ \text{GeV} \tag{3c}$$

$$m_Z = 94.1 \pm 0.6\ \text{GeV} \tag{3d}$$

How do these predictions compare with experiment? Deep-inelastic ν-N scattering and e-D scattering asymmetry measurements, including $O(\alpha)$ radiative corrections yield [8,9]

$$\sin^2\theta_W(m_W) = 0.215 \pm 0.014 \quad (R_\nu \text{ data}) \tag{4a}$$

$$\sin^2\theta_W(m_W) = 0.216 \pm 0.020 \quad (\text{e-D asymmetry}) \tag{4b}$$

These results are in excellent agreement with the prediction in Eq. (3b), and thus provide strong support for minimal SU(5). It would be nice to reduce the experimental errors in Eqs. (4) and then use $\sin^2\theta_W(m_W)$ to predict m_X, $\Lambda_{\overline{MS}}$ etc. Eventually, m_W and m_Z will be determined (at CBA and LEP) to within 0.1-0.2 GeV; then Eq. (1) can provide a precise $\sin^2\theta_W(m_W)$.

Another interesting prediction of minimal SU(5) concerns the ratio m_b/m_τ. Employing only a Higgs 5-plet to provide fermion masses leads to the natural lowest order relation [1] $m_b{}^0/m_\tau{}^0 = 1$. However, that ratio is strongly renormalized to [10,11,12]

$$m_b/m_\tau = 2.9 \pm 0.2 \tag{5}$$

for the $\Lambda_{\overline{MS}}$ range in Eq. (2). (A fourth generation increases this result by $\simeq 0.25$.) For comparison,

$$(m_b/m_\tau)^{exp} \simeq 2.6 - 2.9 \tag{6}$$

The agreement is impressive; however, the same scenario leads to $m_s/m_d \simeq 200$ whereas current algebra implies $m_s/m_d \simeq 20$. Coupling a 45-plet of Higgs scalars to the fermions overcomes this problem. Unfortunately, all fermion masses, including m_b and m_τ are rendered arbitrary.

1.2 GAUGE BOSON MEDIATED PROTON DECAY

The SU(5) model contains a color triplet, $SU(2)_L$ isodoublet of gauge bosons ($X^{\pm 4/3}$, $Y^{\pm 1/3}$) with $m_X \simeq m_Y$ which mediate proton decay. In higher rank groups such as SO(10), a second color triplet, isodoublet ($X'^{\pm 2/3}$, $Y'^{\mp 1/3}$) with $m_{X'} \simeq m_{Y'}$ can also mediate such decays.

The exchange of X, Y and/or X', Y' gauge bosons between quarks and leptons gives rise to the B and L violating (dim. 6) four fermi Hamiltonian [13,14]

$$H = \frac{g^2(m_X)}{2m_X{}^2}\left(\frac{m_X{}^2+m_{X'}{}^2}{m_{X'}{}^2}\right) A\, \epsilon_{ijk}\left[\bar{u}^c_{k_L}\gamma_\mu u_{j_L}\left(\bar{e}^+_R\gamma^\mu d_{i_R} + r_e(\bar{e}^+_L\gamma^\mu d_{i_L})\right)\right.$$

$$\left. - \bar{u}^c_{k_L}\gamma_\mu d_{j_L}\ \bar{\nu}^c_{e_R}\gamma^\mu d_{i_R}\right] + \text{h.c.} + \text{other generations} \tag{7}$$

(in the SU(5) model set $m_{X'} = \infty$.) In this amplitude $g(m_X)$ is the value of the gauge coupling at unification, A is an enhancement factor [11] and $r_e \simeq 2m_{X'}{}^2/(m_X{}^2 + m_{X'}{}^2)$.

In minimal SU(5) with 3 generations of fermions, one finds [3,9] $g^2(m_X)/4\pi \simeq 0.0242$ and $A \simeq 2.9$. Combining these values with m_X from Eq. (3a) determines H. To go from H to partial decay rates and lifetime predictions requires the evaluation of hadronic matrix elements which interpolate from an initial proton to the final state decay products. Unfortunately, the evaluation of such matrix elements is model dependent.

Consider the decay $p \to e^+\pi^0$ induced by H. That mode is the primary quest of the IMB experiment [15]. The matrix element for that process receives contributions from two sources, two quark annihilation and three quark fusion. These amplitudes are proportional to $1/R_p^{3/2}$ and $1/R_p^3$ respectively, where R_p is the effective proton radius. The relative importance of each contribution as well as the overall amplitude is sensitive to R_p. Most calculations find that the two parts are roughly equal and add constructively. Results form a variety of very different calculations are illustrated in Table II.
It is difficult to assess the uncertainty in such calculations. I will employ the results of Isgur and Wise [20], allowing for a factor of 5 uncertainty due to the model dependence of matrix elements. Then, including a factor of 10 uncertainty for the $\Lambda_{\overline{MS}}$ spread in Eq. (2), one finds ($10^{1.7} = 50$)

$$\tau_p = 2 \times 10^{29\pm1.7} \text{ yr.} \tag{8a}$$

$$1/\Gamma(p \to e^+\pi^0) = 4.5 \times 10^{29\pm1.7} \text{ yr.} \tag{8b}$$

Group	Method	τ_p(yr)	$1/\Gamma(p\to e^+\pi^0)$(yr)
Tomozawa [16]	PCAC	$(1\text{-}6)\times10^{29}$	$(1.5\text{-}10)\times10^{29}$
Berezinsky et al. [17]	QCD Sum Rule	4×10^{28}	8×10^{28}
Donoghue & Golowich [18]	MIT Bag	2.2×10^{29}	5.5×10^{29}
Lucha [19]	B-S Eq.	3.2×10^{29}	7×10^{29}
Isgur & Wise [20]	N.R.Q.M.	2×10^{29}	4.5×10^{29}
Thomas & McKeller [21]	Cloudy Bag	3×10^{28}	6×10^{28}

TABLE II: Results of different calculations, normalized to $\Lambda_{\overline{MS}} = 0.16$ GeV, in the minimal SU(5) model.

Note that the partial lifetime in Eq. (8b) is well below the IMB bound [15]

$$1/\Gamma(p \to e^+\pi^0) > 6 \times 10^{31} \text{ yr.} \quad \text{(IMB Exp.)} \tag{9}$$

reported at this meeting. That bound appears to rule out minimal SU(5) with a great desert unless $\Lambda_{\overline{MS}} > 0.3$ GeV and the $p \to \pi^0 e^+$ matrix elements are actually smaller than the estimates in Table II.

The simplest SO(10) models with symmetry breaking patterns that leave m_X unchanged predict a shorter τ_p, i.e. Eq. (8) is reduced by

$$\frac{5}{4}\left[\frac{r_e^2}{1 + r_e^2}\right] \tag{10}$$

where $r_e = 1$ for $m_X = m_{X'}$ (see (Eq. (7)). Of course, there are symmetry breaking schemes in SO(10) which allow intermediate mass gauge bosons to populate the desert and render m_X arbitrary [22]. In those cases τ_p may be much longer.

Baryon number violating neutron decay can also be induced by H. Isospin implies $\Gamma(n \to e^+\pi^-) \simeq 2\Gamma(p \to e^+\pi^0)$; so one expects

$$\tau_n \simeq 1.5 \times 10^{29\pm1.7} \text{ yr.} \tag{11a}$$

$$1/\Gamma(n \to e^+\pi^-) \simeq 2.2 \times 10^{29\pm1.7} \text{ yr.} \tag{11b}$$

Cerenkov radiation from $n \to e^+\pi^-$ should be readily discernable in the IMB experiment.

What other decay modes are induced by H? The branching ratios for $p \to e^+\omega^0$ or ρ^0 should be about 30 ~ 35%. In addition the relation

$$\Gamma(p \to \pi^+\bar{\nu}_e) \simeq \frac{2}{1+r_e^2}\,\Gamma(p \to \pi^0 e^+) \tag{12}$$

implies about a 16% B.R. for $p \to \pi^+\bar{\nu}_e$. A few of the anticipated branching ratios are given in Eq. (13)

$$p \to e^+\pi^0,\quad e^+\omega^0 \text{ or } \rho^0,\quad \bar{\nu}_e\pi^+,\quad \mu^+K^0,\quad \bar{\nu}_\mu K^+$$
$$0.40:\quad 0.30:\quad 0.16:\quad 0.03:\quad 0.03$$

$$n \to e^+\pi^-,\quad e^+\rho^-,\quad \bar{\nu}_e\pi^0,\quad \bar{\nu}_\mu K^0 \qquad (13)$$
$$0.80:\quad 0.05:\quad 0.08:\quad 0.02$$

1.3 UNCERTAINTIES IN τ_p

The proton lifetime prediction in Eq. (8) has about a factor of 10 uncertainty, primarily from the allowed range in $\Lambda_{\overline{MS}}$ [3,9]. In addition, the model dependence of matrix elements implies about another factor of 5 uncertainty. These uncertainties are not sufficient to reconcile minimal SU(5) and the IMB bound in Eq. (9). What other sources of uncertainty exist? The Higgs sector introduces the largest degree of uncertainty; it will be elaborated on separately in Section 2.

What about the uncertainty in the top quark mass, m_t? Fortunately, m_t enters only at the next-to-leading log level; so, ignorance of its value induces less than a 10% uncertainty in τ_p [3,9].

What about added fermion generations? For charged fermions with masses $\simeq m_W$, a fourth generation increases τ_p by about 30%, while a fifth and sixth generation increase τ_p by factors of about 2 each [9,23]. Additional larger representations (technifermions?) cause similar modifications.

Nuclear physics effects can enhance or diminish the proton decay rate depending on the mode [24]. Such effects are about a factor of 2. They are taken into account in the IMB bound.

2.1 HIGGS SCALAR EFFECTS

The only Higgs scalar multiplets that can couple to the known fermions in the SU(5) model are $\underline{5}$, $\underline{10}$, $\underline{15}$, $\underline{45}$ and $\underline{50}$ plets. All of these are included in the $\underline{126}$ of SO(10), but only the $\underline{5}$ and $\underline{45}$ are reqired in SU(5). Some of these scalars induce exotic effects such as proton decay [13], Majorana neutrino mass, neutrinoless double beta decay, n-$\bar{n}$ oscillations [25], H-$\bar{H}$ oscillations [26] etc. In addition, lack of mass degeneracy in the scalar spectrum can change the renormalization group analysis and SU(5) predictions [27,28,29]. To illustrate the latter effect, I first give the $SU(3)_c \times SU(2)_L \times U(1)$ decomposition of the above multiplets and assign their components arbitrary masses m_i, $i = 1,2,\ldots 21$.

$$\underline{5} = \underbrace{(1,2,1)}_{m_1} + \underbrace{(3,1,-2/3)}_{m_2}$$

$$\underline{10} = \underset{m_3}{(1,1,2)} + \underset{m_4}{(\bar{3},1,-4/3)} + \underset{m_5}{(3,2,1/3)}$$

$$\underline{15} = \underbrace{(1,3,2)}_{m_6} + \underbrace{(3,2,1/3)}_{m_7} + \underbrace{(6,1,-4/3)}_{m_8}$$

$$\underline{45} = \underbrace{(1,2,1)}_{m_9} + \underbrace{(3,1,-2/3)}_{m_{10}} + \underbrace{(3,3,-2/3)}_{m_{11}} + \underbrace{(\bar{3},1,8/3)}_{m_{12}} + \underbrace{(\bar{3},2,-7/3)}_{m_{13}}$$
$$+ \underbrace{(\bar{6},1,-2/3)}_{m_{14}} + \underbrace{(8,2,1)}_{m_{15}}$$

$$\underline{50} = \underbrace{(1,1,-4)}_{m_{16}} + \underbrace{(3,1,-2/3)}_{m_{17}} + \underbrace{(\bar{3},2,-7/3)}_{m_{18}} + \underbrace{(6,1,8/3)}_{m_{19}} + \underbrace{(\bar{6},3,-2/3)}_{m_{20}}$$
$$+ \underbrace{(8,2,1)}_{m_{21}}$$

Some of these scalars can mediate proton decay, hence they must be rather massive [13] (m_2, m_{10}, m_{11}, m_{12}, $m_{17} > 10^{10}$ GeV). The other masses are more or less arbitrary. In the "great desert" scenario one would assume $m_1 \simeq m_W$ while all other scalar masses $\simeq m_X$, then the results in Section 1 follow. Allowing arbitrary m_i, one finds [30]

$$\sin^2\theta_W(m_W) = 0.210 - 7 \times 10^{-5}$$
$$\times \ln\left[\frac{m_1^2\, m_5^4\, m_6^7\, m_7^4\, m_9^2\, m_{11}^{24}\, m_{10}^{33}}{m_2^2\, m_3\, m_4^3\, m_8^{11}\, m_{10}^2\, m_{12}^7\, m_{13}^4\, m_{14}^9\, m_{15}^4\, m_{16}^4\, m_{17}^2\, m_{18}^4\, m_{19}^{19}\, m_{21}^4}\right] \quad (14a)$$

$$\tau_p = 1 \times 10^{30\pm1.7} \left(\frac{m_1\, m_3\, m_6^5\, m_9\, m_{11}^3\, m_{12}^4\, m_{13}^7\, m_{16}^4\, m_{18}^7\, m_{19}^4}{m_2\, m_5\, m_7\, m_8^4\, m_{10}\, m_{14}^6\, m_{15}^8\, m_{17}\, m_{20}^6\, m_{21}^8}\right)^{2/33} \text{ yr.} \quad (14b)$$

Predictability is lost; however, some general features can be seen: 1) Relatively light color singlets decrease τ_p. 2) Light m_5, m_7, m_{20} lead to increases in both $\sin^2\theta_W(m_W)$ and τ_p. 3) Assuming a hierarchy condition $0.1 \leq m_i/m_j \leq 10$, $i,j = 2,\ldots 21$ amd $m_1 = m_W$, one finds a maximum increase of a factor of 150 in τ_p while correspondingly $\Delta\sin^2\theta_W(m_W) = +0.001$. Uncertainty in the Higgs sector provides additional motivation for pushing up the bound on τ_p and continuing the search for $p \to e^+\pi^0$.

2.2 PROTON DECAY VIA HIGGS SCALARS

Proton decay can be mediated by the following $SU(3)_c \times SU(2)_L \times U(1)$ Higgs scalar multiplets [13]: (3,1,-2/3), (3,3,-2/3), ($\bar{3}$,1,8/3) Assuming only Higgs 5-plets, Golowich [30] found that the (3,1,-2/3) induced proton decay amplitudes have the following ratios due to couplings alone

$$\begin{array}{lllll} p \to K^+\bar{\nu}_\mu, & K^0\mu^+, & \pi^0\mu^+, & K^0e^+, & \pi^0 e^+ \\ m_s m_d: & m_s^2\sin^2\theta_c: & m_s m_d \sin\theta_c: & m_s m_d \sin\theta_c: & m_d^2: \end{array} \tag{15}$$

where θ_c is the Cabibbo angle ($\sin^2\theta_c \simeq 0.05$). Phase space suppresses the K modes by $\simeq 1/2$, in addition one expects 3 quark fusion to enhance the π^0 and K^+ decay modes but not the K^0 (because the $pK^0\Sigma^+$ coupling is small [31]). Incorporating these effects with Golowich's results [30], I expect

$$\begin{array}{lllll} p \to K^+\bar{\nu}_\mu, & K^0\mu^+, & \pi^0\mu^+, & K^0e^+, & \pi^0 e^+ \\ 0.75: & 0.18: & 0.07: & 0.007: & 0.004: \end{array} \tag{16}$$

for Higgs (3,1,-2/3) mediated proton decay. Although the $K^+\bar{\nu}_\mu$ mode is dominant, the more easily observable $K^0\mu^+$ and $\pi^0\mu^+$ decay rates are significant. Already the IMB experiment has given the bound [15]

$$1/\Gamma(p \to K^0\mu^+) > 1.4 \times 10^{31}\ \text{yr}. \tag{17}$$

In the case of the neutron, a similar analysis suggests that $n \to K^0\bar{\nu}_\mu$ dominates and $\Gamma(n \to \pi^-\mu^+) \simeq \Gamma(p \to K^0\mu^+)$. In this scenario $m_2 \simeq 10^{10}$ GeV corresponds to Higgs mediated decay lifetimes $\simeq 10^{30}$ yr

($\tau_p \propto m_Z^4$); however, there is no compelling reason for introducing a 10^{10} GeV mass scale in the SU(5) model.

3.1 SUPERSYMMETRIC SU(5)

The basic idea of supersymmetry [32] is that each known boson (fermion) has a fermion (boson) partner. Assuming that all such partners of light particles have mass $\approx m_W$ while partners of superheavies have mass m_X (i.e. retaining the "great desert" idea), one obtains the predictions [33,34] illustrated in Table III. Notice that the predictions are very sensitive to the number of light Higgs isodoublets, N_H. In this scenario proton decay is still mediated by the X and Y; so the branching ratios in Section 1 hold.

N_H	m_X (GeV)	τ_p (yr)	$\sin^2\theta_W(m_W)$	m_W (GeV)	m_Z (GeV)
2	7.7×10^{15}	3×10^{35}	0.236	79.3	90.8
4	4.2×10^{14}	5×10^{30}	0.259	75.7	88.0

TABLE III. Supersymmetric SU(5) predictions for $\Lambda_{\overline{MS}} = 0.16$ GeV.

The most striking supersymmetry modification is that $\sin^2\theta_W(m_W)$ increases. That appears to be ruled out by the results in Eq. (4); however, the definitive test lies in the measurement of m_W and m_Z. There are of course ways to lower $\sin^2\theta_W(m_W)$. One popular idea [35] involves adding Higgs $\underline{10} + \underline{\overline{10}}$ multiplets containing light color singlets such that $\sin^2\theta_W(m_W)$ and τ_p are both reduced.

3.2 DIMENSION 5 OPERATORS AND PROTON DECAY

It was noted by Weinberg and Sakai and Yanagida [36] that dimension 5 (fermion-fermion-scalar-scalar) B and L violating operators are induced by Higgsino mixing in supersymmetric GUTS. Through loop effects, these operators give rise to dimension 6 four-fermi amplitudes which lead to proton decay at a rate proprotional to $1/m_X^2$ rather than $1/m_X^4$. Subsequent analyses [37] found that such effects lead to $\tau_p \simeq 10^{30}$ yr through the dominant decay mode $p \to \bar{\nu}_\tau K^+$ (for the neutron, $n \to \bar{\nu}_\tau K^0$). Such short lifetimes appear to be in conflict with the bound

$$\tau_p > 3 \times 10^{31} \text{ yr} \quad (\mu^+ \text{ in final state}) \tag{18}$$

obtained by the Homestake experiment [38]. (64% of the K's should produce a μ^+.) However, there are uncertainties in the theoretical analysis. I should note that the dimension 5 operators can be exorcised by symmetries [36].

3.3 HIGGS SCALARS AND PROTON DECAY

It has been noted that the geometric mass scale $(m_W m_P)^{1/2} \simeq 10^{10}$ GeV arises quite naturally in locally supersymmetric models [39]. (m_P is the Planck mass $\simeq 10^{19}$ GeV.) Such a mass scale in the Higgs sector can lead to scalar mediated decays with $\tau_p \simeq 10^{30}$–10^{32} yr. In that case the analysis of Section 2.2 regarding branching ratios is applicable.

4. COMMENTS

The initial results of the IMB experiment [15] appear to be in conflict with minimal SU(5) and the "great desert" hypothesis. To be more definite, we need better determinations of $\Lambda_{\overline{MS}}$, $\sin^2\theta_W(m_W)$, m_W and m_Z. In any case, a variety of possibilities can populate the desert. They range from new fermions and scalars to additional gauge bosons which occur in larger symmetry groups. All things considered, it is worth pushing the bound on $p \to e^+\pi^0$ as far as possible.

Higgs mediated proton decay exists at an arbitrary rate in the SU(5) model. Some supersymmetric models naturally suggest a lifetime of $\tau_p \simeq 10^{30}$–10^{32} yr. from such decays. In such scenarios the decays $p \to K^+\bar{\nu}_\mu$, $K^0\mu^+$, $\pi^0\mu^+$ and $n \to K^0\bar{\nu}_\mu$, $\pi^-\mu^+$ tend to dominate. All should be searched for. I must remark that if supersymmetric partners exist at mass scale m_W, they will be quite easily discovered and studied at CBA. Underground and accelerator experiments are clearly complementary probes of the fundamental laws of nature.

ACKNOWLEDGMENTS

I wish to thank Maurice Goldhaber for enlightening conversations regarding the IMB experiment and Carl Dover for providing updated meson-baryon couplings.

This work was supported by Department of Energy Contract No. DE-AC02-76CH00016.

REFERENCES

[1] H. Georgi and S. Glashow, Phys. Rev. Lett. 32, 438 (1974).

[2] H. Georgi, H. Quinn and S. Weinberg, Phys. Rev. Lett. 33, 451 (1974).

[3] Cf. W. Marciano, in "Field Theory in Elementary Particles", Jan. 1982, Eds. B. Kursunoglu and A. Perlmutter, (Plenum, NY) p. 71.

[4] W. Marciano and A. Sirlin, Phys. Rev. Lett. 46, 163 (1981).

[5] W. Marciano, in Orbis Scientiae, Recent Developments in High Energy Physics", Jan. 1980, Eds. A. Perlmutter and L. Scott (Plenum, NY) p. 121.

[6] W. Marciano, Phys. Rev. D20, 274 (1979);
A. Sirlin, Phys. Rev. D22, 971 (1980);
W. Marciano and A. Sirlin, Phys. Rev. D22, 2695 (1980).

[7] A. Buras, in Proc. of the Bonn Lepton-Photon Symp. Aug. 1981.

[8] A. Sirlin and W. Marciano, Nucl. Phys. B189, 442 (1981);
C. Llewellyn Smith and J. Wheater, Phys. Lett. 105B, 486 (1981).

[9] W. Marciano and A. Sirlin, In Proc. of the Second Workshop on Grand Unification, Ann Arbor, 1981, Eds. J. Leveille, L. Sulak, and D. Unger (Birkhauser, Boston).

[10] M. Chanowitz, J. Ellis and M.K. Gaillard, Nucl. Phys. B128, 506 (1977).

[11] A. Buras, J. Ellis, M.K. Gaillard and D. Nanopoulos, Nucl. Phys. B135, 66 (1978).

[12] M. Fischler and J. Oliensis, Phys. Lett. 119B, 385 (1982).

[13] S. Weinberg, Phys. Rev. Lett. 43, 1566 (1979);
F. Wilczek and A. Zee, Phys. Rev. Lett. 43, 1571 (1979).

[14] P. Langacker, Phys. Rep. 72, 185 (1981).

[15] B. Cortez describes the Irvine-Michigan-Brookhaven (IMB) experiment in these proceedings.

[16] Y. Tomozawa, Phys. Rev. Lett. 46, 463 (1981).

[17] V. Berezinski, B. Joffe and Ya. Kogan, Phys. Lett. 105B, 33 (1981).

[18] J. Donoghue and E. Golowich, Phys. Rev. D26, 2888 (1982).

[19] W. Lucha, Phys. Lett. 122B, 381 (1983).

[20] N. Isgur and M. Wise, Phys. Lett. 117B, 179 (1982).

[21] A. Thomas and B. McKellar, CERN preprint 3376 (1982).

[22] Cf. Y. Tosa, G. Branco and R. Marshak, VPI preprint (1983).

[23] M. Fischler and C. Hill, Nucl. Phys. B193, 53 (1981).

[24] C. Dover, M. Goldhaber, T.L. Trueman and L.-L. Chau, Phys. Rev. D24, 2886 (1981)

[25] Cf. L.N. Chang and N.P. Chang, Phys. Lett. 92B, 103 (1980).

[26] G. Feinberg, M. Goldhaber and G. Steigman, Phys. Rev. D18, 1602 (1978);
L. Arnellos and W. Marciano, Phys. Rev. Lett. 48, 1708 (1982).

[27] G. Cook, K. Mahanthappa and M. Sher, Phys. Lett. 91B, 369 (1981).

[28] L. Ibanez, Nucl. Phys. B181, 105 (1981).

[29] W. Marciano and Z. Parsa, unpublished.

[30] E. Golowich, Phys. Rev. D24, 2899 (1981).

[31] C. Dover, private communication.

[32] Cf. P. Fayet and S. Ferrara, Phys. Rep. 32C, 249 (1977).

[33] S. Dimopoulos, S. Raby and F. Wilczek, Phys. Rev. D24, 1681 (1981);
S. Dimopoulos and H. Georgi, Nucl. Phys. B193, 150 (1981);
N. Sakai, Z. Phys. C11, 153 (1981).

[34] L. Ibanez and G. Ross, Phys. Lett. 105B, 439 (1981);
M. Einhorn and D.R.T. Jones, Nucl. Phys. B196, 475 (1982);
W. Marciano and G. Senjanovic, Phys. Rev. D25, 3092 (1982);
L. Ibanez and F. Yndurain, Phys. Lett. 113B, 367 (1982).

[35] U. Berezinsky and A. Smirnov, INR preprint (1982);
Y. Igarashi et al., Phys. Lett. 116B, 349 (1982);
A. Masiero et al., Phys. Lett. 115B, 298 (1982).

[36] S. Weinberg, Phys. Rev. D26, 287 (1982);
N. Sakai and T. Yanagida, Nucl. Phys. B197, 533 (1982).

[37] S. Dimopoulos, S. Raby and F. Wilczek, Phys. Lett. 112B, 133 (1982);
J. Ellis, D. Nanopoulos and S. Rudaz, Nucl. Phys. B202, 43, (1982).

[38] M.L. Cherry et al., Phys. Rev. Lett. 23, 1507 (1981).

[39] Cf. L. Ibanez, Phys. Lett. 118B, 73 (1982).

THE K. G.F. NUCLEON DECAY EXPERIMENT

MR Krishnaswamy, MGK Menon, NK Mondal, VS Narasimham, BV Sreekantan
Tata Institute of Fundamental Research, Bombay 400005, India
Y Hayashi, N Ito, S Kawakami
Osaka City University, Osaka, Japan
and
S. Miyake
Institute for Cosmic Ray Research, University of Tokyo, Japan

A 140 ton nucleon decay detector is in operation in the Kolar Gold Mines in India at a depth of 2300 metres since November 1980. The experimental details, the method of analysis and the results obtained after different periods of observation are available in a series of earlier publications[(1)].

Basically the detector comprises 34 layers of proportional counters with 1.2 cm thick iron plates in between the counters with a total weight of 140 tons. The counters have a cross sectional area of 10 cm x 10 cm and in the direction parallel to the tunnel, they are 6 metres long and in the perpendicular direction, 4 metres long. For each event the ionisation is measured in all the hit counters, provided it exceeds the threshold of $\sim 1/3\ I_{min}$. While the ionisation measurement is linear upto $\sim 100\ I_{min}$, data are available upto the level of saturation set by electronics. Since December 1982, timing information accurate to 0.5 μsec is also available from each of the triggered counters. The basic trigger is a 5-layer coincidence of pulses from counters in any of the successive 11 layers, with the top and bottom-most layers as well as the end counters in each layer being eliminated from the trigger. An additional 2-layer trigger has also been introduced in 1981 to record events that have tracks confined to less than 5 consecutive layers.

Table 1 summarises the events that have been recorded from the beginning (November 1980) upto 21st March 1983 in an effective running

Table 1

A detailed break-up of events recorded in the KGF experiment

	Run time: 1.9 years		Observed	Expected
1.	Single penetrating muons (atmospheric origin)		1215	∿ 1190
2.	Multiple penetrating muons (atmospheric) n = 2: 13 events n = 3: 2 events		15	-
3.	Muon induced showers ($E_{vis} \gtrsim 10$ GeV)		8	
4.	Side showers		6	
5.	Neutrino interaction in rock			
	(a) With secondary muon penetrating the detector ($\theta > 55°$)		38-3* } =39	37±5
	(b) Without penetrating tracks		4 }	
6.	Neutrino interactions inside the detector as estimated from the following data:		13	< 20
	(a) Multiprong events with vertex inside (ν-induced)	7**		
	(b) Single tracks penetrating the sides or bottom of the detector (ν-induced)	5		
	(c) Single tracks penetrating the top (ν-induced and stopping atm. muons)	6		
7.	Kolar events		2	
8.	Nucleon decay candidates Partially confined: 3 Fully confined : 3		6	

* In the angular bin of 55-60°, 3 events are estimated as due to atmospheric muons.

** Includes one of the Kolar events with vertex at the edge of the detector.

time of 1.9 years. The criteria for classification of events have been discussed in the earlier publications.

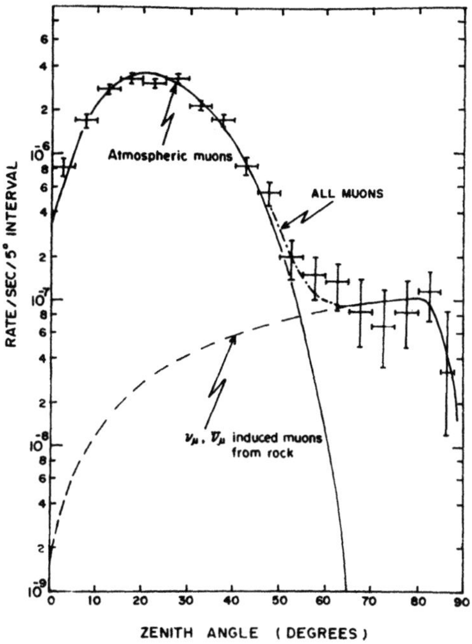

Fig. 1

The zenith angle distribution of the penetrating tracks (muons) is shown in Fig.1 and is fully consistent with what is to be expected on the basis of the contribution of atmospheric muons ($\theta < 55°$) and from the ν, $\bar{\nu}$ induced muons for larger angles. The fluxes of ν and $\bar{\nu}$ of atmospheric origin and of energy larger than a few GeV that contribute to the observed muons are fairly well determined. It can be seen from Table 1 that the observed number of 39 events with $\theta > 55°$ agrees very well with the expected number of 37±5 calculated from the spectrum of neutrinos using cross sections determined from accelerator experiments.

The detailed break-up of the neutrino interactions observed inside the detector are given in Table 1. These include (a) multiprong events with vertex inside the detector; (b) single tracks which either start or stop in the detector and correspondingly either leave or come from the bottom or side of the detector and (c) single tracks that either start or stop in the detector and correspondingly either leave or enter from the top of the detector. It is estimated that in category (b), 3 out of the 5 cases are produced in the detector and the muons move out and in category (c) 3 out of 6 cases are estimated as due to atmospheric muons stopping in the detector. Thus out of 18 events, 13 are to be attributed to secondaries produced in neutrino interactions in the detector itself. The expected number, however, is $\sim$ 20. The discrepancy may be, apart from statistical fluctuations, due to

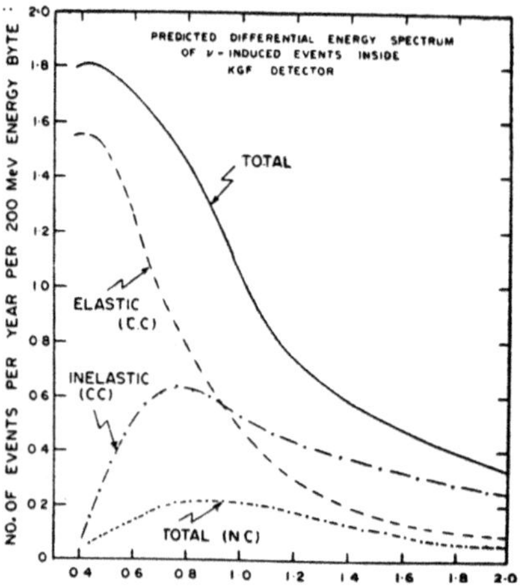

Fig. 2

uncertainties in the fluxes of low energy neutrinos (<2 GeV) particularly from the latitude effect, as discussed in detail in earlier publications.

In Fig. 2, the predicted differential energy spectrum of neutrino induced events inside the KGF detector is given for the energy range 0.5 - 2 GeV. It is seen that the main contributions around 1 GeV, relevant to proton decay background estimates, is from the inelastic charged current processes.

In Table 1, six events have been classified as candidates for nucleon decay from the KGF experiment thus far. Out of these, in three cases (Fig. 3) secondaries are fully confined in the detector and constitute rather strong evidence for nucleon decay. The detailed profiles of these events are shown in Figures 4, 5 and 6. Table 2 summarises the most probable interpretation of these events,

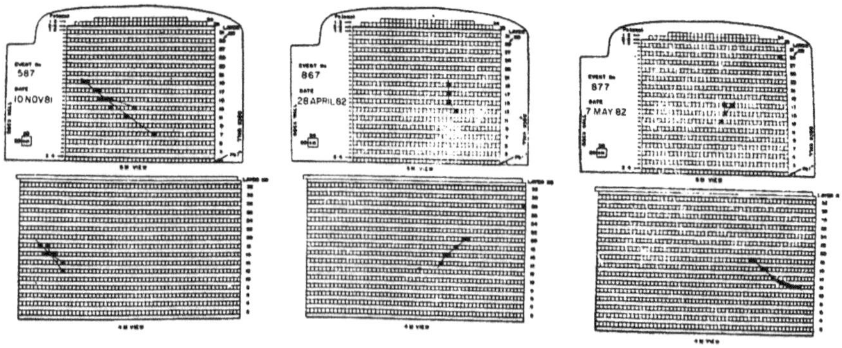

Fig. 3. Fully confined decay candidates.

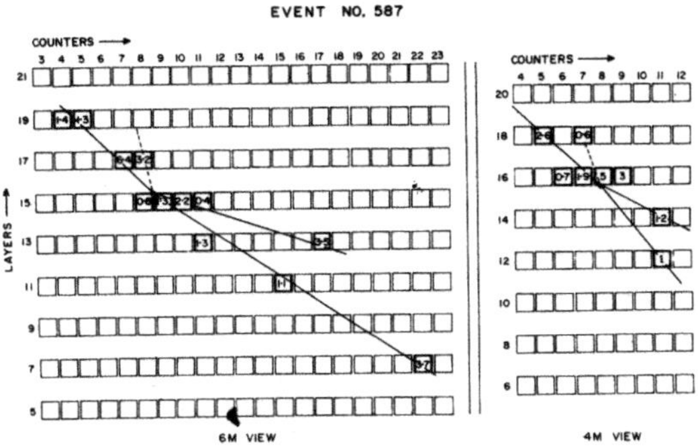

Fig. 4

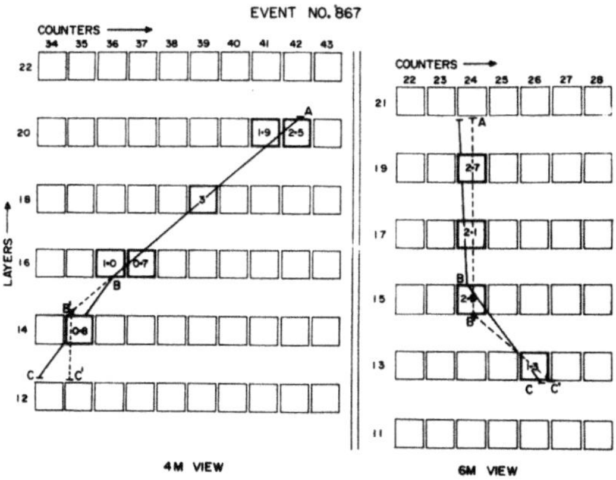

Fig. 5

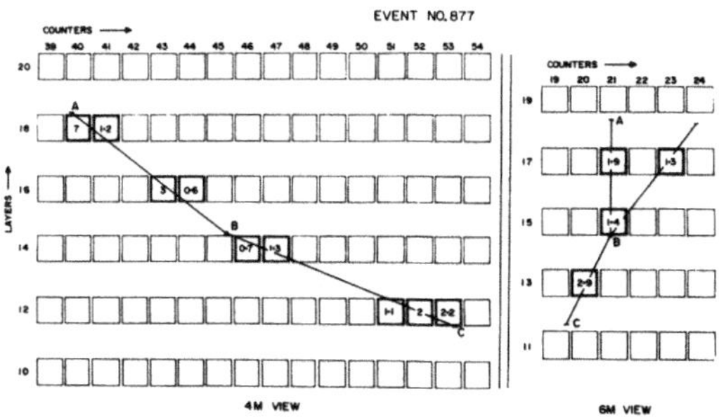

Fig. 6

the energy of the secondaries, the total visible energy, alternative modes of decay and the calculated neutrino background for each of the events.

Among these, event 587 has the highest probability of being a neutrino interaction, if it is interpreted as a *single* cascade of *one* GeV energy. However, the main objections for this being a single cascade are (1) a 1 GeV shower very rarely extends upto and beyond 20 radiation lengths as observed here; (2) the position of the shower maximum is around 6-7 radiation lengths and not at 2-3 radiation lengths as is to be expected for a single cascade of energy 1 GeV; (3) there is a suggestion of a second maximum at ~ 12 radiation lengths; and (4) the direction of motion will then be downwards

Table 2

Details of fully confined nucleon decay events in KGF detector

Event No.	Decay Mode	Particle range (g/cm²)	Energy estimates (MeV)[a]	Other plausible modes	ν-background events/1.9 yrs.
587	$P \to e^+ \pi^o$	e^+:115	$U_e \sim 500$	-	<0.7[b]
		π^o:150	$U_{\pi^o} \sim 500$		
867	$P \to \bar{\nu} \pi^+$	ABC(π^+):182	$U_\pi \sim 435$	$P \to \bar{\nu} K^+_{\to \mu\nu}$	<0.07
877	$N \to e^+ \pi^-$	AB(e^+):104	$U_e \sim 400$	$P \to \mu^+ + K^o_s \to \pi^+ \pi^-$	<0.07
		BC(π^-):135	$U_\pi \sim 450$	U_μ:340 MeV	
				$U_{\pi^\pm}$:320,220 MeV	

a) U refers to total energy.

b) Estimated rate for a single cascade. It would be much lower for the preferred interpretation of 2 cascades starting at the mid-point of the event (see Fig. 4).

with zenith angle of $\sim 58°$ and azimuth of $\sim 35°$ East of North. In this direction the flux of $\lesssim 1$ GeV neutrinos is highly suppressed due to large geomagnetic cut-off rigidity - a fact not considered so far explicitly for different azimuths in evaluating the ν-background.

The event 876 is clearly a case of a single track originating and stopping well inside the detector. The probability of this track being a muon is ruled out from the very large angle ($\sim 37°$) scattering observed, particularly when the residual range is ~ 45 g.cm^{-2}. The track has the highest probability of being a pion of energy around 450 MeV and fits well with the interpretation $P \to \bar{\nu} + \pi^+$. While the event is consistent with the decay mode $P \to \bar{\nu} + K^+_{\to \mu\nu}$, it has a lower probability since the kaon will have to decay in flight rather than interact in the iron absorber.

The event 877 is the strongest candidate for nucleon decay. It has one non-showering track of range 135 g.cm^{-2} and in the opposite direction it could either be a shower or a pair of charged particles. In the first case it is consistent with the interpretation $N \to e^+ + \pi^-$ and in the second case with the interpretation $P \to \mu^+ K^o_s$ with $K^o_s \to \pi^+ \pi^-$.

The reason why all these three events are not cases of neutrino interactions has been discussed earlier.

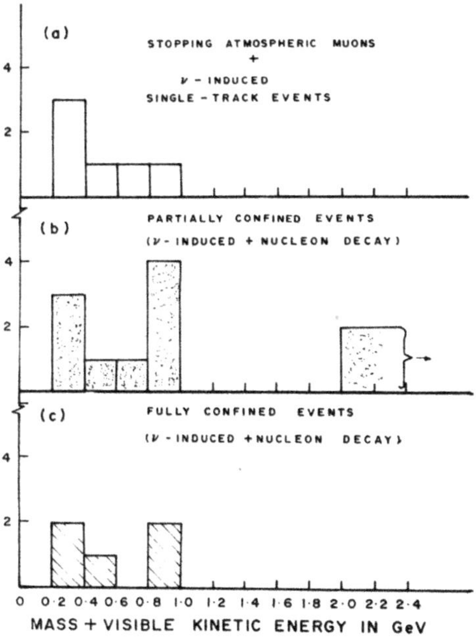

Fig. 7

In Figs. 7a,b,c, we plot the energy distribution (visible energy + mass) of all events considered as decay candidates, the ν-interaction inside the detector as well as the stopping muons of atmospheric origin. The confined events shown in Fig. 7c* clearly demonstrate the low background encountered in this experiment. It is remarkable that, though the statistical weight is meagre, the three decay candidates stand out in this plot with events 587, 877 around proton mass and event 876 at half of that energy.

Another way to distinguish between ν-interactions and the decay events is to study the distribution of the maximum opening angle in multitrack events. This is plotted in Fig. 8 using only the events for which the spatial angles are measured. A clear distinction is apparent with the ν-events predominantly populating the smaller opening angles and the events identified as decay candidates crowding beyond 140°, as is to be expected.

Summary of the KGF results

The main evidence for nucleon decay (based on total energy and configuration considerations) comes from events 877 and 587. Event 877 is characteristic of the decay mode $e^{+}\pi^{-}$ or $\mu^{+}K_{s}^{o}$ and event 587 of $e^{+}\pi^{o}$. Supporting evidence comes from event 876 with decay mode $\pi^{+-}\nu$ or $K^{+-}\nu$.

---

* A recently observed confined event is not plotted here as the energy estimates vary in the range 1-2 GeV due to complexity in the configuration of individual tracks.

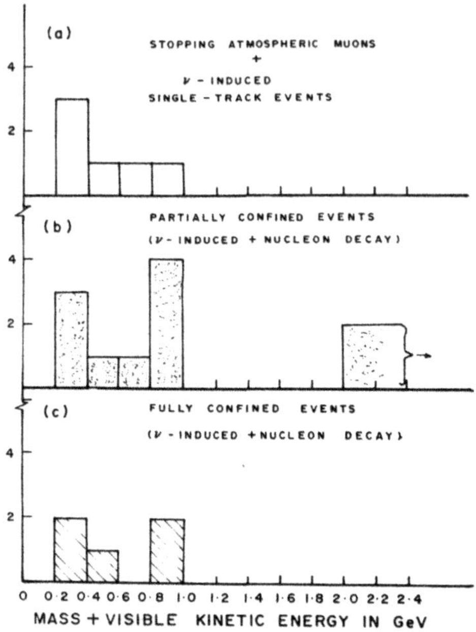

There is a suggestion of a peak around 1 GeV in the energy spectrum measurements amidst the very low background from ν-interactions at the KGF latitude.

Estimate on the lifetime of bound nucleons: We consider here only the confined events and use an effective fiducial weight of 60 tons with a livetime of 1.9 years. The estimated efficiency for detection is 0.5 inclusive of hadron absorption in the iron nucleus. On the basis of 3 confined events the lifetime estimate is $\tau/\text{B.R.} \sim 1.1 \times 10^{31}$ years. However, the lifetime based on individual decay modes is $\tau/\text{B.R.} \sim 1.7 \times 10^{31}$ years for the modes

$$P \to e^+\pi^o \text{ or } \mu^+K^o_s, \; \bar{\nu}\pi^+ \text{ or } \bar{\nu}K^+ \text{ and}$$
$$N \to e^+\pi^-$$

where B.R is the branching ratio for decays into the modes considered.

Limits on the fluxes of primordial magnetic monopoles: The KGF detector is sensitive to GUT monopoles having velocities in a certain range. Two methods viz. (a) ionisation and (b) progressive time delay have been used to set limits on the flux of these monopoles. In the ionisation method, a search is made for particles that cross at least 1.5 m of the detector and show an ionisation greater than 2.5 I_{min} on a statistical basis. This criterion and the coincidence resolution time of 2 μsec limits the velocity range to values shown in Table 3. From December 1982, timing (with a resolution of 0.5 μsec) of the proportional counter pulses has been introduced. The present data acquisition system permits the study of time of flight of slow particles upto a time delay of $\sim$ 5 μsec. No event was recorded so far satisfying the criteria of either (a) or (b). The relevant details

Table 3

Experimental details for the study of magnetic monopoles

Area x solid angle: $\sim$250 m^2 sterad.

Method of analysis	Range of β	Period of operation (hours)	Upper limit of fluxes (90% C.L) ($cm^{-2}.Sec^{-1}.St^{-1}$)
Ionisation $I > 2.5\ I_{min}$	$1.3 \times 10^{-3*}$-1	10136	2.2×10^{-14}
Timing over > 1.5 m track length	7.5×10^{-4}-4×10^{-3}	2300	9.7×10^{-14}

* Based on the calculation of Ritson (1982).

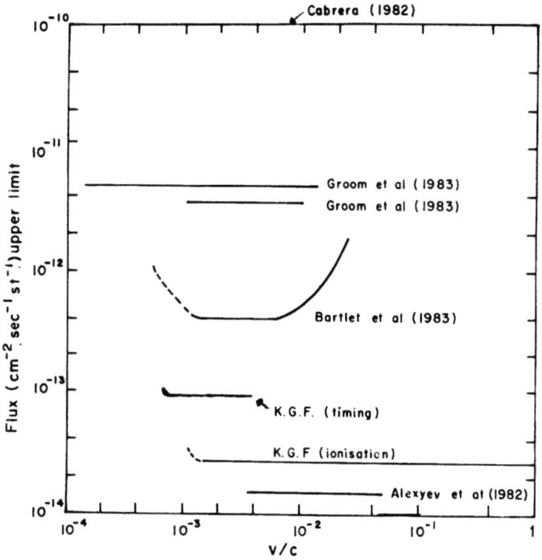

Fig. 9

and the flux limits obtained are shown in Table 3. In Fig. 9 we compare the present results with those of other experiments[3]. Clearly, the KGF data provide one of the lowest limits achieved so far over a large span of monopole velocities ($v/_c = 10^{-3}$-1).

Acknowledgements: We wish to thank Messrs. R.M. Wankar, V.M. Punekar, R.P. Mittal and R. Kulkarni for the technical assistance provided during the course of this experiment. The cooperation of the officers and other staff of Bharat Gold Mines Ltd. at all stages of our experiment is gratefully acknowledged. We are thankful to the Ministry of Education, Japan for partial financial support to this experiment.

References:

1. Krishnaswamy, M.R. et al., Phys. Lett., B106, 339, 1981
   Phys. Lett., B115, 349, 1982
   Proc. Int. Colloq. on Baryon Nonconservation, Bombay, Pramana Suppl., 115, 1982
   Pramana, 19, 525, 1982
2. Ritson, SLAC Report, SLAC-pub-2950, 1982 (unpublished)
3. Alexeyev, N. et al., Lett. Nuovo Cimento, 35, 413, 1982
   Bartelt, J. et al., Phys. Rev. Lett., 50(9), 655, 1983
   Cabrera, B., Phys. Rev. Lett., 48, 1378, 1982.
   Groom, D.E. et al., Phys. Rev. Lett., 50(8), 573, 1983.

NEW RESULTS FROM THE SOUDAN 1 DETECTOR

E. Peterson

The Soudan 1 detector[1] began operation in May 1981, and has produced reliable data since October 1981. We present here results (preliminary in some cases) from three classes of data that have been obtained, on nucleon decay, magnetic monopoles and multiple-muon events.

The Soudan 1 nucleon decay detector, a 31.5 metric-ton tracking calorimeter, is instrumented with 3,456 gas-filled proportional tubes. It is located 625 m underground in northeastern Minnesota and is rectangular, 2.9 m x 2.9 m x 1.9 m high. It consists of 48 layers of 72 proportional tubes each, with the tube axes turned by 90° in alternate layers, in order to provide two views of each event. The 2.9 m-long proportional tubes are each 2.8 cm in diameter with an 0.8 mm-thick steel wall. The tube axes are spaced by 4 cm in the horizontal direction and adjacent tubes are staggered up and down by 0.45 cm from the layer center. The vertical distance between layers is 4.1 cm. Most of the detector mass is provided by a matrix of heavy concrete in which the tubes are embedded. This substance was made from a mixture of purified iron ore (taconite), Portland cement and water. The total mass of protons in the detector is 15.1 tons; the neutron mass is 16.4 tons. The average density of the detector is 1.85 g/cm^3; the average radiation length is 9.3 cm. Additional information about events which originate externally is provided by a scintillation counter shield which covers the top and four sides of the detector.

During normal operation, the outputs of all proportional tubes and shield scintillation counters are continually recorded in a buffer memory. A trigger is defined as a coincident signal above threshold in any one of the 72 proportional tubes in any 3 out of 4 adjacent layers. For each trigger with more than 5 proportional tube hits, the output of each proportional tube and each shield counter is permanently recorded for 32 time frames beginning about 370 nsec before the trigger and ending about 8 μsec after the trigger. This detailed time-structure data is used to determine the dE/dx ionization in the proportional tubes from the time over threshold, to detect stopped muon decays, to search for nonrelativistic particles and to diagnose malfunctioning channels. The data sample reported here was collected between October 1981 and March 1983. The actual time that the detector was receptive to triggers during that period was 0.621 years, as measured by a 10 kHz, crystal-controlled clock.

The search for nucleon decay candidates has been described in detail in the literature [2]. Only one event satisfied all of the selection criteria outlined in that paper, and no further candidate events have been found. By assuming that the candidate event shown in Reference 2 is not a nucleon decay, we can set a lower bound for the lifetime of nucleon (bound in iron, oxygen and the other nuclei found in the Soudan 1 detector) decaying to other than neutrino modes as 1.68×10^{30} years at the 90 percent confidence level. The sensitivity to modes which include one neutrino is about one-fifth of this limit. The subsidiary limits on lifetime divided by branching ratio (also 90 percent confidence level) are listed in Table 1. These results are independent of any theoretical assumptions about branching ratios.

Table 1 Nucleon Decay Lifetime Limits

Mode	Acceptance	Lifetime/Branching ratio
$p \rightarrow e^{+}\pi^{0}$	0.34	0.84×10^{30} yr.
$p \rightarrow e^{+}\rho^{0}$	0.31	0.75
$p \rightarrow e^{+}\omega$	0.38	0.93
$p \rightarrow \mu^{+}K$	0.38	0.93
$p \rightarrow \nu K^{+}$	0.07	0.18
$n \rightarrow e^{+}\pi^{-}$	0.32	0.85
$n \rightarrow e^{+}\rho^{-}$	0.35	0.94
$n \rightarrow \mu^{+}K^{-}$	0.27	0.71
$n \rightarrow \nu K_{s}$	0.18	0.48

Several recent experiments, including the one reported here, have searched for slowly moving particles which are capable of penetrating significant amounts of matter, thus signaling the passage of a magnetic monopole. A complication in assessing the sensitivity of these searches is the expected strong dependence of the ionization and energy-loss of monopoles on velocity. Recent work by Ritson [3] on the ionization by slow monopoles in argon indicates that above 0.001c the ionization rate should exceed that of a relativistic muon. If the monopoles have velocities above 0.01c, the ionization rate should exceed this "minimum" rate by a factor of 40.

The experiment utilizes the gas proportional tubes in the detector to measure the ionization caused by the passage of a particle. A relevant feature of the detector that is of interest in a search for slowly moving particles is the time resolution of the system. The information that is recorded for each tube is a time history of whether or not the pulse height in each tube is above a fixed threshold. This is recorded in 16 frames of 185 nsec. (with the trigger time defined as the 3rd frame) followed by 16 additional frames of 335 nsec. (225 nsec., for data recorded after August 1982). With this time information, an ionizing particle travelling at a velocity of less than 0.01c can be clearly distinguished as non-relativistic.

The geometrical acceptance of the detector is a function of angle and of the minimum number of hit tubes required. A requirement of 8 proportional tubes or more produces a very clean sample of muons, and has been adopted for the monopole search. With this requirement the effective area of the detector is 5.7 m^2 averaged over 4π steradians. The limits on the velocity acceptance of the detector are set both by the time resolution of the proportional tubes and by the trigger requirement. The acceptance was determined by a Monte-Carlo calculation which traced the deposition of ionization through the detector and simulated the proportional tube drift times (0 to 300 ns) as well as the trigger. Two fitting procedures were then applied to the model events: a geometrical fit to a straight line through the detector and a constant velocity fit to the simulated arrival times of the ionization. The parameters in the simulation were selected to reproduce the measured characteristics of the muon background events. Criteria were then established for the velocity fit which selected events that had well-determined velocities which were significantly less than c. Events with velocities above 0.01c often failed to meet these criteria. The low-velocity limit of the acceptance is probably set by the velocity dependence of the ionization by a monopole. The detector response is negligible for ionization depositions below half of that of a relativistic muon, and it has been estimated that this level is not reached for monopoles with velocities below 0.0008c. (The trigger is effective for monopoles of velocities as low as 0.0003c.) The velocity acceptance is a smooth function between these limits, and is maximum at v = 0.002c.

Above .01c, the very high ionization rate of the monopoles should be a distinguishing characteristic. Other processes which lead to very high ionization rates are either slow protons or nuclear fragments which cannot penetrate a significant portion of the detector before stopping, or are processes that produce many ionizing particles (electromagnetic showers, for example) which do not appear as single tracks in this relatively fine-grained detector. After eliminating noise events and events that do not meet the velocity-selection or ionization criteria outlined above, no candidate event remains. The 90-percent confidence-level flux limits that are obtained are given in Table 2, below.

Table 2 Magnetic Monopole Flux Limits

Beta	Flux Limit
1×10^{-3}	1.7×10^{-13} /cm^2 sec sr
2×10^{-3}	1.5×10^{-13}
3×10^{-3}	1.7×10^{-13}
5×10^{-3}	1.7×10^{-13}
$>1\times10^{-2}$	1.5×10^{-13}

An extension to lower monopole velocities is possible if the process of nucleon decay catalysis occurs.[4] The cross-section for catalysis is usually parameterized by $\sigma = x\,\sigma_0/\beta$ where σ_0 is a typical strong-interaction value, taken to be $1/m_\rho^2$ in natural units (440 μbarns), β is the monopole-nucleon relative velocity and x is a scale factor of order unity that reflects theoretical uncertainty. If x is larger than 10, monopoles would catalyze so profusely in Soudan I that the criteria for straight highly ionizing single tracks would not apply to monopoles. This situation would invalidate the limit quoted above for monopoles faster than 0.01c. If x is much smaller than 0.1, the probability of 2 or more catalyzed decays in the detector is small and no useful flux limit for monopoles can be inferred. If x is unity, interesting limits can be set for monopole velocities as low as 10^{-5}c.

The acceptance for monopole catalysis is difficult to calculate. There is a possibility of a catalysis event occuring in the rock wall adjacent to the detector which sends a relativistic decay fragment into the detector. This would cause a trigger and subsequent read-out with its associated large dead-time before the slow monopole actually entered the detector. Such events would be lost. Rough estimates indicate that this effect seriously degrades the acceptance for monopoles entering the top or bottom of the detector at velocities below 10^{-3}c, and for monopoles entering the north or south faces below 10^{-4}c.

The existing data has been scanned for catalysis candidates, i.e. events with a distinct time separation between groups of hits. No candidates have been found. A Monte-Carlo calculation of the acceptance is still in progress, but we estimate that flux limits of 10^{-11}/cm^2 sec sr for x=1.0 and 10^{-10}/cm^2 sec sr for x=0.1 will be valid over the velocity range from 10^{-5}c to 10^{-3}c.

The study of muon multiplicities in Soudan I has been pursued for some time as a way of understanding the composition of the primary cosmic rays. This question has recently become more interesting because of data from the SPS collider, which permit a better understanding of the hadron dynamics in the primary cosmic ray collisions at a relevant energy. Although the detector is too small to directly measure the usual decoherence function, it does have two strong assets. The two-track resolution of Soudan 1 is about six times better than the last large scale experiment [5]. Secondly, the Soudan 1 data set, which contains 4650 multiple muon events, provides much better statistical accuracy than most similar experiments. These multiple muon events have shown preliminary evidence for two new effects, in data samples consisting of events with high muon multiplicity and in pairs of multiple muon events that occur within a limited time period. These will be discussed in turn.

While the angular distributions of single muon and the low-multiplicity multiple muon events seem to be smooth, a directional anisotropy can be discerned in the (limited) data sample consisting of events where five or more parallel muons are observed. This sample, which consists of 105 events, shows an enhancement in the region of right ascension from 8 hours to 13 hours, 20 minutes (120° to 200°). This coordinate, the star-fixed longitude, has a smooth acceptance due to the mismatch of the solar and sidereal days. The data is shown in Figure 1. While the peak is impressive (5 standard deviations), the statistics are limited, and the probability that the true distribution is flat is not negligible (7%).

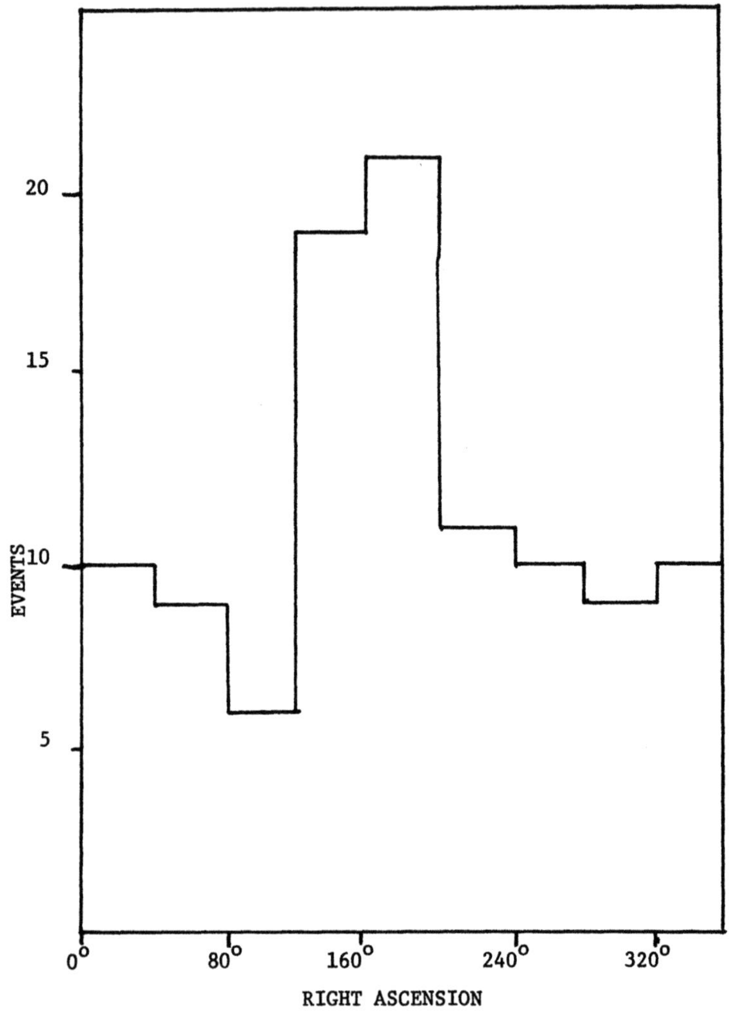

Figure 1 Muon multiplicity of 5 or more

An entirely unexpected observation has been an apparent correlation between time of arrival vs. direction of arrival for multiple muon events. Figure 2 shows the type of angle-time correlations which have been observed. Plotted are the cosines of the star-fixed angular separations between multiple muon events that are time-separated by less than 5000 seconds. A peak is evident for event separations of less than 8°. The smooth dotted curve is a background estimate obtained by selecting events that are separated by an additional period of one or more sidereal days. The enhancement corresponds to a 3 standard deviation effect, which is suggestive but not conclusive.

The implications of these observations are still being assessed. It appears, however, that further study of these effects could be very interesting. To that end, the multiple muon acceptance of the Soudan I detector is currently being expanded by the addition of a two-layer system of large proportional tubes adjacent to the existing detector. These tubes, provided by Purdue University and installed by a group from Tufts University, will extend the area by an approximate factor of three.

We acknowledge the cooperation of the Department of Natural Resources, State of Minnesota, especially the crew at Tower-Soudan State Park. The construction and operation of the Soudan 1 detector represents the efforts of many individuals including J. Blazey, H. Bristol, T. Copie, D. Feyma, J. Greenwalt, W. Heikinnen, S. Heppelmann, S. Heilig, N. Hill, M. Hirsch, H. Hogenkamp, D. Jankowski, C. James, X. Li, D. Logan, Z. Malloi, B. Neace, J. Osen, N. Pearson, J. Povlis, R. Taylor, D. Wahl and D. Wicks. This work was supported by the Department of Energy and by the Graduate School of the University of Minnesota.

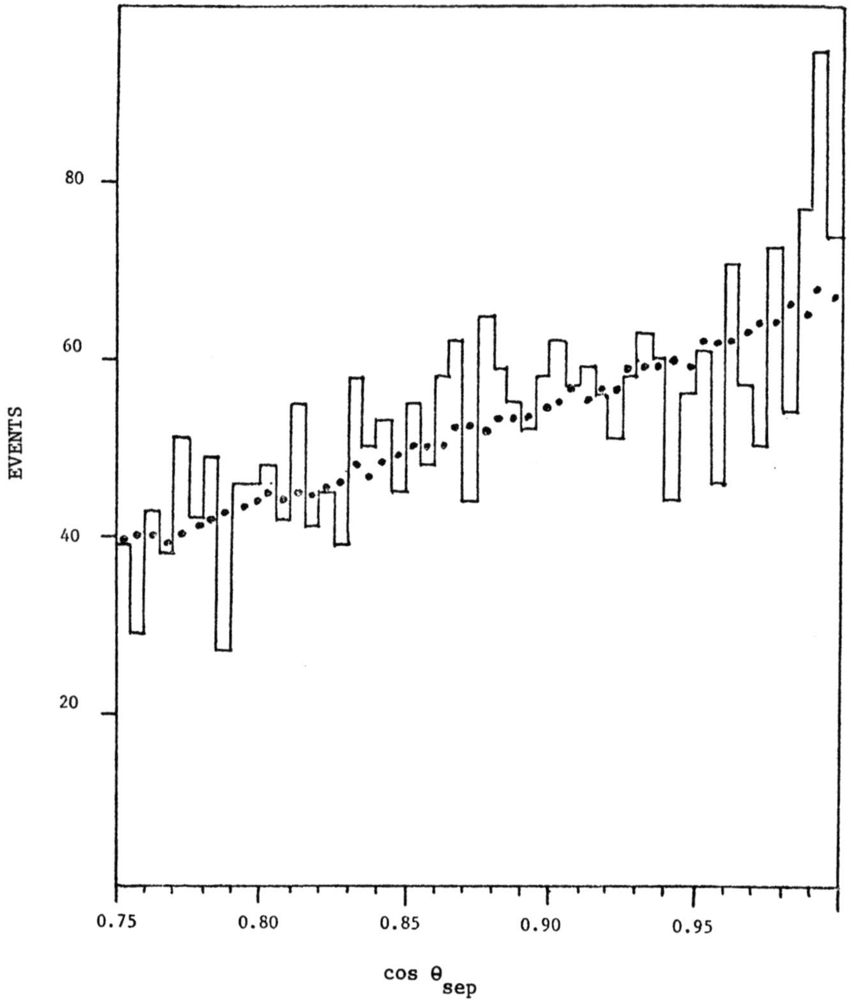

Figure 2 Separation angle distribution for multiple muon events arriving within 1-5000 seconds. The dotted line is a background estimate, normalized to the same number of events.

References

[1] The Soudan I collaboration consists of J. Bartelt, H. Courant, K. Heller, T. Joyce, M. Marshak, E. Peterson, K. Ruddick and M. Shupe, University of Minnesota, and D. Ayres, J. Dawson, T. Fields, and E. May, Argonne National Laboratory.

[2] J. Bartelt, et al., Phys. Rev. Lett. 50,651 (1983); 50,655 (1983).

[3] D. M. Ritson, SLAC-PUB-2950, July 1982, (unpublished).

[4] V. A. Rubakov, Zh. Eksp. Teor. Fiz. 33,658 (1981) [JETP Letters 33,644 (1981)]; C. G. Callan, Princeton University preprint LPTENS 82/20, unpublished.

[5] G. K. Ashley, et al., Phys. Rev. 12D, 20 (1975).

RESULTS FROM THE IMB DETECTOR*

THE IRVINE-MICHIGAN-BROOKHAVEN COLLABORATION

R. M. Bionta[2], G. Blewitt[4], C. B. Bratton[5], B. G. Cortez[2,a], S. Errede[2], G. W. Foster[2,a], W. Gajewski[1], M. Goldhaber[3], J. Greenberg[2], T. J. Haines[1], T. W. Jones[2,7], D. Kielczewska[1,b], W. R. Kropp[1], J. Learned[6], E. Lehmann[4], J. M. LoSecco[4], P. V. Ramana Murthy[1,2,c], H. S. Park[2], F. Reines[1], J. Schultz[1], E. Shumard[2], D. Sinclair[2], D. W. Smith[1,d], H. W. Sobel[1], J. L. Stone[2], L. R. Sulak[2], R. Svoboda[6], J. C. van der Velde[2], and C. Wuest[1].

(1) The University of California at Irvine
Irvine, California 92717

(2) The University of Michigan
Ann Arbor, Michigan 48109

(3) Brookhaven National Laboratory
Upton, New York 11973

(4) California Institute of Technology
Pasadena, California 91125

(5) Cleveland State University
Cleveland, Ohio 44115

(6) The University of Hawaii
Honolulu, Hawaii 96822

(7) University College
London, United Kingdom

ABSTRACT

Results are presented from the first 80 days of the IMB detector, with 69 observed interactions in a 3300 metric ton fiducial volume. The 69 events are consistent with atmospheric neutrinos. Limits are set at the 90% CL for the lifetime/branching ratio τ/B for $p \to e^+\pi^\circ$ at 6.5×10^{31} years and $p \to \mu^+K^\circ$ at 1.4×10^{31} years. For $n \to \bar{n}$, in oxygen, the lifetime is $\tau > 7 \times 10^{30}$ years. Limits on monopole catalysis of nucleon decay up to a flux of $10^{-14}/cm^2$ sec sr are described.

*Presented by B. Cortez

1. The Detector

The Irvine-Michigan-Brookhaven proton decay detector[1] is located 2000 feet underground in the Morton-Thiokol salt mine in Painesville, Ohio. It is a large rectangular tank of water, 22.8 x 17.8 x 16.9 m, with 2048 photomultiplier tubes (PMT's) on all six walls. The fiducial volume of 3300 metric tons or 2.0×10^{33} nucleons begins 2.0 meters in from the tube planes.

Charged particle tracks in the water give off Cherenkov light above the threshold $\beta = 0.75$. A track that goes towards a wall and exits will leave a filled in circle of lit PMT's, while a stopping track will light up a ring. By utilizing the relative timing of the lit PMT's and their geometric pattern, the position of single tracks can be reconstructed to accuracies of better than 1 m. For two track nucleon decay events with opening angles greater than 90° (so that the Cherenkov cones do not overlap), the angle can be reconstructed to within ~ 15°, and the vertex to 40 cm.

The PMT's, five inch diameter hemispherical tubes, and their associated electronics are sensitive to light at the single photon level. The discriminator is set at ~ 0.2 photoelectron (pe) level, giving an average dark noise of 2.5 kHz. For the purposes of triggering, the detector is divided into 32 squares of 64 tubes called patches. A coincidence of three tubes in a patch in 50 ns will fire the patch. A coincidence of any two patches within 150 ns (the time of flight for light across the detector) will trigger the detector, a minimum threshold of 6 tubes or ~ 25 MeV. An independent trigger is made whenever 12 tubes anyplace in the detector fire within 50 ns, setting an unbiased threshold of 50 MeV. The measure of energy is E_C, the visible Cherenkov light yield of a massless, nonshowering particle of that total energy. For an electron, photon, or π°, $E_C \approx E_{TOT}$. For a muon, $E_C \approx E_{TOT} - 250$ MeV, due to the muon rest mass and the energy lost below Cherenkov threshold.

Three 9 bit numbers are recorded for each PMT--T1, T2, and Q. T1 is the time the tube fired relative to the trigger with 1 ns least count. (PMT resolution is 5.5 ns HWHM at low light levels.) T2 is the time that a PMT fired up to 7.5 μsec after the main trigger with 15 ns least count. The T2 scale is used to look for muon decay electrons from particles that stopped in the detector. Q is the integrated pulse height from the PMT, which is proportional to the light incident on the PMT.

There are 2.7 triggers per second due to cosmic ray muons passing through the water. The number of accidental triggers is negligible. All events in the energy range of interest to proton decay (0.05-2 GeV) are written on tape for offline analysis.

2. Calibrations

The raw T1, T2, and Q values are transformed to times and energy using the calibration procedure. This is based on a pulsed nitrogen laser sending 337 nm light into an optical fiber which ends in a diffusing ball in the center of the pool. This forms an isotropic light source whose intensity can be varied over five orders of magnitude by placing neutral density filters in front of the beam under computer control. The time response and corrections as a function of pulse height can be obtained by varying the time of firing of the laser with respect to the trigger and changing the light level.

The pulse height response is linearized in terms of the laser light level. There is an arbitrary constant, the same for all PMT's, that must be derived from cosmic ray events to convert light level to MeV of Cherenkov equivalent light. This is done by taking a sample of muons that go straight through the detector, 18.5 m in path length. Since the mean muon energy is $\sim$ 300 GeV, corrections must be made for electromagnetic processes such as knockons, pair production, etc. which increase the Cherenkov light output relative to a single minimum ionizing track, which would deposit 3.7 GeV. Averaging over the muon energy spectrum, we estimate that the peak is increased by 22%, with a tail at the high energy end. This and other geometric corrections give 4.6 GeV per track, or $\sim$ 3.8 MeV per photoelectron, to be compared to the expected energy deposition of 938 MeV from $p \rightarrow e^+\pi^\circ$. We estimate a systematic error of 15% in our absolute energy determination at 1 GeV.

We have written event simulation programs in order to evaluate our efficiency for detecting various proton decay modes. Track generation in the homogeneous water medium is straightforward. The SLAC EGS program is used to generate electromagnetic showers. Muons are propagated using measured dE/dx values. Pion interactions, including absorption, charge exchange and elastic and inelastic scattering cross sections for both oxygen and free protons are included. Light generation comes from the Cherenkov formula. Each photon is followed through the water with allowances for absorption or Rayleigh scattering. The attenuation length is in excess of 40 meters around 400 nm. Measurements are shown in Figure 1. Photons striking a PMT have ~ 15% probability of producing measured photoelectrons.

The PMT efficiency is normalized by comparing generated straight through muons to the data. Figure 2 shows that probability of a PMT firing as a function of the distance the photon travelled from the track to the tube. The agreement over all distances is good. Figure 3 shows the distribution of photoelectrons for the data and simulation. The tails, due to fluctuations in electromagnetic processes agree remarkably well.

We make the assumption that we can properly simulate events at 1 GeV and below using a program which agrees with the data at 4 GeV. We then determine the proper cuts in numbers of lit PMT's and pe's for different decay modes and their ultimate efficiency to pass through the analysis.

3. Data Reduction

There are three independent analysis programs to find events originating in the detector. These could be atmospheric neutrino interactions, possible proton decay candidates, or other new physics. The details of one analysis will be presented here, but the conclusions from the other analyses are the same.

There are 235,000/triggers per day, predominantly straight through cosmic ray muons. The first cut is to require 30-300 lit PMT's, reducing the data by a factor of 3. This range includes most proton decay modes and even 2 GeV $n\bar{n}$ annihilations. $p \rightarrow e^{+}\pi^{\circ}$ is expected to light up 170 ± 20 PMT's, well within these cuts.

The second step is the point fit. We find the point that minimizes the χ^2 for the hypothesis that the lit tubes come from a single point source of light. By requiring that this point lies within the fiducial volume, we eliminate a factor of 100 on entering events while saving 80% of neutrino interactions originating in the fiducial volume. Over 90% of multitrack events, e.g., proton decay events, will be saved. The main advantages of the point fit are its speed and efficiency.

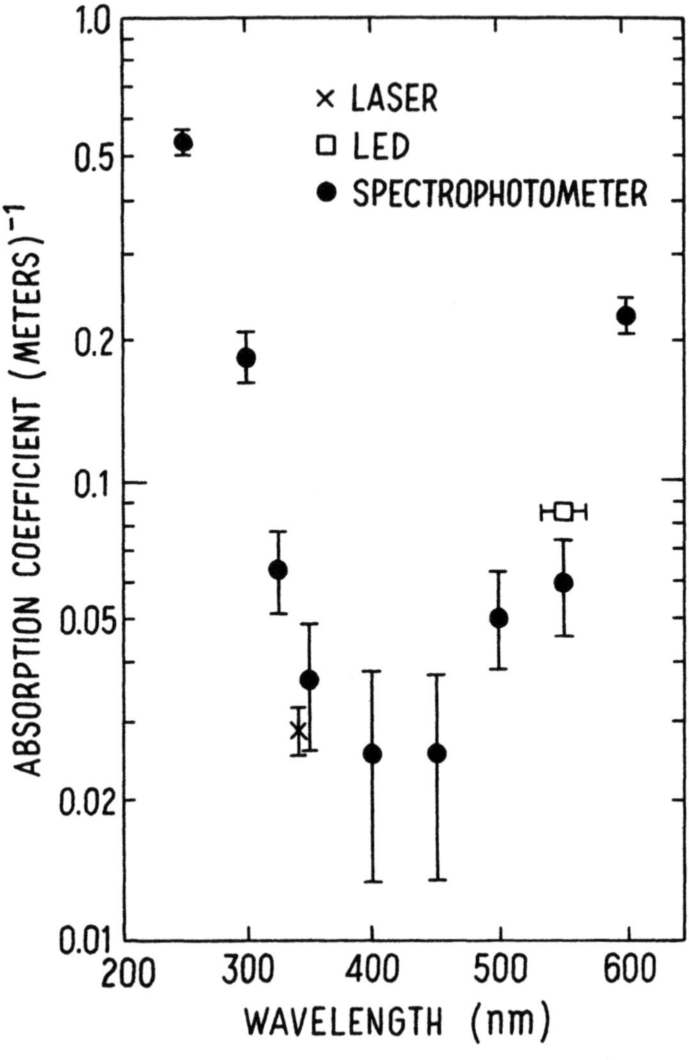

Fig. 1. Attenuation measurements of the water vs wavelength. The points made in the tank with laser and led light are also shown.

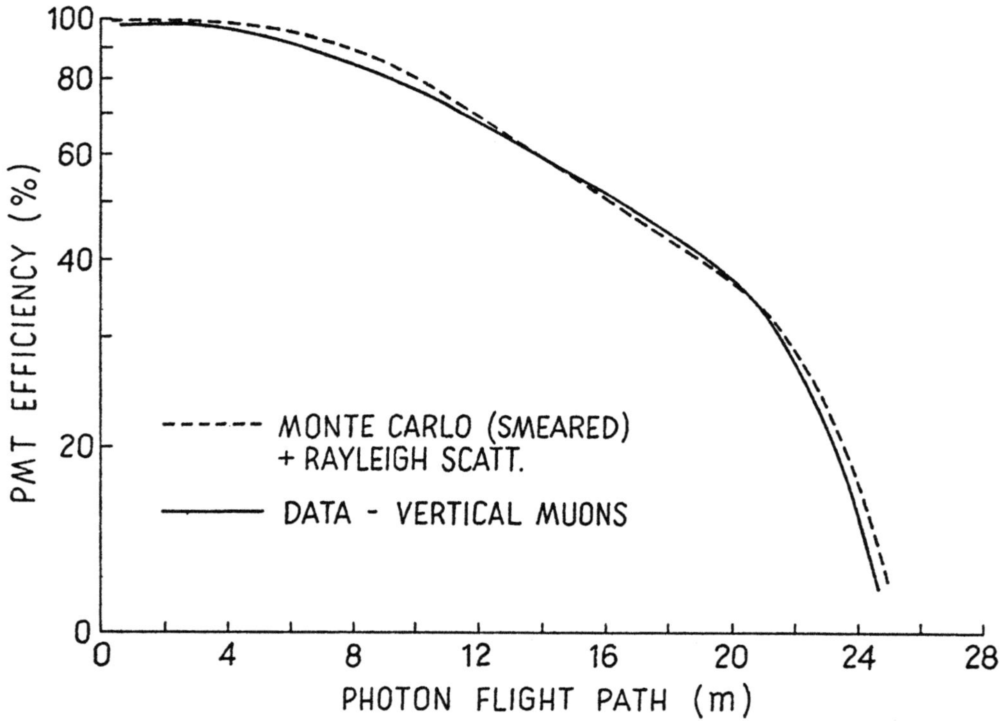

Fig. 2. Occupancy (fraction of events that a PMT fires) as a function of photon flight path from a cosmic ray muon track. The event simulation program is normalized to this curve.

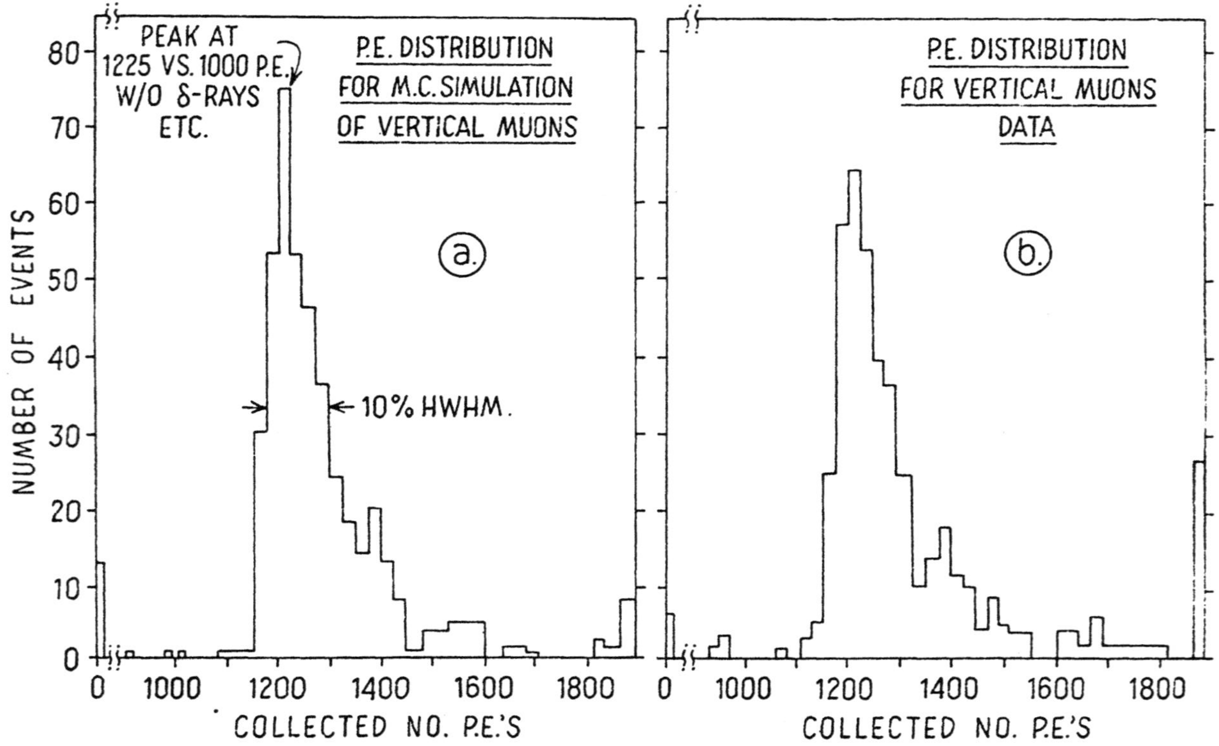

Fig. 3. The distribution of photoelectrons from straight through muon events for both real and simulated events. The peaks are forced to coincide to determine the energy calibration of 3.8 MeV per photoelectron.

The remaining events are primarily short stopping muons and longer tracks that clip the corners. At this stage we employ a more sophisticated program that uses the Cherenkov angle of 41° to find the vertex. First, a guess for the vertex is made by using the earliest tubes to fire. The event is rejected if the χ^2 for this hypothesis is good enough. These tubes typically light up near the entry point of a charged particle into the detector due to direct or scattered light. Then the vertex is allowed to vary throughout the detector, giving ~ 3 events per day with a best fit in or near the fiducial volume, with 40 or more lit tubes. Scanning with a color graphics system reduces this to ~ 1 event per day. It eliminates failures of the fitting program which are primarily top entering and downward going tracks for events with small numbers of PMT's. A neutrino interaction near the top at the edge of the fiducial volume and going downward could be thrown out by such scanning. This is included in the efficiency, averaged over ν types and energies, of 75%.

4. Sample of Events Originating in the Fiducial Volume

There are 69 events in 80 days that remain. According to the recent neutrino flux calculation by Gaisser, et al.,[(2)] and using our detection efficiency we would expect 95 ± 30 events in this period. The older calculation by E. C. M. Young[(3)] predicts 40% fewer events. The data is consistent with both estimates.

To arrive at these estimates, we use the events found by the Gargamelle collaboration in a freon bubble chamber[(4)]. This gives the relative weighting of quasielastic, single pion and multipion production in neutrino interactions in nucleii. In this way the large angle pions that could mimic proton decay are properly put in. The Gargamelle events are weighted to the atmospheric neutrino energy spectrum, and electron neutrino events are generated by changing the observed μ to an electron of the same energy.

The characteristics of these events are as expected from ν interactions. Figure 4 shows the energy distribution. The variable plotted is E_{min}, which is the visible Cherenkov energy corrected by particle type. We add 250 MeV to the visible light yield for events with a muon decay, due to the muon rest mass and kinetic energy lost below the Cherenkov threshold.

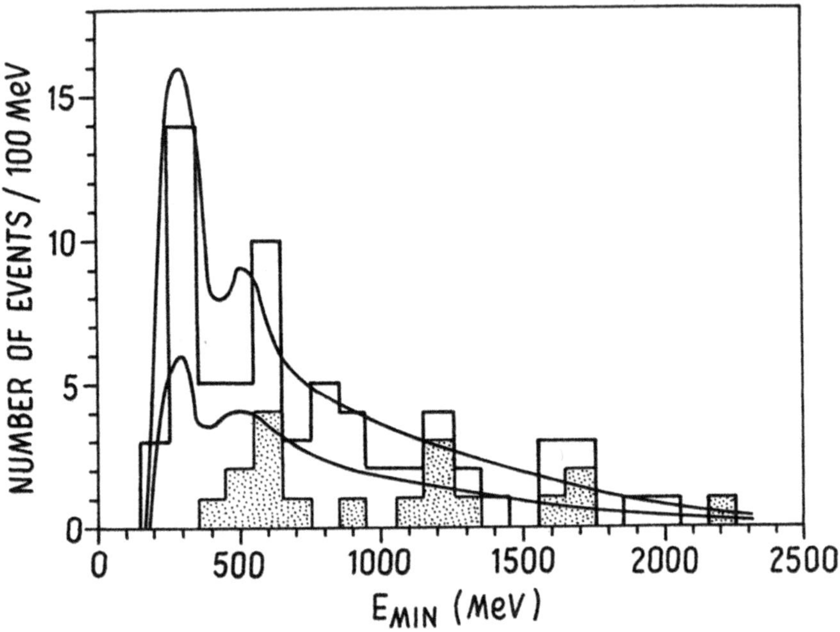

Fig. 4. Distribution of neutrino energy E_{min} for 69 contained vertex events. 250 MeV is added to events with a visible μ decay to account for the loss of Cherenkov light due to the μ mass. The curve shows the range of the expected distribution from E. C. M. Young.

A muon decay electron is identified by a coincidence of 5 or more PMT's in a 60 ns window up to 7.5 μsec after the main event. We have a 62 ± 4% measured efficiency for μ^+ decay detection from a selected sample of stopping tracks. 18 of the 69 events have a muon decay. Although the ratio of ν_μ/ν_e is ~ 2 /1, the ratio of μ events/e events is ~ 1/1 due to the lower neutrino energy threshold for electron events. We expect ~ 30% of the neutrino events to have an identified μ decay, compared to 30 ± 8% observed.

Neutrino events should occur uniformly in the detector and be reasonably isotropic in direction. Figure 5 shows a side view of the detector with each dot representing the vertex of the event and each arrow the unit vector direction projected onto the view. There is no obvious up/down or top/bottom asymmetry. Figure 6 are histograms of the x, y, and z position of the vertex. Figure 7 are histograms of $\cos\Theta_x$, $\cos\Theta_y$ and $\cos\Theta_z$ of the direction of the main track in the event. Superimposed on the $\cos\Theta_z$ distribution is the expected distribution based on the neutrino flux calculations of Gaisser, et al. The calculation includes the effect of the earth's magnetic field to arrive at an angular distribution. There are not enough statistics yet to distinguish between a flat distribution or this distribution.

5. Limits on $p \to e^+\pi^\circ$

The minimal SU(5) theory predicts $\tau/B = 4 \times 10^{29\pm1.7}$ years for $p \to e^+\pi^\circ$(5). At the upper limit, we would expect 7 events in this sample.

To look for $e+\pi^\circ$, we search for events with two tracks, each with more than 40 tubes and an opening angle greater than 100°. Three events remain, described in Table I. The third has too much energy to be a proton decay. The second has a small opening angle and too much energy. It cannot be $e^+\pi^\circ$ because there is a visible μ decay. The first event cannot be $e^+\pi^\circ$ for the same reasons. Figure 8 shows the opening angle vs energy sharing between the two tracks. The dashed line indicates the region where $e^+\pi^\circ$ events lie, when the effects of Fermi motion of protons in oxygen and detector resolution are considered. The number of expected neutrino interaction in ten years are shown as dots. The first event is near the $e^+\pi^\circ$ region, but the visible μ decay with 15 PMT's at 1.2 μsec makes it impossible to be $e^+\pi^\circ$.

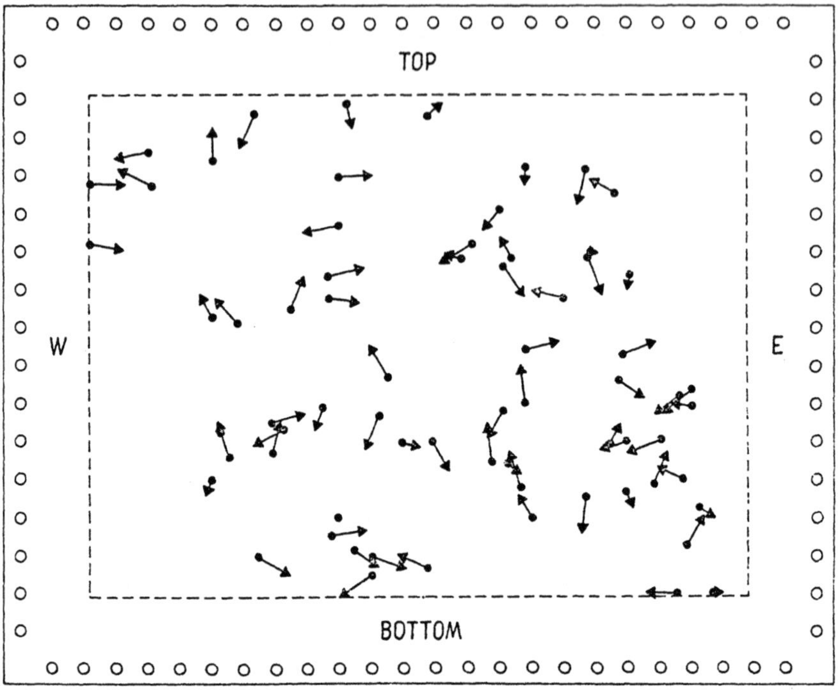

Fig. 5a. Unit vector projection of the 69 events viewed from the side of the detector.

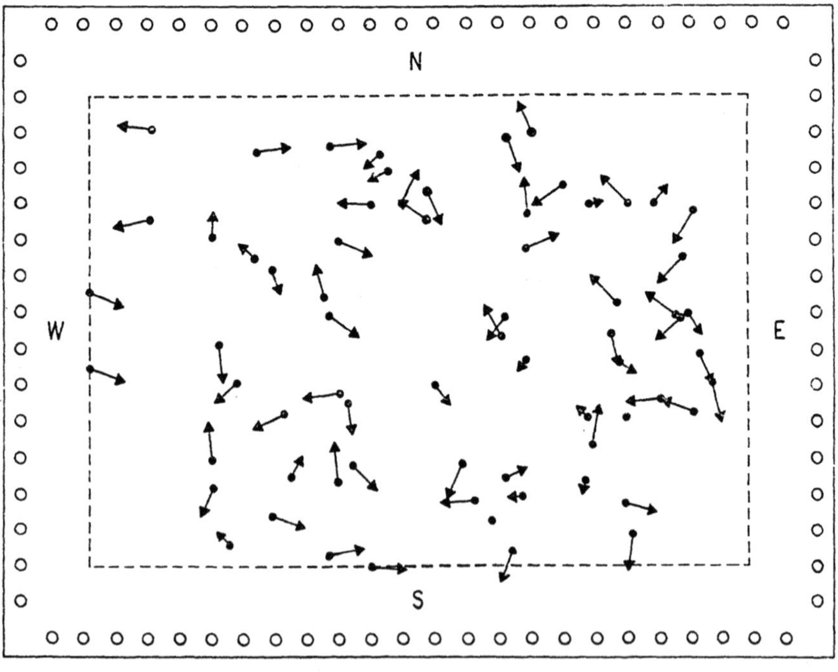

Fig. 5b. Same as (a) but viewed from the top.

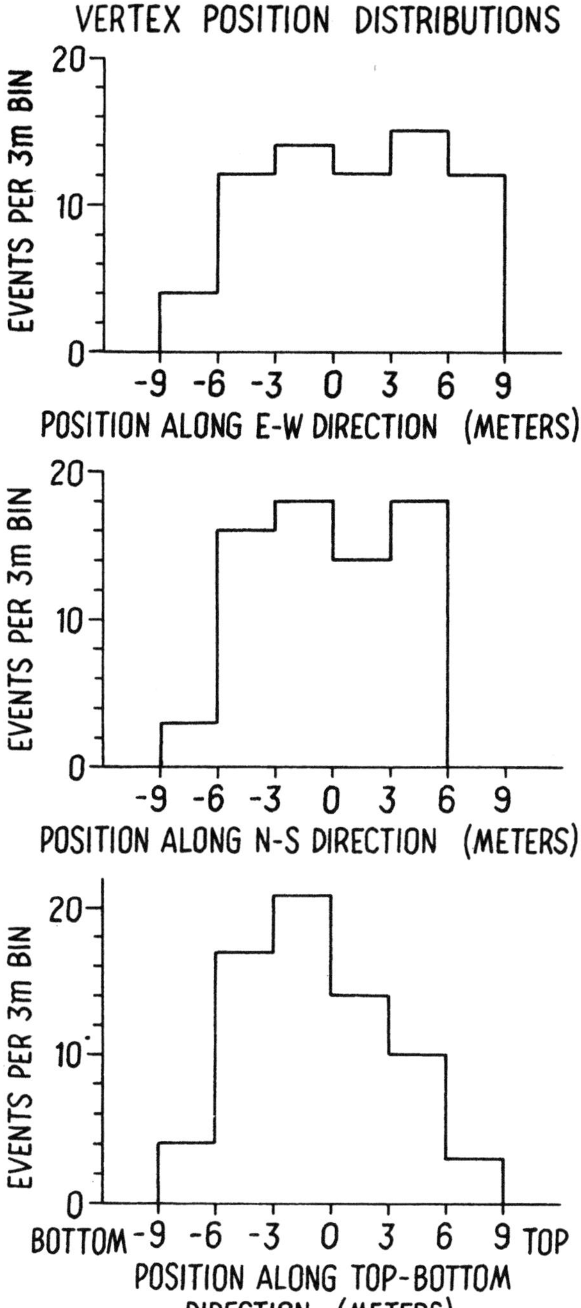

Fig. 6. Histograms of the x, y, and z vertex position of the 69 events.

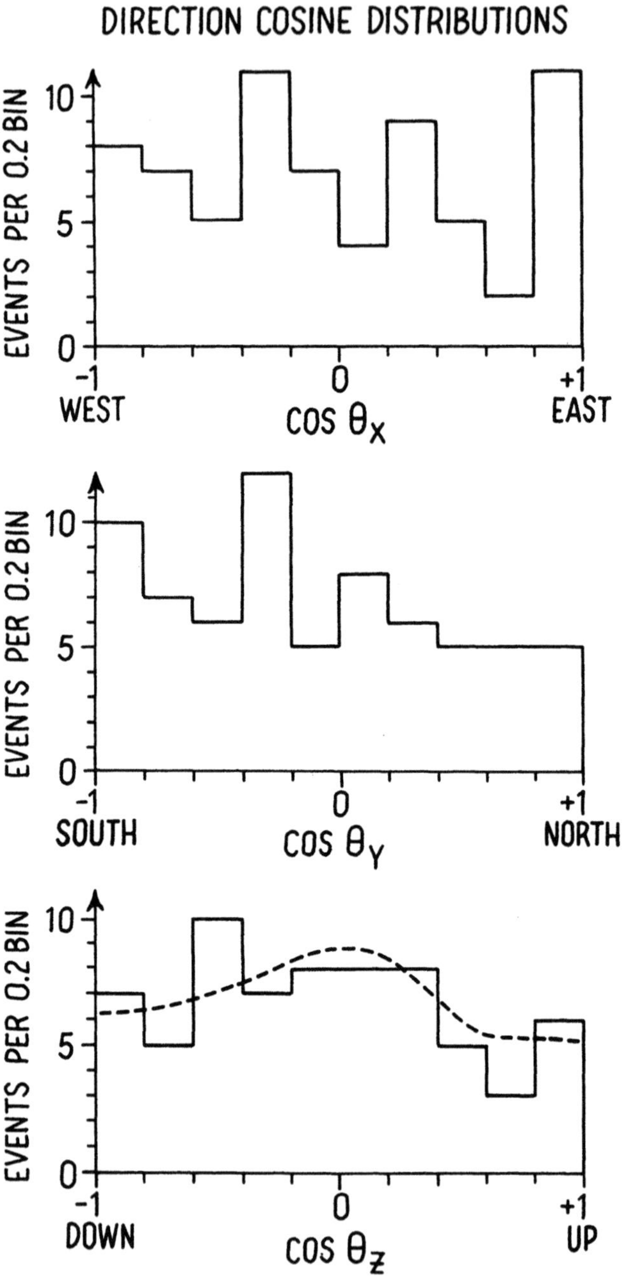

Fig. 7.

Distribution in $\cos\theta_x$, $\cos\theta_y$, $\cos\theta_z$ for the primary track direction in the 69 events. Superimposed on $\cos\theta_z$ is a prediction from Gaisser, et al. normalized to the data.

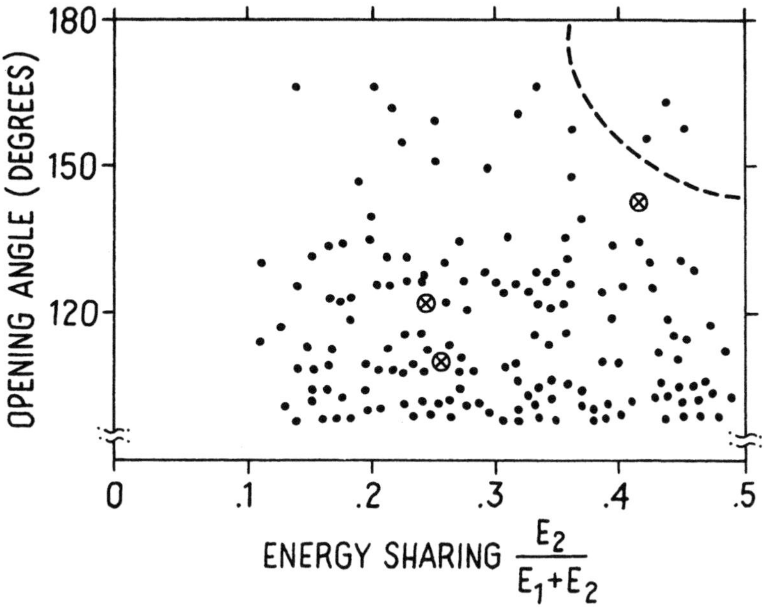

Fig. 8. Scatterplot of opening angle vs energy sharing for two track events. The 3 observed events are shown on a distribution expected from 10 years of data. The dashed region is for $p \to e^+\pi^\circ$ events.

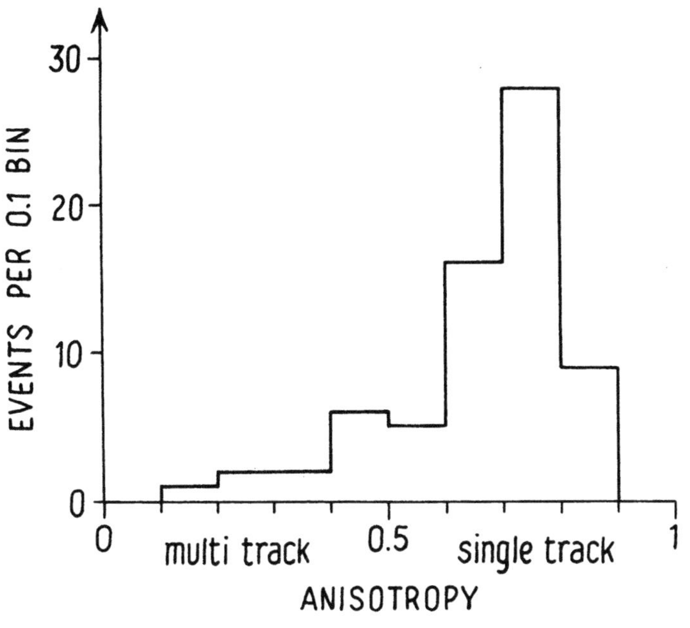

Fig. 9. Distribution of the anisotropy A for the 69 events.

TABLE I. WIDE ANGLE TWO-TRACK EVENTS

EVENT NO.	NO. OF PMT	E_C CHERENKOV ENERGY (MeV)[a]	E_1 (MeV)[a]	E_2 (MeV)[a]	θ_{12} (Deg)	OBS. MUON DECAYS
151-35037	188	1000±100	600±90	400±60	135±7	1
225-07794	166	950±100	700±130	250±80	125±25	1
388-19376	340	1700±170	1050±170	650±100	115±15	0

[a]Does not contain additional possible systematic errors of 15%. E_C is calculated assuming the particles are showering. If one particle is a charged pion or muon, as must be the case in the first two events in the table, 250 MeV must be added to the energy of one track and to the total energy listed.

We can set the following limits with no observed events. First consider free protons. There is no Fermi motion or nuclear effects for the π°. We set the limit at the 90% C.L.

$$\tau/B > (1/2.3) \times (80/365) \times 2\times10^{33} \times (2/18) \times 0.9 = 1.9 \times 10^{31} \text{ years}$$

where 80/365 is the livetime in years, 2/18 is the fraction of free protons and the overall detection efficiency is 90%. In oxygen, our intranuclear cascade program calculates that 40% of the π°'s are lost outside of our cuts. This gives the limit

$$\tau/B > (1/2.3) \times (80/365)\ 2\times10^{33} \times (10/18) \times 0.9 \times 0.68 = 6.5 \times 10^{31} \text{ years}$$

where 0.68 represents the probability for π°'s to survive nuclear interactions averaged over all protons in water.

The KGF experiment reports one possible event of the type $p \to e^+\pi^\circ$.(6) With an order of magnitude more exposure, and no events seen, their interpretation may be in doubt.

6. Limits on $p \to \mu^+ K^\circ$

We have two different ways to look for $p \to \mu^+ K^\circ$. The first is based on the decay $K_S^\circ \to \pi^\circ \pi^\circ$ which gives off 4 γ's and isotropic light. The second method will make use of the decay $K_S^\circ \to \pi^+ \pi^-$ and the eventual decay $\pi^+ \to \mu^+$. This gives two muon decays, a distinctive signal.

6.1 Method 1: $p \to \mu^+ K^\circ$, $K^\circ \to \pi^\circ \pi^\circ \to \gamma\gamma\gamma\gamma$.

The light from the tracks overlap and it is difficult to separate them. Instead, we construct a measure of the isotropy of the event by calculating the norm of the average of all the unit vectors from the vertex to each lit PMT. If this quantity A (the anisotropy) is near 0, then the event is isotropic. For a single track event $A \sim 0.7$. Figure 9 is a histogram of A for the 69 events.

The second requirement is that the visible Cherenkov energy, E_C, lie in the range $550 < E_C < 850$. We do not require the presence of a μ decay. Figure 10 is a scatterplot of E_C vs A for the 69 events. The dashed region is the expected area for this decay mode. Two events fall inside the region. The first can be ruled out because there is 600 MeV in one track, which is not possible in this decay. The second event cannot be simply ruled out. We will consider it as a possible candidate in setting limits. The expected background from ν interactions is 1 ± 0.5 events. Without a background subtraction, the limit is

$$\tau/B > (1/3.9) \times (80/\ 3.65) \times 2\times10^{33} \times (10/18) \times 0.16 \times 0.9 =$$
$$8.8 \times 10^{30} \text{ years at the 90\% C.L.}$$

where 0.16 is the branching ratio $K^\circ \to \pi^\circ \pi^\circ$ (ignoring all $K^\circ{}_L$ interactions) and 0.9 is the detection efficiency. There is no nuclear correction since all K°'s escape the oxygen nucleus unscathed.

6.2 Method 2: $p \to \mu^+ K^\circ$, $K^\circ \to \pi^+ \pi^-$

We require that there be two visible μ decays. To increase the detection efficiency, since the mean light level is only 45 PMTs for $p \to \mu^+ K^\circ$, $K^\circ \to \pi^+ \pi^-$, we include events down to 30 PMTs. There is 65% efficiency for detecting the event and 75% efficiency for the π^+ to stop and decay to a μ. This gives an overall detection efficiency of $0.75 \times 0.65 \times (0.62)^2 = 0.19$. One event has two μ decays. However, it has only one clean visible track with 550 MeV total energy. The second μ decay probably comes from a π^+ below Cherenkov threshold, produced by

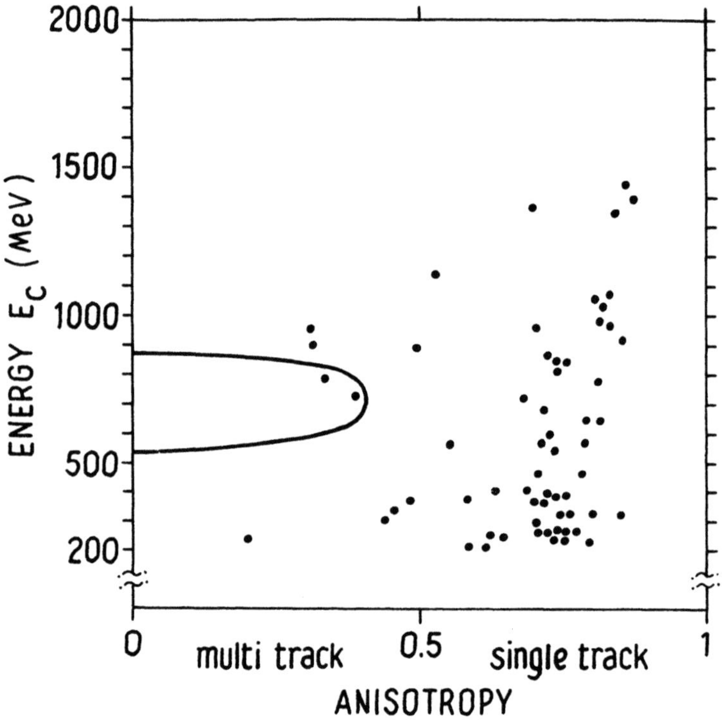

Fig. 10. Scatterplot of Cherenkov energy vs anisotropy for the 69 events. The dashed region is for $p \rightarrow \mu^+ K^\circ$, $K^\circ \rightarrow \pi^\circ \pi^\circ$ events.

a ν interaction. No simulated $p \to \mu^+\pi^+\pi^-$ events out of 250 give only one clean track of this energy. With no signal the limit is

$$\tau/B > (1/2.3) \times (80/365) \times 2\times10^{33} \times (10/18) \times 0.19 \times 0.35 = 7\times10^{30} \text{ years}$$

using the branching ratio of 0.35 for $K^\circ \to \pi^+\pi^-$.

Combining the two independent limits for $p \to \mu^+K^\circ$, we arrive at the final answer $\tau/B > 1.4 \times 10^{31}$ years at the 90% C.L.

The Mont Blanc Collaboration reports a possible μ^+K° event(7). With a factor of six more sensitivity we cannot confirm this interpretation.

7. Neutron-antineutron Annihilation

We also search for $\Delta B = 2$ processes. If a neutron turns into an antineutron, it will immediately annihilate in the nucleus and give off 2 GeV in the form of pions. Many will be π^+, which can stop and decay to μ^+ to e^+. Two μ decays along with an isotropic event will be clean signal for $n \to \bar{n}$.

No events are observed. The calculation of the detection efficiency can be broken down into three steps.

7.1 The initial pion distribution from the annihilation is generated. Both $\bar{n}n$ and $\bar{n}p$ are possible in oxygen. We use final states measured from $\bar{p}p$ and $\bar{p}n$ interactions. The mean pion multiplicity is 4.8. The fraction of events with two or more π^+ is 64%.

7.2 The pions must escape the nucleus. They can be absorbed, charge exchange, or inelastically scatter. Our model shows the pion multiplicity is reduced by 25% and the fraction with two or more π^+ is 40%.

7.3 The π^+ must travel through the water and come to rest, and decay to μ^+. This reduces the fraction of events with two or more μ^+ to 23%.

The limit is

$$\tau > (1/2.3) \times 2\times10^{33} \times (8/18) \times (0.23) \times (0.62)^2 \times 0.9 = 7 \times 10^{30} \text{ years}$$

where 0.62 is the μ-e decay detection efficiency and 0.9 is the efficiency to pass through the analysis.

Using the calculations for the conversion from nuclear lifetimes to a free neutron lifetime[8], this gives $\tau_{free} > \sim 3 \times 10^7$ sec.

8. Monopole Catalysis of Nucleon Decay

It has been suggested that monopoles in grand unification theories could have cross sections in the millibarn range to catalyze proton decay.[9,10] If the cross section is large enough then several proton decays can be catalyzed by a monopole traversing our detector. Our electronics is sensitive to multiple interactions occurring up to 7.5μsec after the event.

We search through the data for events with 50 or more PMTs in a 300 ns coincidence in the T2 time scale.[11] There are 4.6 ± 0.3 events per day passing these cuts. We expect 4.7 per day from random coincidences of two unrelated cosmic rays. None of these events looks anything like a proton decay. At the 90% C.L., we have the flux limits shown in Figure 11 for different catalysis cross sections and monopole velocities. For the moderate β expected for astrophysical monopoles of $\sim 10^{-3}$ and cross sections above 1 mb, the limit is $F_m < 10^{-14}/\text{cm}^2$ sec sr, for 100 days of detector livetime. This is only one order of magnitude above the Parker bound.

If the monopole is moving slowly, or the catalysis cross section is small only one interaction would be observed. All 69 events originating in the fiducial volume are candidates for monopole catalysis, if we do not make assumptions on the decay modes. This gives a limit on the product of flux and cross section as shown in Figure 12 with a minimum limit at $F \times \sigma < 10^{-40}$ sec sr.

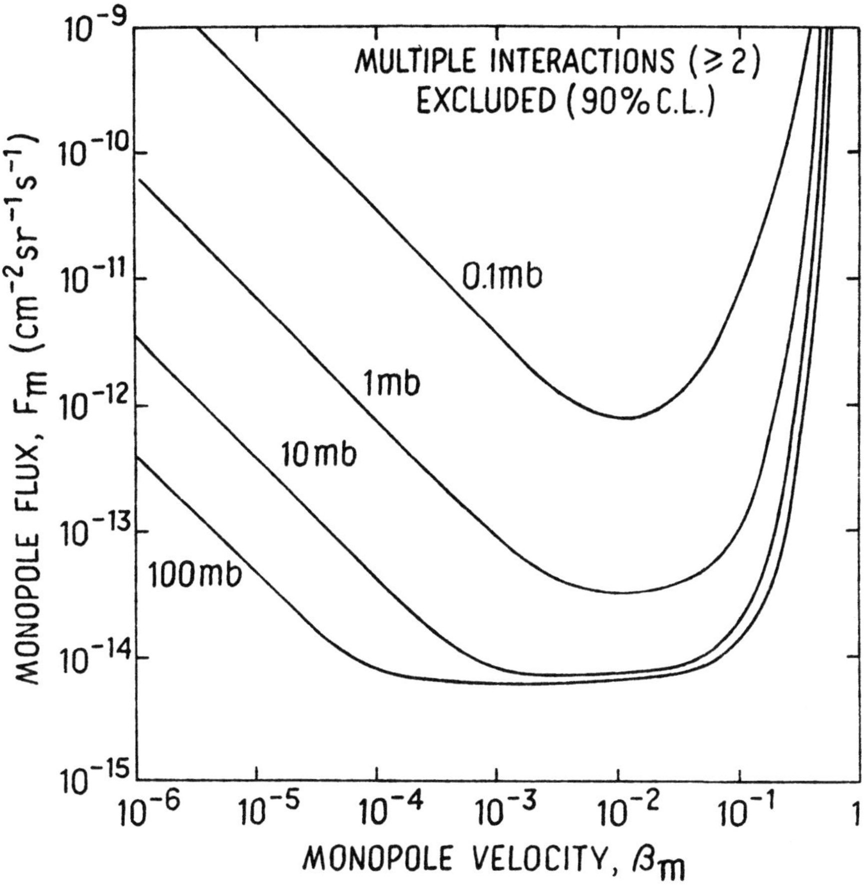

Fig. 11. Limits on monopole flux as a function of catalysis cross section and monopole velocity for multiple interaction.

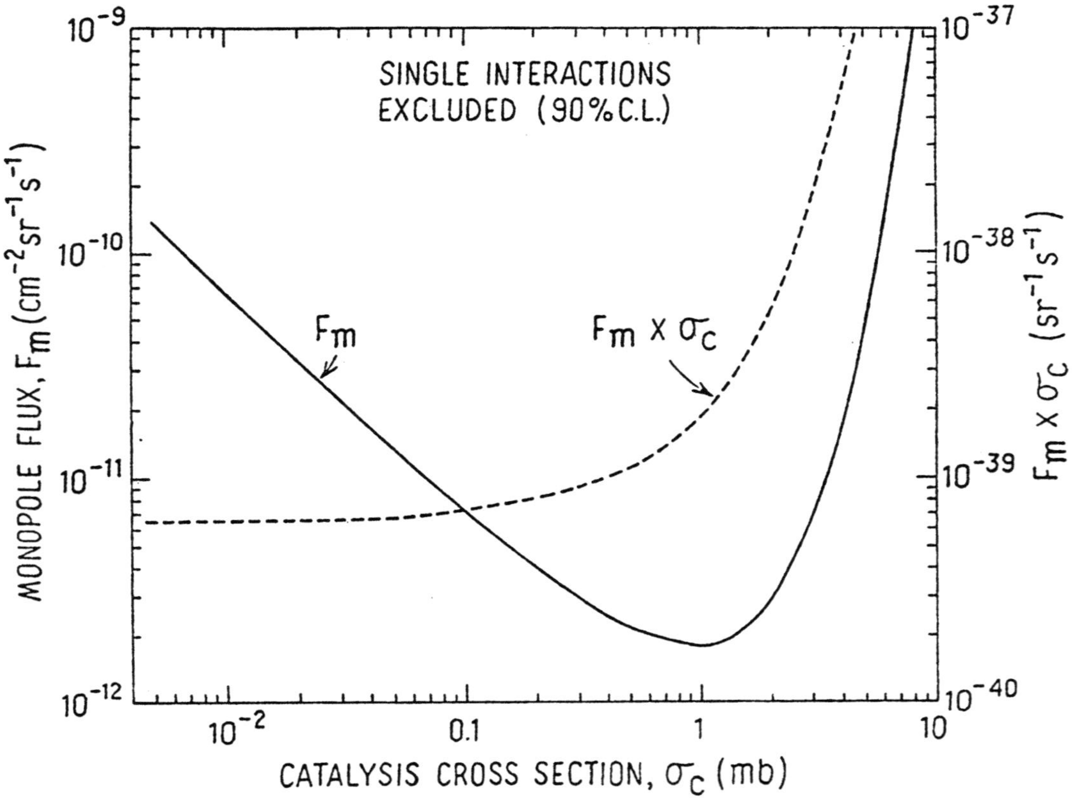

Fig. 12. Limits on monopole flux times cross section from single interactions as a function of cross section.

9. Conclusion

There is no evidence for proton decay, $\bar{n}n$, or monopole catalysis. The current limits are,

$\tau/\beta > 6.5 \times 10^{31}$ years, $p \to e^{+}\pi^{\circ}$

$\tau/\beta > 1.4 \times 10^{31}$ years, $p \to \mu^{+}K^{\circ}$

$\tau > 7 \times 10^{30}$ years, $n \to \bar{n}$ in oxygen nucleus

$F_m < 10^{-14}/cm^2$ sec sr, monopole catalysis for $\sigma > 1mb$, $\beta \sim 10^{-3}$.

Acknowledgements

We would like to thank the many people who made this detector a reality and especially the employees of Morton-Thiokol who operate the Fairport mine. This work is supported in part by the U.S. Department of Energy.

REFERENCES AND FOOTNOTES

a. Also at Harvard University;

Permanent Addresses:

b. Warsaw University, Poland

c. Tata Institute of Fundamental Research, Bombay, India

d. Now at University of California, Riverside, California 92521

[1]R. Bionta et al. PRL (1983, to be published), R. Bionta, et al., Proceedings of the 1982 Moriond Conf., Tran Than Van, ed., (1982).

[2]T. Gaisser, et al. in these proceedings.

[3]J. L. Osborne and E. C. M. Young in "Cosmic Rays at Ground Level" (A. Wolfendale, ed.) 1973.

[4]H. Deden, et al., Nuclear Physics B85, 269 (1975)

[5]P. Langacker, Phys. Rev. 72, 185 (1981) and proceedings of the 1982 Workshop on Proton Decay, ANL-HEP-82-24 (D. S. Ayres, ed.) p. 64; M. A. B. Beg and A. Sirlin, Phys. Rep. 88, 1 (1982). N. Isgur and M. B. Wise, Phys. Lett. 117B, 179(1982); W. J. Marciano BNL 31036, presented at Orbis Scientiae 1982, and private communication.

[6]M. R. Krishnaswamy, et al., Phys. Lett. 115B, 4, p. 349 (1982).

[7]G. Battistoni, Phys. Lett. 118B, p. 461 (1982).

[8]C. B. Dover et al., BNL 32097, V. A. Kuzmin et al., Sov. Phys. JETP 12, 228 (1970), W. M. Alberico et al., Phys. Lett. 114B, 226 (1982).

[9]C. G. Callan, Jr., Phys. Rev. D26, 2058 (1982); Phys. Rev. D25, 2141 (1982); and in "Magnetic Monopoles," ed. by R. A. Carrigan and W. P. Trower (Plenum, New York 1983). To be published.

[10]V. B. Rubakov, Nucl. Phys. B203, 311 (1982); JETP Lett. 33, 644 (1981).

[11]S. Errede, et al., PRL (1983, to be published), S. Errede, et al., Proceedings of the 1983 Moriond Conference, Tran Than Van, ed. (1983).

REVIEW OF FUTURE NUCLEON DECAY EXPERIMENTS

A.L. Grant
CERN, Geneva, Switzerland

ABSTRACT

At a time when the first results are becoming available from the IMB Water Chenenkov, the Kolar Gold Field and the Mont Blanc tracking calorimeters, it is appropriate to review the status and expected performance of the whole range of future nucleon decay experiments; from those under construction, Komioka and Fréjus, to the future dreams of kilotons of liquid argon.

1. INTRODUCTION

At the time when the first round of nucleon decay experiments is beginning to produce results on nucleon decay, it is appropriate to look critically at the status and expected performance of the whole range of second round experiments. The detectors discussed cover those just about to start taking data; Kamioka [1] and those under construction at Fréjus [2] and Soudan [3]. Proposals exist for conventional iron calorimeters for Gran Sasso [4,5] and the Kolar Gold Field, and for fully active liquid scintillator detectors at Homestake [6] and in the "Penn design" [7]. The far future belongs to the extrapolation of the tests done with a few kilos of argon, liquid [8,9] or gas [10], to multi-kilo ton detectors.

The common feature of all these new experiments is their initial large size, greater than 1000 t fiducal volume, and with the exception of the water Cherenkov detector, a modular construction which would allow a further increase in mass.

A second feature of the experiments is the high level of technical innovation in the proposals; from the massive 20" diameter photomultipliers used in the Kamioka experiment to the new concept of resistive plastics in the construction of drift tubes in the Soudan2 Proposal. It is already evident, for example, from the impact of the design of the limited streamer tubes [11] built for the NUSEX detector in the Mont Blanc tunnel on many of the calorimeters in the LEP experiments, that the design of proton decay experiments is at the forefront of detector technology.

The sheer size of these "multi-kiloton" experiments causes new problems, often underestimated, in the scale of production of the components and eventual installation underground. Fortunately, experience from the earlier experiments in the difficulties involved in working underground is now reflected in the modular construction of self-contained units which are built up into the final detector.

In the analysis of the experimental data from the calorimeter detectors, many old, and now rediscovered techniques are being developed, e.g. the use of multiple scattering to determine the direction of tracks. However, the competition of the very massive water Cherenkov detectors is still very strong. Also here the development of new ideas in analysis can give information on decays which at first sight would be undetectable. It is clear that to be competitive the calorimeter experiments must rely on very fine tracking to give a good picture of the possible decay candidates. In addition, they must have a redundancy to detect and differentiate from neutrino induced background the predicted supersymmetric decay modes such as $p \rightarrow K^+\nu$.

2. STATUS OF THE EXPERIMENTS

A brief description is given of the various experiments, either under construction; proposals or future plans. Emphasis is placed on new technical developments in hardware or analysis in each, particularly with reference to their ability to detect the difficult decay modes predicted by supersymmetric grand unified theories. The ordering of the experiments is not random, it reflects the expectations of the groups as to the experimental time scale.

2.1 Kamioka Water Cherenkov

The Japanese water Chenenkov detector [1], situated in the Kamioka mine 300 km west of Tokyo, is now almost complete and ready to be filled with water for the first time.

The site of the experiment is an active lead and zinc mine with good rail access and a mountain rock cover of 2400 to 2700 m.w.e. It is not as deep as the European experiments in their alpine road tunnels, but somewhat better than most North American sites. A schematic drawing of the specially constructed cavity and access tunnels is shown in fig. 1. The steel water tank of 3420 m^3 and the 1050 20" diameter photo tubes are installed ready for use.

These very large photo tubes, the result of a fruitful collaboration of the Japanese groups and the Hamamatsu Corp., qualitatively change the prospects of the detector by allowing the reliable detection of the weak signals from charged hadrons in proton decay. The system will be triggered on the total energy of the event summed over all pulses. There is no attempt to provide individual timing for each photomultiplier, however, the time structure of the total signal is digitized in bins of 100 ns for 5 μs to enable the detection of the delayed electron from one or more muon decays.

The expected signals from the two-body decays of the nucleon are shown in fig. 2, one hemisphere plotted against the other. This level of photon statistics would allow clean separation of different decay modes. Table 1 gives the expected performance for all SUSY decay modes [12]. As always it is difficult to separate the single track and the ν decay modes from ν induced background.

2.2 <u>The Fréjus Experiment</u>

The Fréjus experiment, a collaboration of European laboratories at Orsay, Palaiseau, Saclay and Wuppertal is situated in the Modane Underground laboratory in a new road tunnel east of Grenoble in France. There is a reasonable sized cavity, 4000 m^3, at a depth of 4500 m.w.e.

The design of the Fréjus detector is now fixed, full scale production of the 6 × 1.5 m^2 flash chamber modules has started and is expected to reach a production rate of 5 modules per day. It is expected to have 400-500 t installed by the end of the year and 1500 t fully operating by the end of 1984. The cavity is now ready and installation of power, gas supplies, computer, etc. under way.

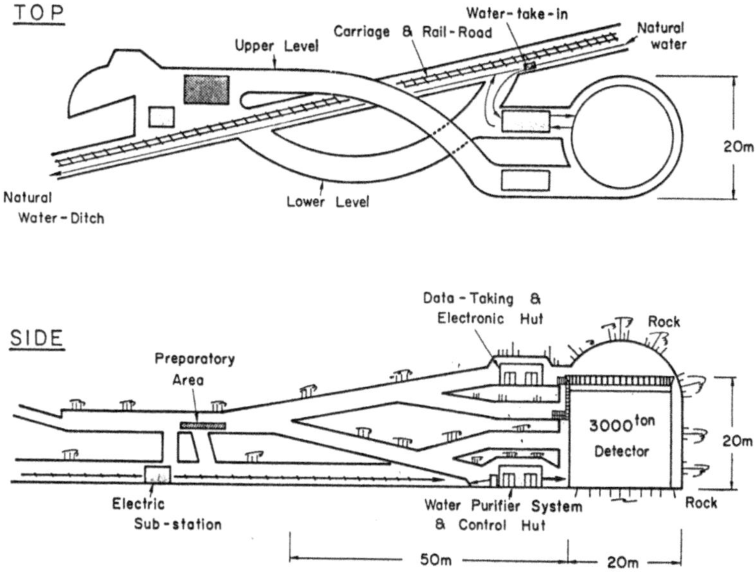

Fig. 1 Layout of the Kamioka proton decay experiment.

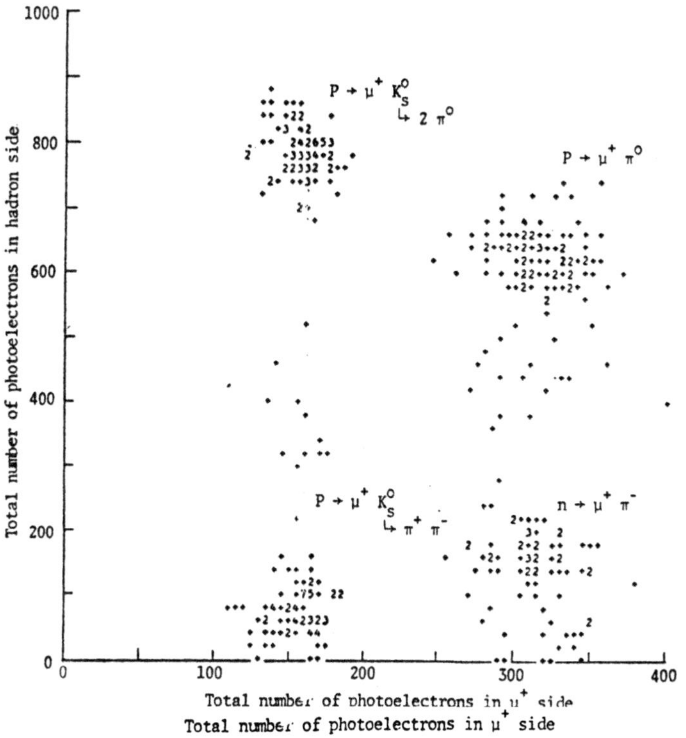

Fig. 2 Monte Carlo simulation of some nucleon decay modes in the Kamioka detector.

TABLE 1

Performance of Kamioka Water Cherenkov for SUSY decay modes

Decay	$\#_{pe}$	ΔE_{total}	ΔE_{μ}	Comment
1) $P \to \mu^+ K_s^o$ $\quad \mapsto \pi^o \pi^o$	2000	3.6%	μ 13% 4 γ's	Easily identified
2) $P \to \mu^+ K_s^o$ $\quad \mapsto \pi^+ \pi^-$	540	90% of dks 17.3%	μ 13% 2 π's	16% 3 rings K_s^o mass 75% 2 rings - difficult
3) $P \to K^+ \nu$ $\quad \mapsto \mu^+ \nu$	130		μ 9%	Background $\bar{\nu}N \to \mu^+$ in same μ energy bin 2-3/year
4) $P \to K^+ \nu$ $\quad \mapsto \pi^+ \pi^o$	700		E_{K^+} 12%	Rarely 3 rings much ν inelastic background
5) $n \to K_s^o\ \nu$ $\quad \mapsto \pi^o \pi^o$	1600		E_{K^o} 7%	Very rare to reconstruct K_s^o, ν background
6) $n \to \nu\ K_s^o$ $\quad \mapsto \pi^+ \pi^-$	230		E_{K^o} 8%	50% of K_s^o reconstruct not much ν background

Fig. 3 shows a schematic drawing of a single super module of 8 detector planes and 2 trigger planes. It is hoped to have the first of these 6 x 6 m^2 units installed and working in the cavity by the end of May 1983.

The detector well satisfies the need of fine sampling, 3 mm iron, and fine lateral cell size, 5 mm, to provide good separation between electrons and hadrons on the basis of visible electromagnetic showers. The flash chamber planes are triggered by planes of Geiger tubes 1.4 x 1.4 cm^2 cross section, between each super module of flash chamber, e.g. 1.4 radiation lengths of material. The intrinsic resolution of these tubes is ~200 ns,

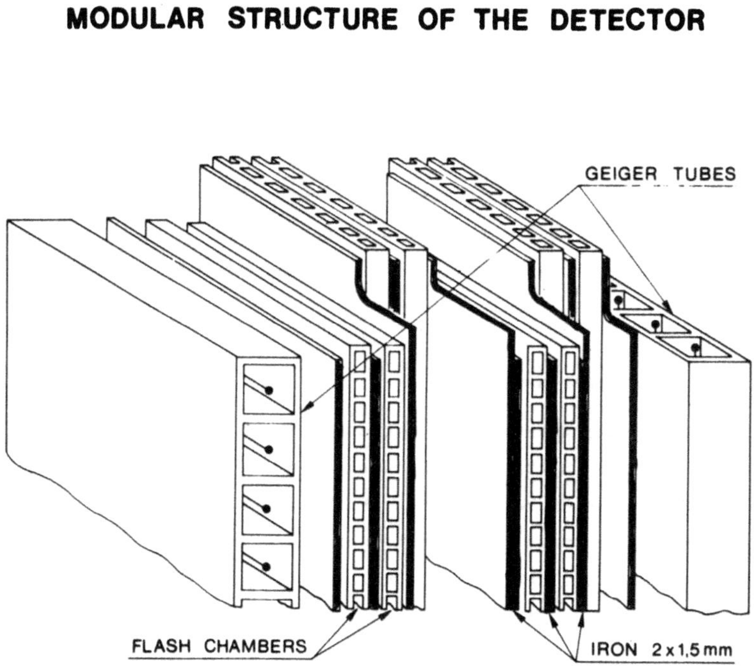

Fig. 3 Detail of the structure of a super module for the Fréjus detector.

adequate for a simple coincidence trigger, but insufficient to give decay at rest delay times for K's or π's or track direction by time-of-flight.

The direction of muons from, for example, kaon decay can be measured by the increase in multiple scattering, 60% useful, and the positron from μ^+ decay can be seen by using a long HT pulse on the flash chambers, 10μs. The last feature has been tested in the laboratory and works at the expense of an increased gas flow. To save cost, the time of the delayed signal is digitised by TDC's connected to groups of 8 channels, 4 cm in space, instead of the single channel of the normal readout electronics.

2.3 Soudan2 Experiment

The Soudan2 detector [3] proposed by the Minnesota, Argonne, Oxford, RAL and Tufts Collaboration to be installed in a mine in the Tower-Soudan Historical State Park, bears many similarities to the Fréjus experiment. Both detectors are fine grained tracking calorimeters, modular in construction, with iron plates sandwiched between gaseous tracking chambers. Both detectors are of similar size, Soudan2 will start at 1000 t with an option to extend to 5000 t. Though the Soudan2 will start a year after Fréjus it is clear that there will be strong competition between them.

The option taken by the Soudan2 Proposal has been to trade the good two track resolution, but digital readout of the flash chambers, for the advantage of analog pulse height measurement of dE/dx in drift chambers. The single track spatial resolution of the two detectors is comparable, although Soudan2 has an advantage in space points rather than the uncorrelated stero views of the flash chambers.

The potential of the dE/dx measurement in identifying tracks and telling their direction of motion is shown in fig. 4. Also the intrinsic timing will allow the detection of the delayed e^+ from μ^+ decays at rest.

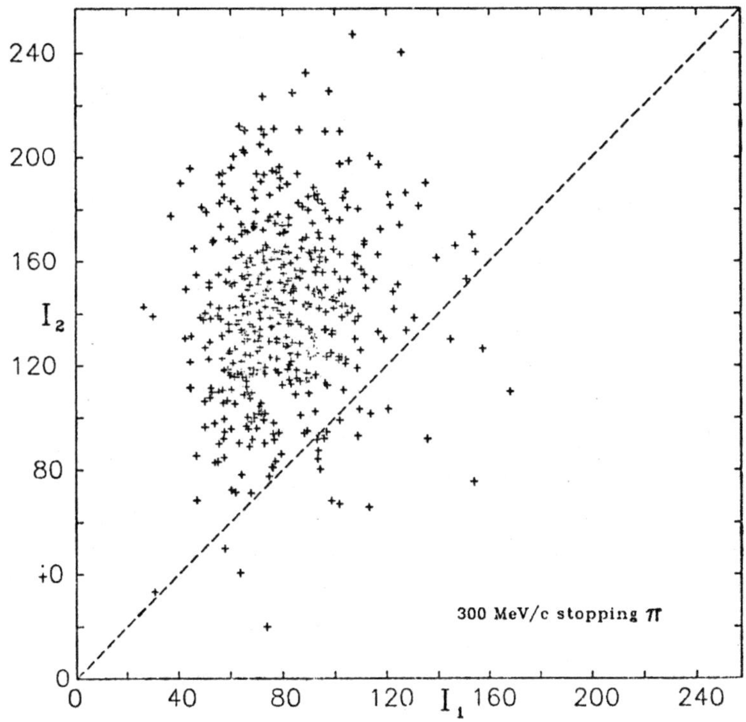

Fig. 4 Possibility of determining the direction of a stopping track using dE/dx measurement. The total ionization of the last third of the track, I_2, is plotted against that of the first third, I_1.

The Soudan2 Proposal is now authorised both by DOE and SERC, and the detailed design of the drift chamber scheme is underway. Two 1-2 t prototypes are being built on rather different lines, both technically inovative and worth describing in some detail.

The proposal of the Argonne Group is to build planar drift units of 5 x 0.5 m^2. The ionized electrons drift over 50 cm down a 1 cm wide gap to an anode wire 5 m long running down one edge of the chamber. A space point is obtained from drift time, an anode and a cathode strip linking several chamber units. The difficult step in the construction is to provide, at reasonable cost, the drift field shaping electrodes. These are made from a mylar sheet with a minimum number of printed copper electrodes connected by a resistor chain to a 10 K high tension source. Irregularities in the field are smoothed by coating the mylar with resistive ink. Chambers have been laboratory tested and a 2 t fully equipped prototype is under construction. The size of the project is best gauged by the fact that 10 000 such chambers have to be built.

A novel alternative for a construction scheme has been proposed by the Oxford Group. Schematic drawings of the detector are shown in figs 5 and 6. In this design the ionized electrons drift down hexagonal resistive plastic tubes held between shaped 2 mm thick iron plates. The advantage of this layout is the small number of electrical connections and hence ease of construction. The HT and insulation from the iron is supplied by a single sheet of mylar, with copper electrodes on the tube side, "woven" between the steel plates.

The problem with the design is to obtain a plastic of the correct resistivity, 10^{10}-$10^{12}\Omega$ per □, and adequate uniformity. Too little resistance and the tubes never reach the delicate dynamical equilibrium between leakage and enough charge induced on the tube surface to give the right field shape; too much resistance and the field quickly becomes unstable.

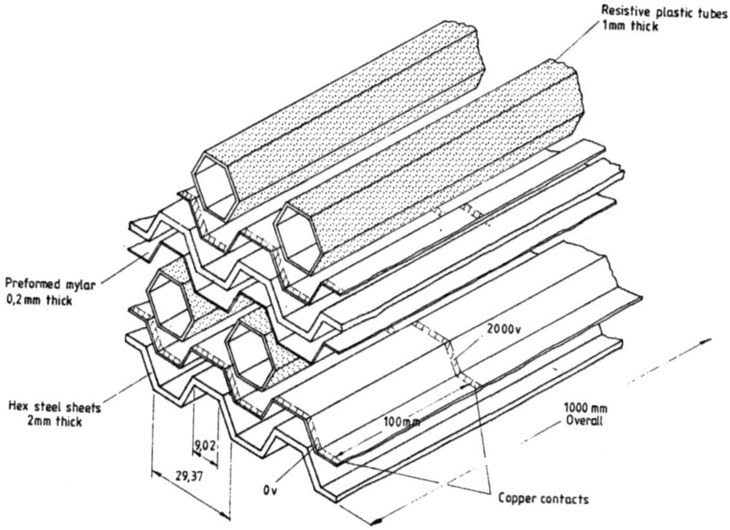

Fig. 5 **Schematic drawing of the structure of the resistive tube calorimeter.**

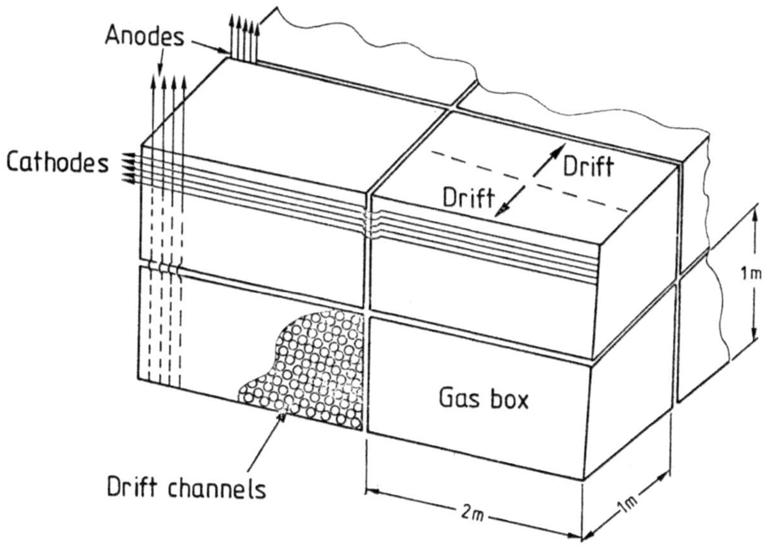

Fig. 6 **Construction of resistive tube calorimeter, showing the layout of the anode wires and cathode strips.**

A one ton prototype detector is under construction and will be compared with the one from Argonne before freezing the design at the end of this year. The proposal calls for the first 50 t detector to be installed in the mine by mid 1984 and the complete detector to be operating by the end of 1985.

2.4 Kolar Gold Field Experiment

As a result of their experience with the small detector in the Kolar Gold mine, the groups involved have proposed an extension of the experiment with a new detector with a mass of up to 1000 t.

A new cavern of 3000 m^3 would be constructed at a depth of 2100 m, large enough to contain three 6 x 6 x 6 m^3 modules. The detector would be based on the same principle as before, i.e. thick walled iron proportional tubes giving position and dE/dx measurements. In the new detector the tubes would be smaller in cross section, 5 x 5 cm^2 giving better spatial resolution.

The proposal is still at the design stage, but in principle a one kilo-ton detector could be ready by the "end of 1984".

2.5 The Gran Sasso Project

The construction of the Gran Sasso underground laboratory [4] has started. The project calls for three large caverns, each the order of 11 x 16 x 100 m^3, complete with good access and user facilities. Fig. 7 shows the layout and cross section of the tunnels. To date, the central cavern and a single access road has been excavated, but still has to be finished. Completion of the concrete lining, installation of power, ventilation etc. awaits further funding.

Two projects exist for detectors to be installed in the finished cavern. GS1, an extension of the NUSEX experiment in the Mont Blanc tunnel, and the GUD flash chamber detector proposed at the Rome Workshop in 1981 [5].

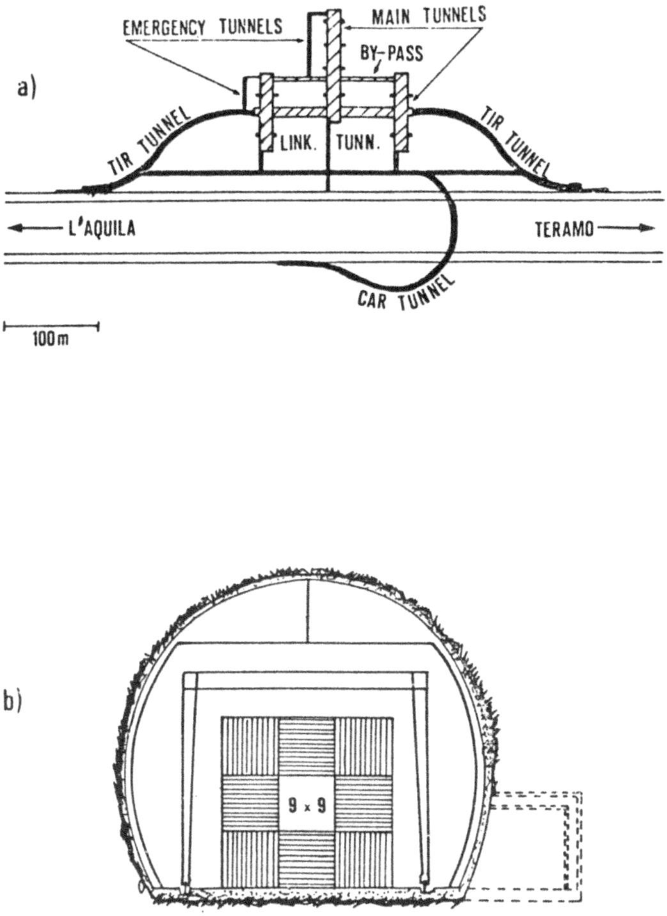

Fig. 7 (a) General plan of laboratory (three "Main Tunnels" connected by a linking tunnel) and access to it;

(b) Cross-sectional view of Main Tunnels.

GS1 is a fine grained tracking calorimeter based on the same structure and limited streamer tubes used as in the NUSEX experiment. Experience has shown that a lower density and finer sampling are desirable, hence iron plates of 5 mm, instead of 10 mm, will be used. A 1000 t detector is foreseen with means of extension to 3000 t. Minor improvement of the readout electronics will increase the detection efficiency of the delayed e^+ from μ decay at rest. The finer sampling will also improve the measurement of multiple scattering to give the direction of flight of the tracks (fig. 8).

The time scale for the proposal would allow the experiment to be taking data with a full 1000 t detector by the end of 1985.

Little has changed in the proposal for the very massive GUD flash chamber detector since the presentation in October 1981. Construction of the large area, 3 x 1 m^2, resistive plate chambers [13] for a current $n\bar{n}$ oscillation experiment by the same group is under way. These chambers allow the very fast trigger and time-of-flight measurements crucial to the success of this very large 10 000 t detector.

2.6 Liquid Scintillator Detectors

Two proposals exist and are being strongly upheld by their authors as the best possible solution for a next generation nucleon decay detector. The advantages of a fully active calorimeter are great. The energy resolution ~ 4%, is at the level of the energy smearing due to Fermi motion and the potential time resolution is excellent ≲ 2ns. This timing would in principle allow the detection of the delayed coincidence from a K meson decay at rest, an unambiguous signal for one of the predicted SUSY decay modes. Disadvantages are however significant; inherently poor spatial resolution, poor containment of neucleon decays and cost, especially in electronics if the timing ability is to be

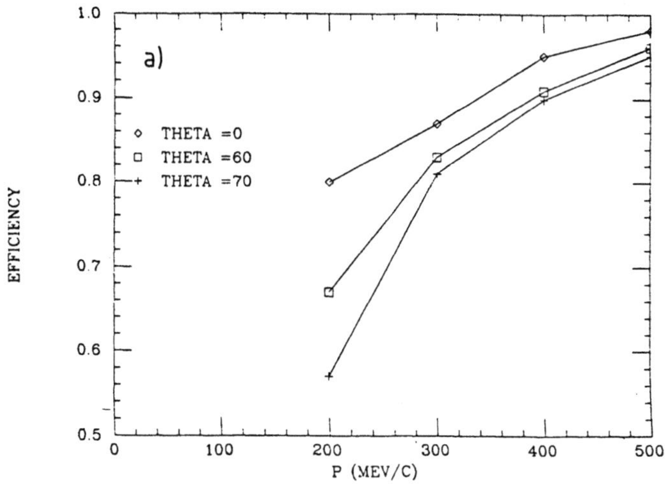

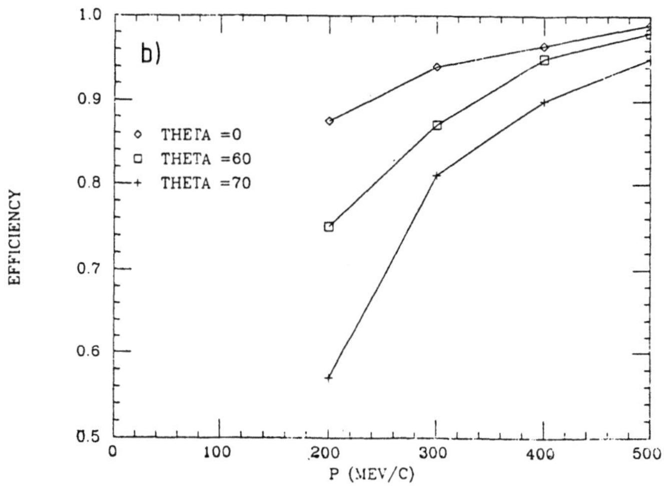

Fig. 8 **Efficiencies for various incidence angles for an apparatus having 0.5 cm of iron, 2.2 cm between the tube planes. The efficiency is the fraction of tracks found in the correct direction:**

(a) refers to σ_x equivalent = 3 mm and σ_y = 3 mm;

(b) refers to σ_x = 3 mm and σ_y = 3 mm.

used to its full potential. The lack of two track spatial resolution is a serious effect, it implies that spurious energy measurements would often occur due to the undetected interactions of hadrons.

The Homestake proposal is for a 1400 t stack of 2000 liquid scintillator modules each .3 x .3 x 8 m^3. At present 200 of these elements are being manufactured to construct a veto plane to surround the existing ^{37}Cl solar neutrino tank. This detector, though of a poor aspect ratio for a nucleon decay detector, will provide an excellent test of working with liquid scintillator deep underground.

The second proposal is the result of a design study by the University of Pennsylvania based on the extrapolation in size of an existing accelerator neutrino detector at BNL. A 2600 t detector is proposed with a fiducal volume of 1300 t. The problem of spatial resolution is overcome by putting xy planes of proportional drift tubes between each liquid scintillator unit. It is clear that the detector proposed is by far the most highly instrumented, with a corresponding high performance and cost.

2.7 Liquid Argon Detectors

There are several ideas for liquid [8] or high pressure gas [10] Argon detectors working in a TPC-like mode. There is even a proposal for a 3000 t liquid argon bubble chamber [9] with a 50% duty cycle to study proton decay. Liquid argon has many beautiful properties as a detector medium: energy measurement by scintillation light in the UV, particle identification from Cherenkov light and fast timing. TPC quality tracks can be obtained after drifting over reasonable distances in the liquid. High pressure gas detectors, 500 atm., have the same properties and the benefit of gas amplification of the signals at the anode wires. At present several small scale tests are under way. However, the main problems of cost and deployment underground still remains to be overcome.

3. CONCLUSIONS

Table 2 shows a summary of the main points of the active and proposed experiments. It is a substantial list corresponding to a massive expenditure of manpower and money. By the end of 1985 there will be at least two large water Cherenkovs and two large fine grained tracking calorimeters in operation. In a rational world this would seem to be an adequate investment, at least until the proton lifetime has been measured!

TABLE 2

Summary of future experiment

Detector	Site	Type	Size k tons	Starting date	Comments
Kamioka	Kamioka mine Japan	water Cherenkov	1.0	early 1983	good p.e. statistics
Fréjus	Modane underground laboratory	flash chamber	1.5	end 1984	track dir. by m.s.
Soudan2	Soudan -Tower mine	drift chamber	1.→5.	end 1985	track dir. by dE/dx
Kolar gold	Kolar mine India	prop tube	1.0	end 1984?	dE/dx
GUD	Gran Sasso	flash chamber	10.0	proposal	fast timing ≲ 2ns
GS1	Gran Sasso	limited streamer tubes	1.→3.	proposal	track dir. by m.s.
"Penn design"	?	liquid scintillator prop. tubes	2.6	proposal	dE/dx and good tracking
Homestake	Homestake mine	liquid scintillator	1.4	proposal	dE/dx and good timing

REFERENCES

[1] T. Suda, Kamioka proton decay experiment, Neutrino 1981, Maui, Hawaii, July 1981;

Kamioka proton decay experiment KEK preprint, 82-7 (1982).

[2] Orsay-Palaiseau-Saclay-Wuppertal Collaboration, Proposition d'une experience pour l'étude de l'instabilité du nucléon au moyen d'un détecteur calorimétrique (December 1979).

[3] Soudan2 Proposal, Minnesota-Argonne-Oxford Collaboration, ANL-HEP-PR-81-12.

[4] The Gran Sasso Project, A. Zichichi, INFN/AE-82/1.

[5] The GUD Project, M. Conversi, CERN/EP 81-13.

[6] Homestake tracking spectrometer, R. Steinberg, 1982 Summer Workshop on proton decay experiments, 7 June 1982, ANL-HEP-PR-82-24.

[7] Design study for a fine grained detector for nucleon decay and neutrino studies, A.K. Mann et al, University of Pennsylvania, July 1981.

[8] H. Chen, 1982 Summer Workshop on proton decay experiment, June 1982, ANL-HEP-PR-82-24.

[9] G. Harigel, A. Hervé and K. Winter, CERN/EP 83-47.

[10] P. Oddone, 1982 Summer Workshop on proton decay experiments, June 1982, ANL-HEP-PR-82-24.

[11] G. Battistoni et al., Nuch. Instr. & Meth. 164 (1979) 57.

[12] M. Koshiba, private communication (March 1983).

[13] R. Santonico and R. Cardarelli, Nuch. Instr. & Meth 187 (1981) 377.

THE ANGULAR DISTRIBUTION AND FLUX OF ATMOSPHERIC NEUTRINOS

T.K. Gaisser and Todor Stanev**

Bartol Research Foundation of The Franklin Institute
University of Delaware
Newark, Delaware 19711

and

S.A. Bludman and H. Lee

Department of Physics
University of Pennsylvania
Philadelphia, Pennsylvania 19104

Abstract

Because of the variation in space of the geomagnetic cutoff of the primary cosmic ray spectrum, the flux of $\lesssim$ GeV secondary neutrinos is not isotropic. This has important consequences for neutrino physics with underground detectors and for calibration and background studies of nucleon decay experiments.

1. Introduction

There are four reasons for interest in the flux of cosmic ray neutrinos at this time. (1) With the advent of large deep underground nucleon decay detectors, contained neutrino interactions are being detected in statistically significant numbers for the first time [1]. Previously, the flux of neutrinos produced by interactions of cosmic rays in the atmosphere was inferred from the observation of upward going muons produced by ν_μ interactions in the surrounding rock [2]. Direct detection of the neutrinos thus constitutes a benchmark cosmic ray measurement and also allows detection of ν_e as well as ν_μ. (2) Neutrino interactions are the most significant background for nucleon decay, so a knowledge of the expected flux is essential for this reason. (3) Comparison between calculations and observations can also serve as an important calibration of the detectors. This is particularly so for the angular dependence, which is relatively free of calculational uncertainty. Comparison of upward and downward fluxes of neutrinos, and particularly of the ν_μ/ν_e ratio, has the potential to extend the search for neutrino oscillation to $\Delta m^2 \gtrsim 10^{-4}$ eV^2 [3,4]. (4) Finally, the atmospheric neutrinos serve as a pedestal for unexpected extraterrestrial sources of neutrinos.

We first review briefly the results of a recent calculation of the neutrino flux [5], giving details of the calculation useful for understanding the important effect of the geomagnetic field. We then discuss the angular dependence of the neutrino flux, resulting from the geomagnetic effects and from zenith angle dependence of neutrino production. In an Appendix we describe briefly the parametrization of meson production in hadron–nucleus collisions that we have used for the calculation. We also compare our results to an analytic calculation [6].

2. Implications for ν-oscillation search

The sketches in Fig. 1 illustrate the geometry of a nucleon decay

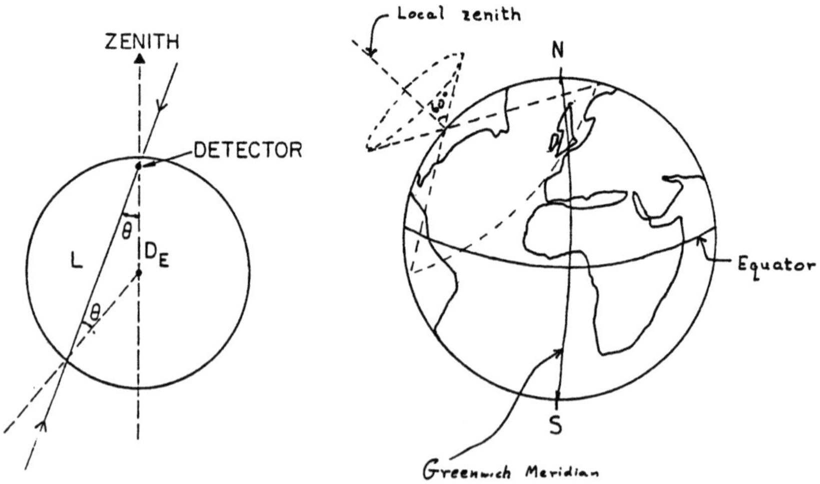

FIGURE 1 Geometry of an underground detector for cosmic ray neutrinos.

detector. Fig. 1a shows that the geometry is symmetric [4]: for every downward trajectory with zenith angle θ there is a corresponding upward trajectory. Thus, if the primary spectrum at Earth were isotropic and there were no neutrino oscillations, the flux of neutrinos produced by cosmic rays in the atmosphere would be up/down symmetric at any location. The geomagnetic field destroys the isotropy of the primary beam by deflecting particles of low magnetic rigidity before they penetrate to the atmosphere. In equatorial regions, the vertical cutoff is typically 15 GeV, with a strong East-West effect. At high geomagnetic latitudes the cutoff is low and the E-W effect small.

Fig. 1b illustrates the geometry of a typical nucleon decay detector in North America. Because of the limited statistics expected in current nucleon decay detectors (~ 100 events per year per kiloton fiducial volume) it is natural to divide the total solid angle into downward ($\theta < 60^\circ$) and upward ($120^\circ < \theta < 180^\circ$) cones of half angle 60° and into upward and downward wide-angle regions with $90^\circ < \theta < 120^\circ$ and $60^\circ < \theta < 90^\circ$ respectively. This binning is oriented to neutrino oscillation searches, where it is desirable to compare upward and downward fluxes. In any case, only crude angular resolution is possible because of the significant loss of directionality between ν and ℓ in $\nu + N \to \ell + N'$ at low energy, which is the dominant neutrino signal around 1 GeV [7].

For typical North American sites with geographic latitude 40-48° N the vertical cutoff is ~ 3-1.5 GeV [8]. (Because the geomagnetic poles are offset, the geomagnetic latitude in the US is higher than the geographic latitude.) Nearly all downward primaries contributing significantly to neutrino production thus penetrate to the atmosphere. Upward trajectories, however are collected from a large portion of the Earth's surface, much of which is at low geomagnetic latitude with high cutoffs. At high geomagnetic latitude, the median cutoff for the upward 60° cone is 10 GV. Thus the flux of upward neutrinos is significantly lower than the flux of downward neutrinos at the U.S. underground sites [9]. (The opposite is true at KGF where the average upward cutoff is similar but the vertical cutoff for downward primaries is about 16 GV.) Table I summarizes the results of Ref. 5 for flux ratios at the Cleveland site.

TABLE I. Upward (↑) and downward (↓) neutrino fluxes (in cones of half angle 60°) at Cleveland for solar maximum (max) and solar minimum (min). Here $\nu = \nu_e + \bar{\nu}_e + \nu_\mu + \bar{\nu}_\mu$.

	$E_\nu \geq 200$ MeV	$E_\nu \geq 1$ GeV
$\frac{\nu\uparrow}{\nu\downarrow}$ min	0.57	0.84
$\frac{\nu\uparrow}{\nu\downarrow}$ max	0.66	0.88
$\frac{\nu max}{\nu min}$ ↓	0.70	0.90
$\frac{\nu max}{\nu min}$ ↑	0.80	0.94

In Ref. 5 we emphasized that, particularly for low energy neutrinos, the geomagnetic suppression of the upward neutrino flux mimics a neutrino oscillation signal. Our calculation employs Cooke's cut-off functions $\Omega_i(E_0,\theta)$ for absolute suppression of primary particles, averaged over azimuthal angle ψ and computed for a dipole magnetic field at 20 km above the Earth's surface [8]. Details of the Earth's magnetic field and of the azimuthal dependence of actual particle trajectories which we have not incorporated will limit the ultimate sensitivity of any atmospheric neutrino experiment. Because geomagnetic effects decrease with increasing energy, the exact geomagnetic suppresion also depends on detector response to neutrinos. The size of the geomagnetic effect will therefore vary from detector to detector; however, we expect the up/down ratios to lie within the range shown in Table I.

For a neutrino oscillation involving three flavors including ν_μ and ν_e, there can be no more than a 33% suppression of upward relative to downward neutrinos [3], which is in the same direction and of the same order of magnitude as the geomagnetic effect, at a high latitude site. Geomagnetic effects will therefore need be very well understood to establish the existence of such an oscillation. The ν_e/ν_μ ratio is largely independent of the geomagnetic effect, so, if neutrinos of different flavor can be distinguished, a mixing scheme that leads to $(\nu_e/\nu_\mu)\uparrow \neq (\nu_e/\nu_\mu)\downarrow$ will be easier to establish than one in which ν_μ/ν_e = constant.

3. Details of the calculation

Electron and muon type neutrinos and anti-neutrinos are produced by the chain

$$N \rightarrow \pi^{\pm} K \rightarrow \mu \, \nu_\mu \qquad \mu \rightarrow e \, \nu_\mu \, \nu_e \tag{1}$$

Typical production heights are 10–50 km for nucleons incident near the vertical. The muon flux is well measured at sea level but, because of significant energy loss in the atmosphere, cannot be used directly to calibrate the neutrino flux. (Old measurements of muons at high altitude [10], are related to the low energy neutrinos [11].) In order to incorporate this energy loss and the geomagnetic effects, we start our calculation directly from the primary cosmic ray spectrum. We compute the neutrino flux by constructing the integral

$$dN/dE_\nu = \int_{E_\nu}^{\infty} dE_0 \; Y(E_\nu, E_0, \theta) \; \Omega_i(E_0, \theta, \psi) \; dN_i/dE_0 \tag{2}$$

Here dN_i/dE_0 is the primary spectrum of nucleons of energy E_0 per nucleon. The subscript i, which is implicitly summed, denotes proton or alpha. These dominant components of the primary spectrum must be treated separately because their cutoff energies differ. $Y(E, E_0, \theta)dE_0$ is the yield of neutrinos per primary nucleon incident at zenith angle θ. (Both Y and the neutrino flux are computed separately for each flavor.) $\Omega_i(E_0, \theta, \psi)$ is the fraction of nucleons at energy E_0 arriving within solid angle $\Delta\psi\,\Delta\cos\theta$ which pass the geomagnetic cutoff [8].

The yield is computed with a Monte Carlo cascade program for a variety of primary energies sufficient to perform the integral (2). We take into account muon energy loss in the atmosphere, secondary interactions of nucleons, pions and kaons, and all relevant decay channels. Charges are followed separately so that ν and $\bar{\nu}$ can be computed separately. However, because we have not finished a detailed momentum-dependent treatment of π^+/π^- and K^+/K^- ratios at production, in this paper we quote only the combined fluxes, $\nu_e + \bar{\nu}_e$ and $\nu_\mu + \bar{\nu}_\mu$. Another limitation of the calculation at present, in addition to the simplified treatment of the magnetic field mentioned above, is its one dimensional nature . This prevents us, for example, from carrying the present calculation to still lower neutrino energies.

The effective primary spectrum $\Omega dN/dE_0$ depends on θ and ψ through the geomagnetic cutoff factor $\Omega_i(E_0, \theta, \psi)$. It also varies with an 11 year period due to solar modulation. Fig. 2 shows solar activity [12], which precedes by about one year the anti-correlated variation in the low energy cosmic ray flux [13]. Fig. 3 shows $\Omega_i dN_i/dE_0$ for the upward and downward 60° cones (Fig. 1b) at solar maximum and at solar minimum.

FIG. 2. Monthly Mean Zürich Sunspot Numbers January 1944 – July 1982

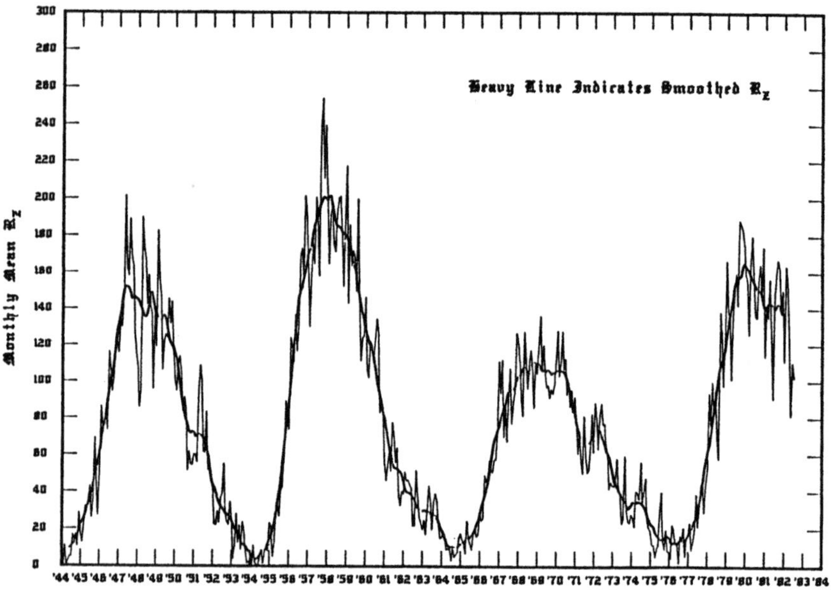

To illustrate the procedure we compute the flux of $\nu_e + \bar{\nu}_e$ above 400 MeV and the flux of $\nu_\mu + \bar{\nu}_\mu$ above 600 MeV at solar maximum averaged over the 60° cones of Fig. 1b at Cleveland. Fig. 4 shows the integral yields, i.e.

$$Y(>E, E_0) = \int_E^{E_0} Y(E', E_0)dE' \tag{3}$$

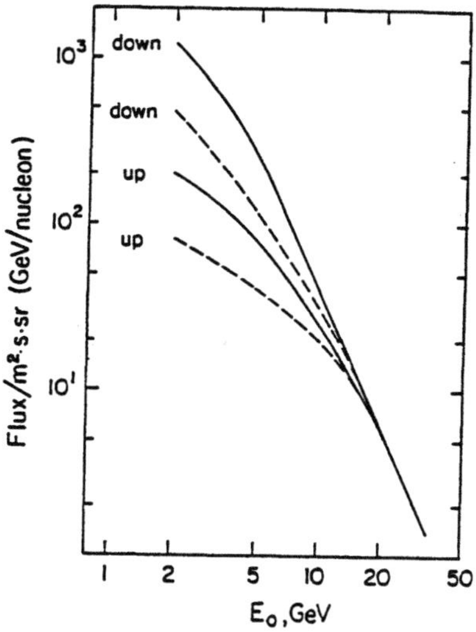

FIGURE 3 Primary spectrum times geomagnetic cutoff averaged over 60° upward and downward cones at Cleveland (Ref. 5). Full lines: solar minimum; dashed lines: solar maximum.

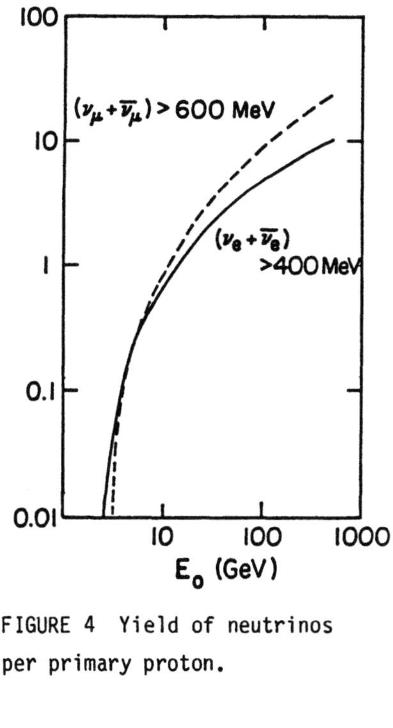

FIGURE 4 Yield of neutrinos per primary proton.

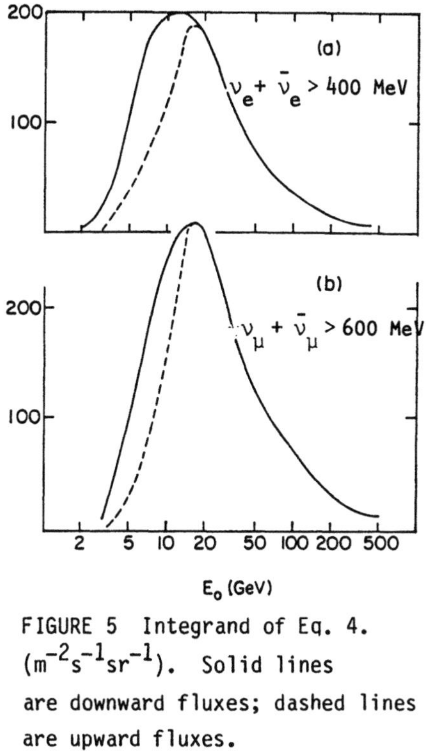

FIGURE 5 Integrand of Eq. 4. ($m^{-2}s^{-1}sr^{-1}$). Solid lines are downward fluxes; dashed lines are upward fluxes.

The integral neutrino flux is then

$$N_\nu(>E) = \int [Y(>E, E_0)(\Omega_i \, dN_i/dE_0)E_0] d\ln E_0, \qquad (4)$$

where the change to a logarithmic variable has been made so that the integral is proportional to the area in the plot of the integrand (square brackets in Eq. 4) vs. $\ln E_0$. Fig. 5 shows plots of this integrand. The difference between the two curves in each plot is just the effect of Ω averaged over the upward 60° cone. The effect of a step function cutoff can be seen by simply taking the area above the cutoff energy E_c.

4. Angular distribution

Since, for downward primaries at high geomagnetic latitudes, the cutoff is low in all directions, the angular dependence of downward muons at Cleveland is dominated by the angular dependence of the yield function and is azimuthally symmetric about the local zenith. (For nearly horizontal zenith, $\theta > 70°$, it becomes important to take into account the curvature of the Earth, which we have done.) This zenith angle dependence is essentially a $\sec\theta$ effect in neutrino production, and is in agreement with that found e.g. by Volkova [14].

For upward going neutrinos at high latitude sites (and for all neutrinos at low latitude sites) the angular dependence of the neutrino flux depends on the angular dependence of the cutoff as well as on that of the yield. Fig. 6 shows the results averaged over azimuth as a function of $\cos\theta$ at Cleveland and at KGF. (These results differ slightly from those shown at the conference because a finer grid is used here for the angular dependence and because here we use solar maximum spectrum rather than solar minimum.) The calculation could in general also include the azimuthal dependence of the cutoff function in order to maximize the angular dependence.

Angular dependence of the type shown in Fig. 6 should be useful for calibrating the nucleon decay detectors and establishing that interactions of atmospheric neutrinos are in fact being detected. Seeing the effect in detectors at quite different geomagnetic sites would be particularly convincing, as illustrated by the difference between Cleveland and KGF in Fig. 6 [15]. In this connection we note that the Kamioka experiment is also at a rather low geomagnetic latitude where the angular dependence will be quite different from that at Cleveland.

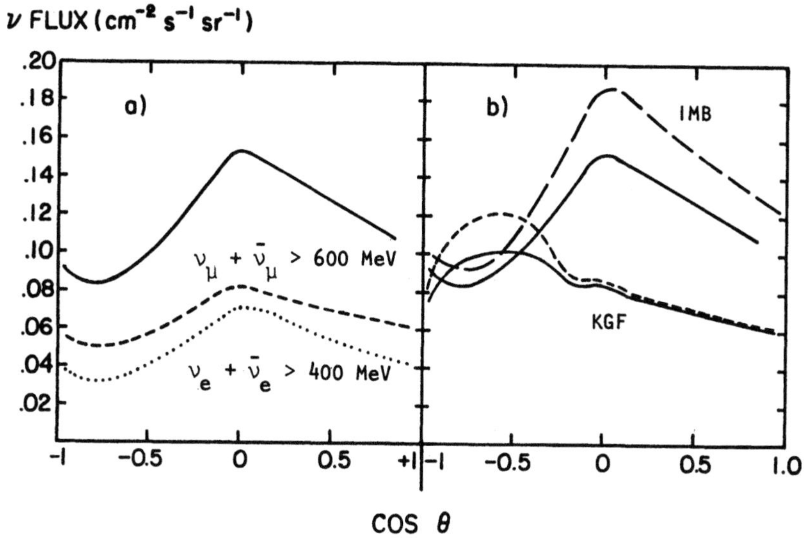

FIGURE 6 Angular distribution of neutrino flux. a) Solar maximum flux at Cleveland: dashed line – $\nu_\mu + \bar{\nu}_\mu$, E > 600 MeV; dotted line – $\nu_e + \bar{\nu}_e$, E > 400 MeV; solid line – total. b) Comparison of angular dependence at Cleveland (upper pair) and KGF (lower pair). In each case dashed line is solar minimum flux and solid line the flux at solar maximum of all neutrino types with thresholds as in (a).

Appendix: Model of meson production and comparison to analytic calculation

A. Treatment of Hadron Interaction

The yield of neutrinos and muons depends on the multiplicity and the momentum distributions of pions and kaons produced in collisions of hadrons (primarily nucleons) on nitrogen and oxygen. The strong interaction model we have used was originally constructed for calculation of high energy air showers [16]. It is based on recent studies of hadron nucleus interactions at FNAL [17] and CERN [18] in the 50-200 GeV range. From Fig. 5 we see that the energy region important for low energy neutrino production is $E_0 \sim 3$–100 GeV. Accordingly, we have tuned the low energy behavior of the model to reproduce the data on light nuclear targets at lower energy [19,20]. Near threshold ($E_0 < 2.5$ GeV) we add up individual channels for resonance production from the particle data tables. We are still making adjustments in the x-dependence of π^+/π^- and K^+/K^- ratio and in the K/π ratio as well as making extensive comparisons to low energy data. Such refinements could lead to some change (probably a reduction) in the predicted neutrino fluxes of perhaps 30% around 1 GeV. We do not expect up/down ratios, angular dependence or solar maximum/solar minimum ratio to be significantly affected by these refinements because such effects largely cancel in taking ratios and constructing the angular dependence.

Because no experiment covers the full kinematic range, and because there are differences in normalization between various experiments of as much as 50%, residual uncertainties in the total yields of secondaries and hence in the calculated neutrino flux from this source will remain. Other surveys [21] of this type of data have also encountered such discrepancies between experiments and ambiguities of interpolation and extrapolation into unmeasured kinematic regions. One of our objectives will be to assess the size

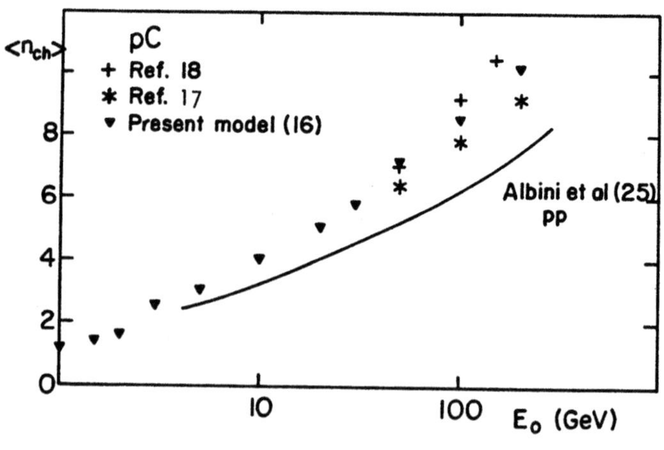

FIGURE 7 Mean charged multiplicity in p Carbon collisions. Solid line shows mean charged multiplicity in pp collisions. E_0 is total lab energy of the incident nucleon.

of this uncertainty. A new measurements of the flux of low energy muons at high altitude would also be of help with the normalization problem [22].

The actual Monte Carlo program for choosing secondaries is based on a modified version of the parametrization of hadron nucleus data due to Stenlund and Otterlund [23]. In a nucleon-nucleus collision the energy of the leading nucleon is chosen first. Then lab rapidities and transverse momenta of mesons are chosen randomly from an appropriate distribution. An x-dependent choice of meson type is made, and the energy of the secondary computed. This procedure continues until all remaining energy is used, with the energy of the last particle adjusted to give exact energy conservation. This procedure gives KNO scaling with the correct shape for the multiplicity distribution [24].

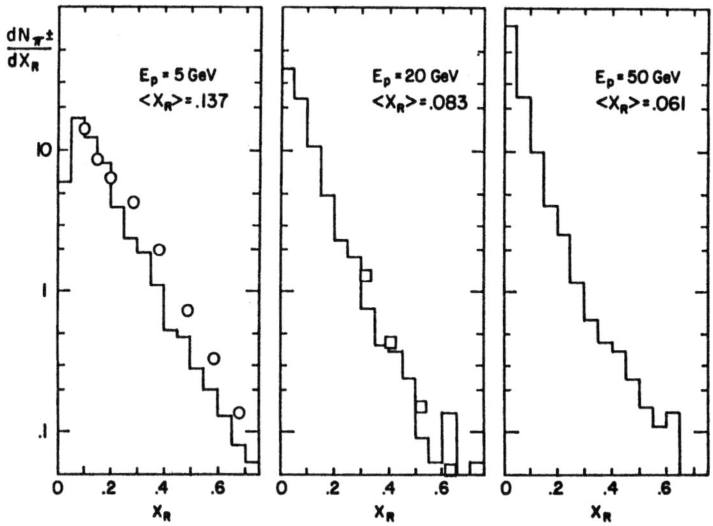

FIGURE 8 Distribution of charged pions produced at three incident nucleon total energies. Histograms show results of the model we use here [16]. Open circles are from the data of Papp [Ref. 20] and open boxes from the data of Allaby et al. [Ref. 19]. We have integrated the data over transverse momentum at fixed $x_R = E_\pi(\text{lab})/E_p$ to obtain the points shown.

In Fig. 7 we show the average charged multiplicity produced by our model as compared to data on p-nucleus [19,20] and on pp collisions [25]. Fig. 8 shows the distribution of $x_{Lab} = E/E_0$ for three important energies. Our procedure is flexible so that it can be adjusted to give non-scaling and threshold behavior correctly as required by the data. Moreover, conservation of energy is ensured.

B. Comparison to an analytic calculation

At this workshop A. Dar originally presented results of an analytic calculation of the neutrino flux in the same energy range as our Monte Carlo calculation. The fluxes of low energy atmospheric neutrinos he presented then were significantly lower than those we find.

Upon working through the derivation of the analytic result, we found that this discrepancy arose from two factors. 1) In the analytic approximation with a cutoff it is necessary to assume a definite, simple analytic form for the inclusive cross section for meson production as a function of x_{Lab}. Dar had assumed $dn/dx = (1-x)^3$, which fails to give any increase of multiplicity with energy. If a specific analytic form must be used, a better approximation is $(1-x)^n/x$. This is approximately valid for $x = x_{Lab}$ over the kinematic range relevant here, and it gives a logarithmically increasing multiplicity. 2) Secondary interactions of nucleons in the atmosphere had been neglected by Dar when those interactions occurred with energy below the cutoff energy. (The cutoff should be applied only to the primaries, which can cascade down to lower energies after entering the atmosphere.)

To compare the two calculations we assumed $dn/dx \propto (1-x)^3/x$ and added the contribution of secondary interactions of nucleons. We then rederived the analytic result for ν_μ from π-decay and found a flux in reasonable agreement with our Monte Carlo result for ν_μ from meson decay. Dar's written contribution to these proceedings [6], now incorporates the above two corrections and apparently agrees substantially with our original Monte Carlo results.

* Work supported in part by the U.S. Dept. of Energy (TKG, SAB, HL) and by the U.S. National Science Foundation (TS).

** On leave from Institute for Nuclear Research and Nuclear Energy, Sofia, Bulgaria.

References

[1] B. Cortez et al., this conference. See also B.V. Sreekantan et al., this conference.

[2] M.F. Crouch et al., Phys. Rev. D18, 2239 (1978); M.M. Boliev et al.,Proc. 17th Int. Cosmic Ray Conf. (ICRC) Paris, 7, 106 (1981); P.V. Ramana Murthy, Proc. 17th ICRC (Rapporteur paper) 13, 381 (1981).

[3] P. Frampton and S.L. Glashow, Phys. Rev. D25, 1982 (1982).

[4] D.S. Ayres, T.K. Gaisser, A.K. Mann and R.E. Shrock, Proc. Snowmass Summer Study, 590 (1982).

[5] T.K. Gaisser, Todor Stanev, S.A. Bludman and H. Lee, Bartol preprint BA-83-20, April 1983.

[6] A. Dar, this conference.

[7] B. Cortez and L.R. Sulak in Unification of Fundamental Particle Interactions (ed. S. Ferrara, J. Ellis, P. van Nieuwenhuizen, Plenum, New York and London) p. 661 (1981).

[8] D.J. Cooke, Utah preprint, 1983.

[9] T.K. Gaisser, in Science Underground (A.I.P. Conf. Proc. No. 96., ed. M.M. Nieto et al.) p. 203, 1982.

[10] M. Sands, Phys. Rev. 77, 180 (1950): M. Conversi, Phys. Rev. 79, 749 (1950). See also S. Olbert, Phys. Rev. 96, 1400 (1954).

[11] A.C. Tam and E.C.M. Young, Proc. 11th ICRC (Budapest), Acta Physica Hungarica 29, Supp. 4, 307 (1970) compute $\nu_\mu+\bar{\nu}_\mu$ and E.C.M. Young in Cosmic Rays at Gound Level p. 105 (ed. A.W. Wolfendale, Institute of Physics, London, 1973) computes $\nu_e-\bar{\nu}_e$. These calculations start from the paramtrization of the high altitude muon flux by Olbert (Ref. 10).

[12] Solar-Geophysical Data prompt reports, Feb. 1983, Number 462-Part 1, page 4 (National Geophysical Data Center, NOAA/NESDIS, E/GC2, 325 Broadway, Boulder, CO 80303, ed. Helen E. Coffey).

[13] M.A. Pomerantz and S.P. Duggal, Rev. Geophysics and Space Physics 12, 343 (1974).

[14] L.V. Volkova, Sov. J. Nucl. Phys. 31, 784 (1980).

[15] We are grateful to A.K. Mann for emphasizing this point.

[16] T.K. Gaisser, R.J. Protheroe and Todor Stanev, paper submitted to 18th ICRC (Bangalore, India) August 1983.

[17] D.S. Barton et al., Phys. Rev. D27, 2580 (1983) and J.E. Elias et al., Phys. Rev. D22, 13 (1980).

[18] K. Braune et al., Z. Physik C17, 105 (1983) and M.A. Faessler et al. N.P. B157, 1 (1979).

[19] See e.g. J.V. Allaby et al., CERN Report 70-12 (unpublished).

[20] James Papp LBL-3633 (Ph.D. Thesis) 1975. See also J. Papp et al. Phys. Rev. Letters 34, 601 E991 (1970).

[21] C.L. Wang, Phys. Rev. Letters 25, 1068, E1536 (1970) and references therein.

[22] See e.g. D.H. Perkins, Oxford University Preprint 60/82.

[23] E. Stenlund and I. Otterlund, CERN-EP/82-42 (April 1982).

[24] T.K. Gaisser, Proc. Workshop on Very High Energy Cosmic Ray Interactions (ed. M.L. Cherry, K. Lande and R.I. Steinberg, U. of Pennsylvania) p. 492 (1982).

[25] E. Albini et al., Nuovo Cimento A32, 101 (1976).

ATMOSPHERIC NEUTRINOS AND ASTROPHYSICAL NEUTRINOS IN PROTON DECAY EXPERIMENTS

Arnon Dar†

Department of Physics, University of Pennsylvania
Philadelphia, Pennsylvania 19104

ABSTRACT

We present simple analytical estimates of the fluxes of atmospheric and interstellar neutrinos. We show that the geomagnetic cutoff strongly suppresses the fluxes of atmospheric neutrinos below 1 GeV from directions of low magnetic latitudes and may open a window for neutrino astronomy. We demonstrate the validity of our estimates through indirect tests.

When high energy cosmic rays propagate through matter (or radiation) they produce π's and K's that decay in flight and produce high energy ν's. At earth one may expect to detect a diffuse background of high energy ν's produced by cosmic ray collisions with galactic interstellar gas and dust, and perhaps discrete ν-sources where high energy ν's are produced when the cosmic rays propagate through matter present at, or near the source[1] (such "beam dump" arrangements may be formed e.g. by ejected or falling matter near supernovae, pulsars, active galactic nuclei, quasars and black holes). Cosmic rays also produce neutrinos in the atmosphere.[2] Usually the flux of atmospheric ν's masks completely the signals of extraterrestial ν's. There are however two domains where the flux of atmospheric ν's is strongly suppressed and thus may open a window for high energy ν-astronomy:

(i) At sufficiently high energies time dilation strongly suppresses the production of atmospheric ν's[3] (the mean decay lengths of π's and K's become larger than their mean free path for inelastic collisions in the atmosphere). In this region ν-astronomy requires extremely large detectors such as the Fly's eye and DUMAND because of the very small

† On leave from the Technion-Israel Institute of Technology, Haifa, Israel

ν fluxes and cross sections.

(ii) At energies below 1 GeV the geomagnetic field strongly suppresses the fluxes of atmospheric ν's from directions of low magnetic latitudes. (The geomagnetic field prevents primary cosmic rays with energy below the geomagnetic cutoff from reaching the atmosphere and produce atmospheric ν's.) If there are significant fluxes of astronomical ν's they may be detected by the massive underground proton decay detectors[4] in directions of low magnetic latitudes (large geomagnetic cutoffs).

In this paper I will present analytical estimates of the fluxes of both atmospheric and interstellar neutrinos. These estimates are based on simple analytical formulae that relate directly the underlying physics to the final numerical results. In particular these analytical formulae for the fluxes of atmospheric muons and neutrinos display explicitly their dependence on energy, zenith angle and geomagnetic cutoff (which depends on geographic location, zenith and azimuth). If the ν-background in proton decay experiments is due to atmospheric ν's, it should exhibit these dependences and in particular a significant asymmetry between the total fluxes of upgoing and downgoing ν's at all the sites of the major proton decay experiments, which result from the strong geomagnetic suppression of the ν-fluxes from directions of low magnetic latitudes. The reliability of our estimates is demonstrated by showing that (i) the predicted spectra of atmospheric π's and μ's, which are the main source of low energy atmospheric ν's, agree well with the measured spectra, and (ii) the predicted ν-fluxes agree very well with recent numerical estimates of these fluxes in the absence of geomagnetic effects.[5] My plan is as follows: First I will summarize production of atmospheric ν's neglecting geomagnetic effects, then I will summarize production of interstellar ν's and finally I will estimate the effects of the geomagnetic field on the fluxes of atmospheric ν's.

<u>Leptons from Meson Decays:</u> The differential fluxes of leptons, $\ell=\mu,\nu$, from $M\to\mu\nu$ decays are related to the differential fluxes of their parent mesons, $M = \pi^{\pm}, K^{\pm}$, through:

$$\frac{dF_\nu}{dE} = \sum_M \alpha_M \int_0^{\alpha_M^{-1}} dx \int_{\alpha_M E}^{\infty} \delta(E-xE_M) \frac{dF_M}{dE_M} dE_M , \tag{1}$$

$$\frac{dF_\mu}{dE} = \sum_M \alpha_M \int_{\beta_M^{-1}}^{1} dx \int_{E}^{\beta_M E} \delta(E-xE_M) \frac{dF_M}{dE_M} dE_M , \tag{2}$$

where $\alpha_M \equiv m_M^2/(m_M^2-m_\mu^2)$ and $\beta_M \equiv m_M^2/m_\mu^2$. Eqs. (1), (2) are valid for energies that satisfy $m_M^2/E^2 \ll 1$. The differential fluxes of the parent mesons (i.e. the mesons that decayed into the respective leptons and photons) which are produced by a primary flux of $dF_p/dE = cE^{-p}$ nucleons/cm^2•sec•sr•GeV at the top of the atmosphere, can be easily estimated assuming that the cross sections for inclusive production of hadrons in nucleon - air nucleus collicions obey Feynman scaling: $dn_h/dx = f_h(x)$; h = N, M, where N denotes either a proton or a neutron and f_h depends on x, the fraction of the momentum of the projectile carried by the produced hadron h, and not on energy. They are given by[6]:

$$\frac{dF_M}{dE} = \frac{B_M}{1+\gamma_M E} \; g_M^{at} \; \frac{dF_p}{dE} \quad . \tag{3}$$

In eq. (3) B_M is the branching ratio for the decay $M \to \mu\nu$ ($B_\pi = 1$, $B_K = 0.632$) and $\gamma_M^{-1} \cong m_M c h_o \sec\theta/\tau_M$ where τ_M is the proper lifetime of M, h_o = 6.5 km is the scale parameter of the upper atmosphere and θ is the zenith angle of the incident cosmic ray. (For zenith angles close to the horizon, such that $\sec\theta \gtrsim \sqrt{R/h_o}$ where R is the earth's radius, $\sec\theta$ has to be replaced by $\lambda(\theta)/\lambda(0)$ where $\lambda(\theta)$ is the atmospheric depth at zenith angle θ). The production coefficients g_M^{at} are defined by $g_M^{at} \equiv g_M/(1-g_N)$ where $g_h \equiv \int_0^1 x^{p-1} f_h(x)dx$. In Table I we list the values of g_h that were calculated by Liland[7] from accelerator data on inclusive production of hadrons in p-hydrogen collisions. (They display explicitly the approach to scaling with increasing energy). Because of the nuclear enhancement of the multiplicity they have to be multiplied by approximately 1.15 when applied to p-air nuclei collisions. When Eq. (3) is substituted into Eqs. (1) and (2) we find that

$$\frac{dF_\nu}{dE} \cong \Sigma \; g_M^{at} \; B_M \; \alpha_M \; K_M(\alpha_M E) \quad , \tag{4}$$

$$\frac{dF_\mu}{dE} \cong \Sigma \; g_M^{at} \; B_M \; \alpha_M \; [K_M(E) - K_M(\beta_M E)] \quad , \tag{5}$$

where $K_M(E) \equiv cE^{-p}/(p+(p+1)\gamma_M E)$. At low energies Eq. (5) has to be modified to include the effects of μ-decay and energy losses in the atmosphere. The energy losses are given approximately by $\rho^{-1}dE/dx = \alpha \sim 2.06$ MeV cm^2/gm, consequently muons at zenith angles θ with energy E at sea level (final atmospheric depth $\lambda_F \cong 1030 \cdot \sec\theta$ gm) are born with initial energy E_o which is given approximately by $E_o \cong E + \alpha(\lambda_F - \lambda_o)$. Their average energy $\bar{E}$ in the atmosphere is $\bar{E} \cong E + \alpha(\lambda_F - \lambda_o)$, and their

M	Decay	B_M	γ_M^{-1} (GeV)	g_M
π^o	$\pi^o \rightarrow 2\gamma$	1	3.63×10^{10}	$.03602 - .00364/\sqrt{E}$
π^+	$\pi^+ \rightarrow \mu^+\nu_\mu$	1	1.16×10^2	$.04190 + .00655/\sqrt{E}$
π^-	$\pi^- \rightarrow \mu^-\bar{\nu}_\mu$	1		$.03014 - .01019/\sqrt{E}$
K^+	$K^+ \rightarrow \mu^+\nu_\mu$	.632	8.70×10^2	$.007354 - .006456/\sqrt{E}$
	$K^+ \rightarrow \pi^o e^+\nu_e$	.0482		
K^-	$K^- \rightarrow \mu^-\bar{\nu}_\mu$	.632	8.70×10^2	$.002633 - .003179/\sqrt{E}$
	$K^- \rightarrow \pi^o e^-\bar{\nu}_e$	.0482		
K_L	$K_L \rightarrow \pi^\pm e^\mp \bar{\nu}_e(\nu_e)$	.387	2.08×10^2	$.003962 - .003912/\sqrt{E}$
	$K_L \rightarrow \pi^\pm \mu^\mp \bar{\nu}_\mu(\nu_\mu)$	.271		
μ^+	$\mu^+ \rightarrow e^+\nu_e\bar{\nu}_\mu$	1	1.04×10^0	
μ^-	$\mu^- \rightarrow e^-\bar{\nu}_e\nu_\mu$	1		$g_N = .3175 + .1816/\sqrt{E}$

Table I: The production coefficients g_M and g_N as calculated by Liland [7].

Site	Magnetic Latitute	Average Stormer Cutoffs (GeV)	
		Downgoing	Upgoing
Soudan Iron Mine, MN	56.3 N	1.7	8.6
Homestake Gold Mine, SD	52.2 N	2.3	8.4
Morton Salt Mine, OH	52.0 N	2.3	8.4
Silver King Mine, UT	47.6 N	3.2	8.6
Mont Blanc & Frejus Tunnels France/Italy	44.4 N	3.9	8.7
Kamioka, Japan	27.0 N	10.8	9.6
Kolar Gold Fields, India	2.2 N	16.0	8.6

Table II: The magnetic latitude and typical Stormer Cutoffs (protons) for the Major proton decay experiments, as calculated by Cooke [11].

probability to reach sea level before decay is given approximately by[6] $S_\mu(\bar{E}) \cong (\lambda_o/\lambda_F)^{1/\gamma_\mu \bar{E}}$ where $\lambda_o \cong 120$ gm/cm^2 and $\gamma_\mu^{-1} \equiv m_\mu c h_o \cdot \sec\theta/\tau_\mu$ $\simeq 1.04$ GeV. Eq. (5) corrected for μ-decay and energy losses in the atmosphere yields for sea level muons:

$$\frac{dF_\mu}{dE} \cong \Sigma\, g_M^{at}\, B_M \alpha_M\, S_\mu(\bar{E})\,[K_M(E_o)-K_M(\beta_M E_o)], \quad (6)$$

where g_M^{at} are evaluated at energy E_o. In Fig. 1 Eq. (6) is compared with measured fluxes of sea level μ's at zenith angles $\theta = 0°$[8] and $\theta = 75°$[9] assuming a primary cosmic ray flux[10] of $1.6\ E^{-2.67}$ nucleons/cm^2·sec·sr·GeV. Fig. 1 demonstrates excellent agreement between theory and experiment.

<u>The ν_μ and ν_e Fluxes from $\mu \to e\nu_e\nu_\mu$ Decays:</u> The ν spectra from μ decays are given by

$$\frac{dI_\nu}{dE} = \int_0^1 dx \int_E^\infty \sum_{i=0}^{3} A_i x^i\, \delta(E-xE_\mu)\, \frac{dF_\mu}{dE_\mu}\, dE_\mu, \quad (7)$$

where A = (5/3 , 0, -3, 4/3) for ν_μ's and A = (2, 0, -6, 4) for ν_e's and where we have neglected small corrections due to μ polarization. The probability that a muon will decay in flight is given approximately by

$D_\mu(E) = 1-S_\mu(E) \cong \delta/(\delta+\gamma_\mu E)$ where $\delta \cong \ell n\,(\lambda_F/\lambda_o)$. Eqs. (6) and (7) yield then the following ν-fluxes produced by μ-decays in the atmosphere:

$$\frac{dI_\nu}{dE} \cong \Sigma \alpha_M B_M g_M^{at} (1-\beta_M^{-p}) \sum_{i=0}^{3} \delta A_i/(p_i\delta+p_{i+1}\gamma_\mu E)\, \frac{cE^{-p}}{p} \quad , \quad (8)$$

where $p_i \equiv p + i$. Note that the ν-fluxes from μ-decays depend on zenith angle through the dependence of δ and γ_μ on zenith angle. In Figs. 2, 3 we compare our analytical estimates as given by eqs. (4) and (8) for the total fluxes of atmospheric ν_μ's + $\bar{\nu}_\mu$'s and ν_e's + $\bar{\nu}_e$'s in the absence of geomagnetic effects (full lines) and the numerical estimates of these fluxes by Volkova.[5] Figs. 2, 3 demonstrate excellent agreement between our simple analytical formulae and the numerical estimates of Volkova.[5]

<u>The ν_e Fluxes from $K \to \pi e \nu_e$ Decays:</u> Time dilation suppresses μ decays in the atmosphere. Consequently at high energies K_{e3} decays become the most important source of atmospheric ν_e's. The ν_e spectra from K_{e3} decays are given by

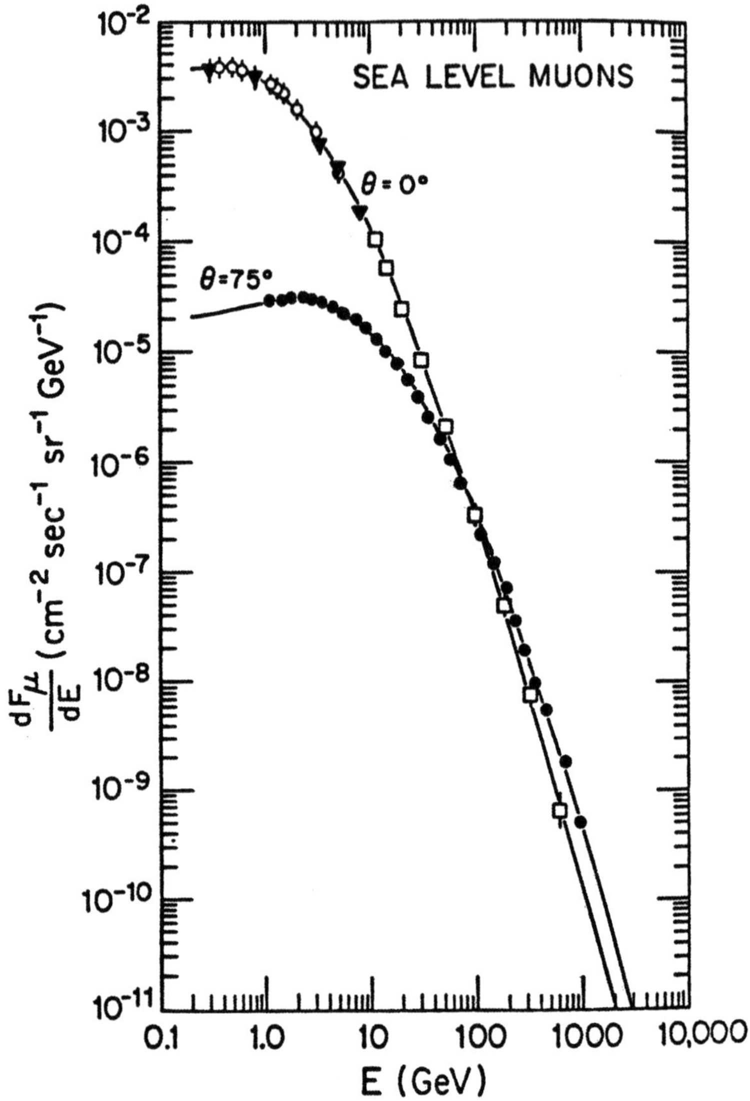

Fig. 1. Comparison between our theoretical predictions and measured fluxes of sea level muons at zenith angles $\theta=0^{\circ}$ and $\theta=75^{\circ}$.

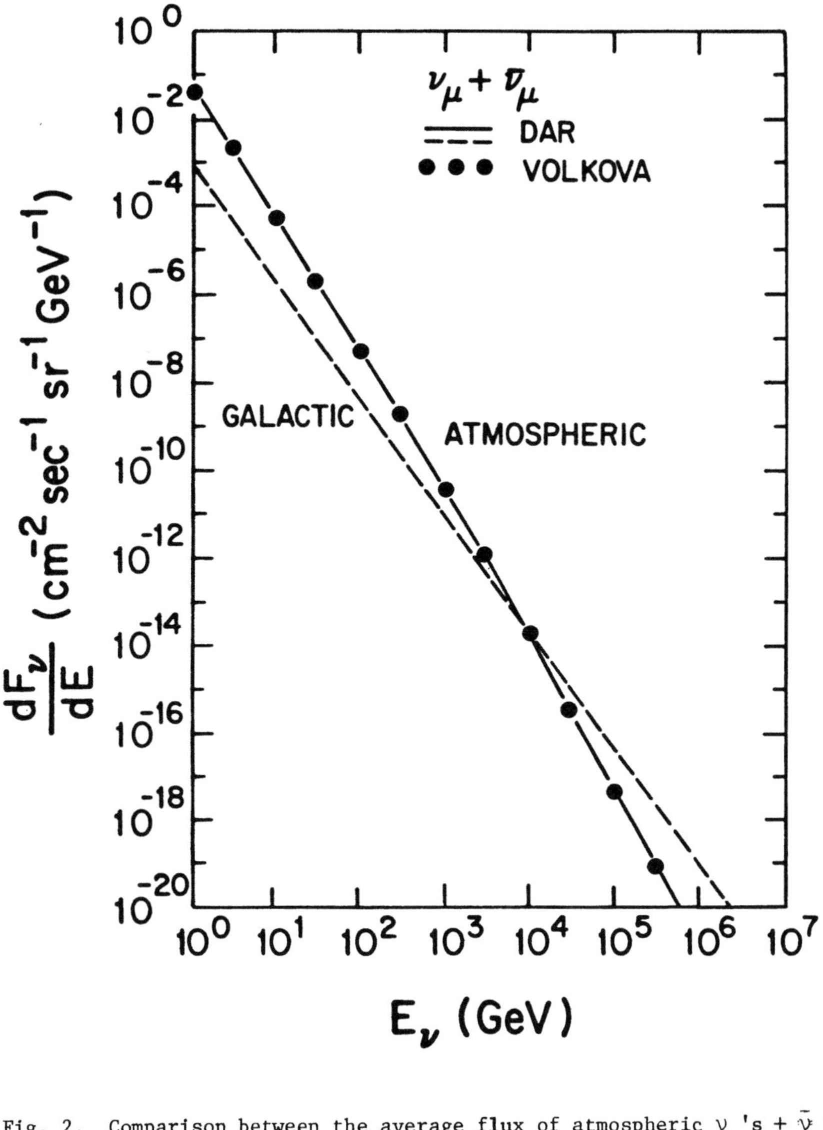

Fig. 2. Comparison between the average flux of atmospheric ν_μ's + $\bar{\nu}_\mu$'s and the flux of interstellar ν_μ's + $\bar{\nu}_\mu$'s from directions close to the galactic center assuming $\lambda \sim 1$ gm/cm^2.

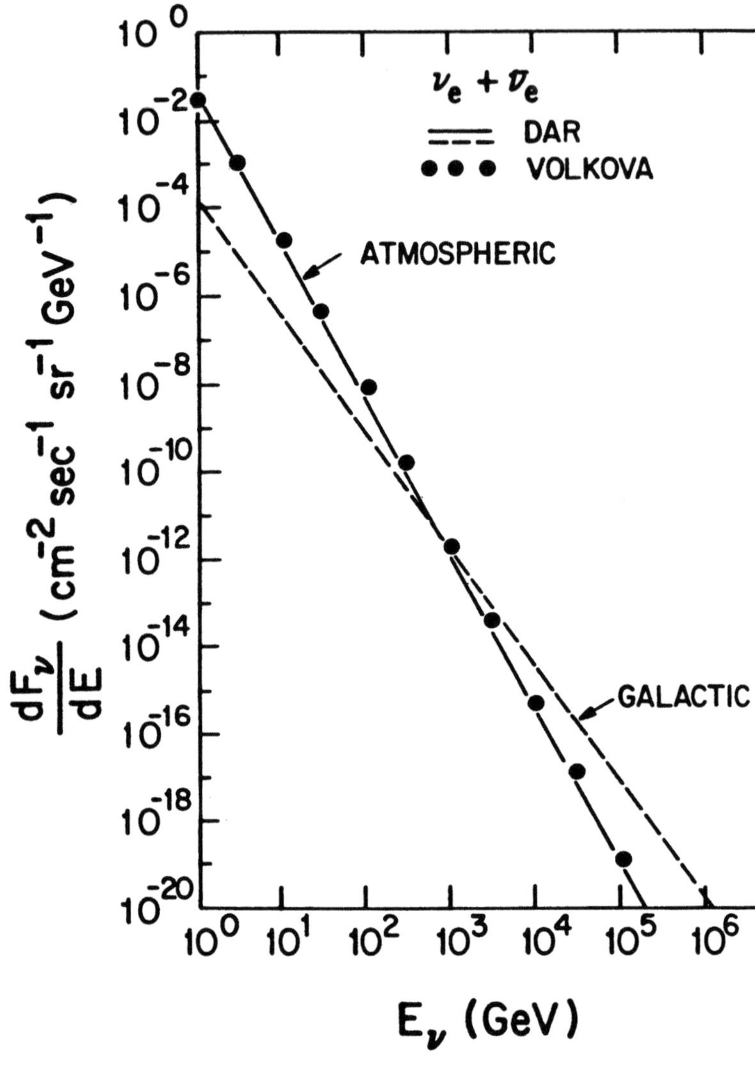

Fig. 3. Comparison between the average flux of atmospheric ν_e's + $\bar{\nu}_e$'s and the flux of interstellar ν_e's + $\bar{\nu}_e$'s from directions close to the galactic center assuming λ~1 gm/cm^2.

$$\frac{dF_{\nu_e}}{dE} = \sum_{K^\pm, K_L} \eta_K \int_0^{1/\eta_K} dx \int_{\eta_K E}^{\infty} 1.72 \left[\sum_{i=0}^{3} C_i x^i + .08 \ln(1-x) \right] \delta(E - xE_k) \frac{dF_K}{dE_K} dE_K \tag{9}$$

where $\eta_K = m_K^2/(m_K^2 - m_\pi^2)$ and $C = (1.33, .08, 4.15, -5.25)$. When eq. (3) is substituted into eq. (9) one finds that the ν_e flux from K_{e3} decays is given approximately by

$$\frac{dF_{\nu_e}}{dE} \cong \sum_{K^\pm, K_L} g_K^{at} B_{K_{e3}} \eta_K^p \, cE^{-p} \, 1.72 \left[\sum_{i=0}^{3} \frac{C_i}{\eta_K^i p_i + \eta_K^{i+1} p_{i+1} \gamma_K E} + \frac{.08 \ln(1-1/\eta_K)}{p_o + \eta p_1 \gamma_K E} \right] \tag{10}$$

Interstellar Neutrinos: The production rate of interstellar ν's depends on the matter density and the cosmic ray flux at the site of production. If the flux of primary cosmic rays is the same everywhere then the ν-fluxes from $M_{\mu 2}$ and μ_{e3} decays that arrive from a given galactic direction are given respectively by

$$\frac{dF_\nu}{dE} = \sum_M g_M^{gl} B_M \alpha_M^{1-p} cE^{-p}/p \tag{11}$$

$$\frac{dI_\nu}{dE} = \sum_M g_M^{gl} B_M \alpha_M (1-\beta_M^{-p}) \sum_{i=0}^{3} A_i/p_i \, cE^{-p}/p \tag{12}$$

where $g_M^{gl} = (1-e^{-\lambda/\lambda p}) g_M$, λ is the amount of interstellar matter (gm/cm^2) along the incident direction and $\lambda_p \cong 40$ gm/cm^2. For directions close to the galactic center $\lambda \sim 1$ gm/cm^2. In Figs. 2 and 3 we plot the fluxes of $\nu_\mu + \bar{\nu}_\mu$ and of $\nu_e + \bar{\nu}_e$, respectively, as given by eqs. (11) and (12) for $\lambda \sim 1$ gm/cm^2. From Figs. 2 and 3 one can see that the fluxes of interstellar ν_e's + $\bar{\nu}_e$'s and ν_μ's + $\bar{\nu}_\mu$'s at energies respectively above 10^3 GeV and 10^4 GeV become larger than the corresponding fluxes of atmospheric ν's, i.e. the windows for high energy ν_e and ν_μ-astronomy open around 10^3 GeV and 10^4 GeV respectively.

The Effects of the Geomagnetic Cutoff: The earth's magnetic field prevents primary cosmic rays with energy below the geomagnetic cutoff energy E_c from reaching the atmosphere. For protons, the cutoff momentum p_c is given approximately by the Stormer Formula:

$$p_c = 59.4 \cos^4\lambda/R^2 (1 + \sqrt{1 - \cos^3\lambda \cdot \sin\theta \cdot \sin\phi}\,)^2 \text{ GeV/c} \tag{13}$$

where λ is the magnetic latitude, θ is the zenith angle, ϕ is the azimuth measured clockwise from the magnetic north and R is the distance from the dipole center of earth in units of earth radii. Table II lists typical Stormer cutoffs at the locations of the major proton decay experiments. The measured geomagnetic cutoffs are significantly larger than those predicted by the Stormer Formula, typically by about 25% for vertical directions, by about 35% for inclined directions and by up to 100% for horizontal directions. Since the real field cutoffs are not well measured all over the globe, in this paper we shall base our estimates on the Stormer Formula.

One can easily show that for a thick atmosphere and $dn_N/dx \sim$ const the effective flux of nucleons with energies below E_c that are produced in air showers is given by $dF_N/dE = g_N c E_c^{-(p-1)}/E$. For the choice* $dn_M/dx \sim (1-x)^3/x$ we then find that the atmospheric fluxes of mesons with energy below E_c (i.e. $x_c = E/E_c < 1$) are given by:

$$\frac{dF_M}{dE} \cong \frac{g_M^{at}(E_c)B_M \; cE_c^{-p}}{(1+\gamma_\mu E)B(p-1,4)} \left[G_1(x_c,p) + g_N G_2(x_c)\right] \quad , \tag{14}$$

$$G_1(x,p) \equiv 1/(p-1)x - 3/p + 3x/(p+1) - x^2/(p+2) \quad ,$$

$$G_2(x) \equiv (11/6 - \ell n x + 3x - 3x^2/2 + x^3/3)/x \quad .$$

In eq. (14), G_1 is due to nucleons with energy above E_c while G_2 is due to nucleons with energy below E_c. (Except for extremely low values of x_c, $G_1 \gg g_N G_2$). When we substitute eq. (14) into eq. (1) we find that the atmospheric fluxes of ν's obtained directly from $M\to\mu\nu$ decays have the following form at ν-energies below E_c/α_M

$$\frac{dF_\nu}{dE} \cong g_M^{at}(E_c)B_M\alpha_M \frac{cE_c^{-p}}{p} \left[1 + \frac{1}{B(p-1,4)} \left(W_1(p,z_M) + g_N W_2(z_M)\right)\right] \quad , \tag{15}$$

$$W_1(p,z) = (1/z-1)/(p-1) + 3\,\ell n z/p + 3(1-z)/(p+1) - (1-z^2)/2(p+2) \quad ,$$

$$W_2(z) = 3/2 - \ell n z/z - 17/6z - 3\ell n z + 3z/2 - z^2/6 \quad .$$

where $z_M = \alpha_M E/E_c < 1$. Similar analytical expressions can be derived[6] for the fluxes of atmospheric μ's from $M\to\mu\nu$ decays and for atmospheric ν_μ's and ν_e's from subsequent $\mu\to e\nu_e\nu_\mu$ decays.

---

*This parametrization was suggested to me by T.K. Gaisser after the conference. It bounds dn_M/dx from above since it is singular at $x=0$; while for $x\to 0$, dn/dx actually falls when $x < m_T/\sqrt{s} \sim 0.4$ GeV$/\sqrt{S}$. My original choice, $dn_M/dx \sim (1-x)^3$, which is finite when $x\to 0$ and which has the correct large x behaviour, leads to a much stronger geomagnetic suppression of the ν-fluxes at $E_\nu \ll 1$ GeV, as was shown in the figures presented in my talk and included in ref. [6].

Results: Using our analytical formulae and the geomagnetic cutoffs that were calculated by Cooke[11] we calculated the atmospheric ν-fluxes at the sites of the major proton decay experiments. We found that the geomagnetic field strongly suppresses the fluxes of atmospheric ν's below 1 GeV from directions of low magnetic latitudes (large geomagnetic cutoffs) and produces a strong asymmetry between upgoing ($\theta>90^o$) and downgoing ($\theta<90^o$) neutrinos:stAt the northern sites the geomagnetic field has only a little effect on downgoing ν's while it strongly suppresses the total fluxes of upgoing ν's which are approximately the same ($<E_c> \cong 8.5$ GeV) at all the sites including the Kolar Gold Field in India, where the geomagnetic field strongly suppresses also the total flux of downgoing ν's ($<E_c>\sim16$ GeV). In Figs. 4, 5 we demonstrate our results for the Morton Salt Mine site in Ohio and the Kolar Gold Field site in India. For reference we also plot there the ν-fluxes in the absence of a geomagnetic field and the predicted flux of horizontal ν's ($\theta=90^o$) from the east at the Kolar Gold Field ($E_c = 54.5$ GeV). Since $\sigma_\nu \sim E_\nu$ the ratio of ν-interactions by upgoing and downgoing ν's is given by the ratio of $\int E(dF_\nu/dE)dEd\Omega$ for $\theta>90^o$ and $\theta<90^o$. For a threshold visible ν-energy between 0.2 to 0.3 GeV this ratio for the northern sites is between 0.54 ± 0.07 and 0.64 ± 0.06, while it is 1.56 ± 0.22 at the Kolar Gold Field in India.

In conclusion, if the ν background in proton decay experiments is mainly due to atmospheric neutrinos, it should exhibit the anticipated dependences on zenith, azimuth, and geographic location for both upgoing and downgoing ν's. For directions of low magnetic latitudes the geomagnetic cutoff strongly suppresses the fluxes of atmospheric neutrinos and may open a window for ν-astronomy with proton decay detectors. Although we believe that our analytic model predicts reasonably well the fluxes of atmospheric ν's, neither our analytic model nor Monte Carlo Codes (one dimensional codes in particular) can be sufficiently reliable to determine whether or not neutrinos oscillate[12] by comparing calculated and measured fluxes of atmospheric ν's, because both calculations are only approximate, because the uncertainties in the experimental input data are large and because the results at low E_ν are very sensitive to the input and to the approximations.

The author would like to thank R. Amado, S. Bludman, M. Cherry, T. Gaisser, P. Langacker, H. Lee and T. Stanev for useful discussions and suggestions.

This research was supported by the U. S. Department of Energy and the National Science Foundation.

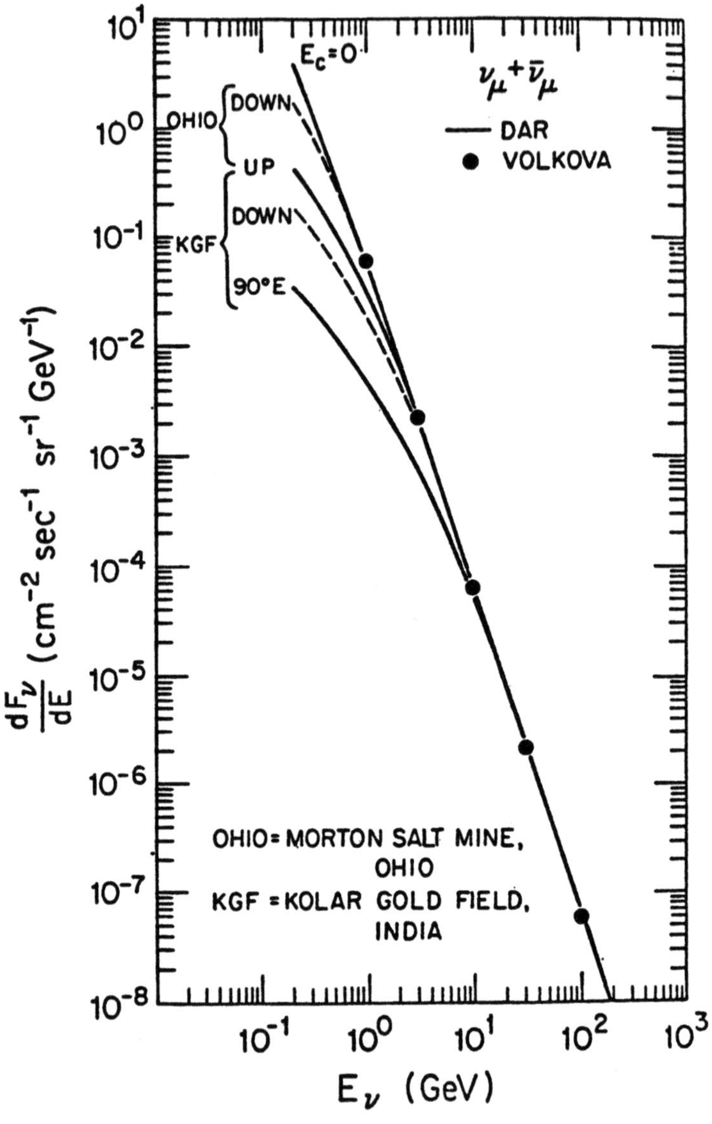

Fig. 4. Predicted fluxes of atmospheric ν_μ's + $\bar{\nu}_\mu$'s from meson and muon decays at the sites of the major proton decay experiments.

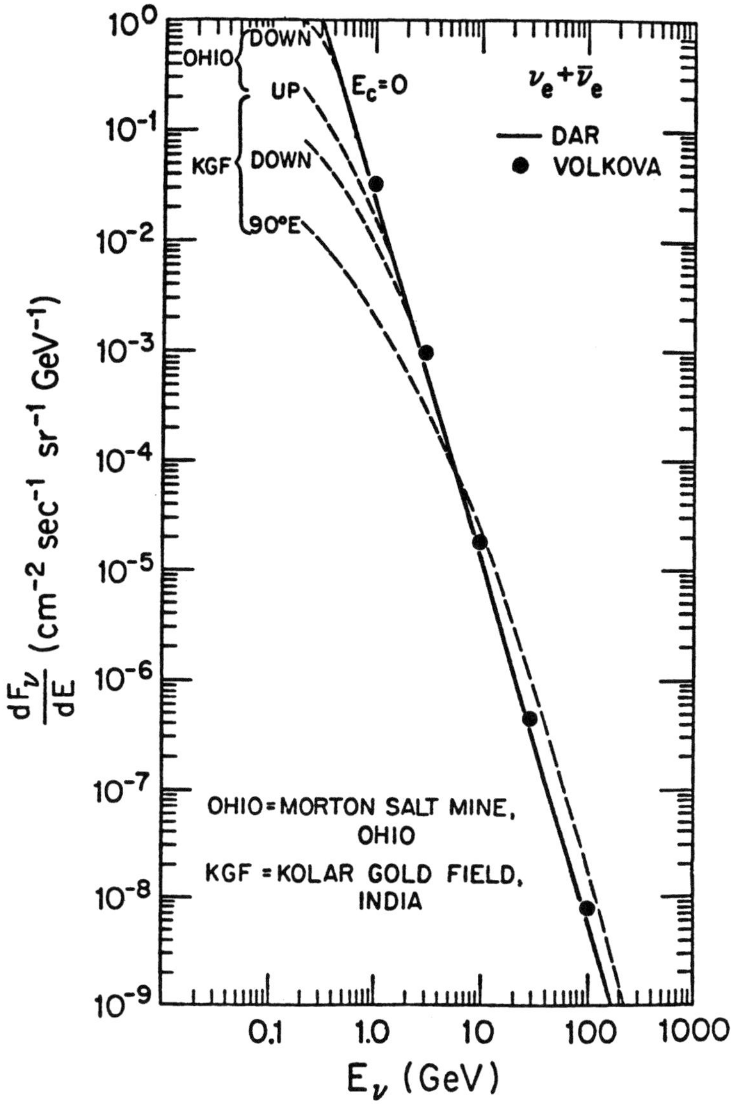

Fig. 5. Predicted fluxes of atmospheric ν_e's + $\bar{\nu}_e$'s from muon decays at the sites of the major proton decay experiments.

References and Footnotes

[1] Berezinsky, V. S. and Zatsepin, G. T., Soviet J. Nucl. Phys. 11, 111 (1970). For recent discussions see P. W. Stecker, AP 228, 919 (1979) and D. Eichler and D. N. Schramm, Nature 275, 709 (1978).

[2] Zatsepin, G. T. and Kuz'min, V. A., Soviet Phys. JETP 14, 1294 (1962); Cowsik, R., et al., Proc. Intl. Conf. Cosmic Rays, Jaipur 6, 211 (1962); Osborne, J. L. et al., Proc. Phys. Soc. 86, 93 (1965); Tamm and Young, Proc. Intl. Cosmic Ray Conf. Budapest (1969); Volkova, L. V., Proc. 1978 DUMAND Workshop, Vol. 1, p. 75 (DUMAND Publications, Ed. A. Roberts), and reference 5.

[3] See for instance, Cline, D., Proc. 1976 DUMAND Workshop, p. 265 (DUMAND publications, Ed. A. Roberts) and references therein.

[4] For a recent theoretical review of proton decay see, for instance, Langacker, P. L., Phys. Reports 72C, 185 (1981) and references therein. For a recent experimental review of proton decay see, for instance, Perkins, D. H., Proceedings of the XXI Int. Conf. on HEP, Paris, July 26-31, 1983 and references therein.

[5] Volkova, Sov. J. Nucl. Phys. 31, 784 (1980).

[6] Dar, A. and Cherry, M., "Atmospheric and Interstellar Neutrinos" paper in preparation.

[7] Liland, A., Proc. 16th Intl. Cosmic Ray Conf. Kyoto 13, 353 (1979).

[8] Allkofer, O. C., Proceedings of the 1978 DUMAND Workshop, Vol. I, p. 13 (DUMAND Publications, Ed. A. Roberts).

[9] Jokisch, H., et al., PR D19, 1368 (1979).

[10] Ormes J. and Freir, P., Ap. J. 222, 471 (1978).

[11] Cooke, D. J., Preprint, University of Utah (March, 1983).

[12] Bilenky, S. M. and Pontecorvo, B., Phys. Rep. 41, 225 (1978).

MONOPOLES, GAUGE FIELDS, AND ANOMALIES

by

Alfred S. Goldhaber
Institute for Theoretical Physics
State University of New York at Stony Brook
Stony Brook, Long Island, New York 11794

The possible catalysis of baryon disintegration by magnetic poles is a beautiful example of the way in which monopoles illuminate fundamental concepts of physics. This hypothetical process is followed step by step, with an effort to minimize arbitrary hypotheses and to state clearly those which remain, as well as the consequences of changing them. The conclusions are that there should be a large cross section for catalysis in collisions of slow monopoles with nucleons, though not with large nuclei, but the intrinsic rate for destruction of a nucleon captured by a pole could be many orders of magnitude below the 10^{23} per second which characterizes strong interactions. It is argued that Rubakov's and Callan's pictures of catalysis are compatible but distinct, and either permits a strong rate but requires at least a weak rate.

1. Introduction: The concept of monopole catalysis of nucleon disintegration, like the concept of a monopole itself, has many layers, which must be peeled off one after another if we hope to end with a reliable conclusion about what really would happen if a nucleon met a monopole.

Let us begin with the unrealistic case of quarks scattering on a classical soliton field, the Dokos-Tomaras [1] SU(5) version of the SU(2) monopole invented by 't Hooft [2] and Polyakov [3]. The Dirac equation for Fermions interacting with this monopole conserves helicity, but for the lowest angular momentum state helicity is proportional to the product of the sign of a particle's charge and the sign of its radial velocity. Thus, to conserve helicity, particles must change charge. This is possible because the classical soliton field does not have well-defined electric charge or color. Indeed the latter fact can be exploited to give an argument for confinement of color to make quantum chromodynamics consistent [4].

At this stage, it seems obvious that monopoles should catalyze baryon-lepton transmutation, since the classical monopole does this automatically. However, we know that nature requires systems with definite electric charge, and if a monopole initially uncharged were to acquire charge by flipping the flavor of a Fermion which collided with it, the resulting energy cost would be overwhelmingly large compared to the incident Fermion energy. Thus, the second thought, valid for the SU(2) monopole, is that charge conservation prevents the monopole from flipping Fermion flavor [5]. Now before attacking this question for the SU(5) monopole, let us examine the long-range interactions of nucleons and poles.

2. The Approach - Bounce or Capture? It is well known that a spinless charged particle feels a centrifugal potential even if it approaches a monopole in the lowest partial wave, while a spin 1/2 particle with a Dirac gyromagnetic ratio feels no repulsion. The result is that a proton or neutron, because of the anomalous moment, feels a centripetal attraction to the pole, while a heavy nucleus is repelled. These long-range interactions determine whether the colliding objects get close enough together for capture to occur at low relative velocities [6].

The most interesting case to consider for laboratory experiment is the collision of a slowly moving pole with a hydrogen molecule. I am unaware of any study on the penetrability of the molecule, but Parke [7] has shown, using a variational electron wavefunction in the Born-Oppenheimer approximation, that there is no barrier to penetration of an isolated hydrogen atom by a pole.

On the plausible assumption that the molecule as a whole is penetrable, we are entitled to focus on the case of a free proton falling into the pole. At radii of the order of the electron Compton wavelength, the vacuum becomes a highly polarizable plasma of electron-positron pairs [5,8]. It is nearly certain that there would be some dissipation of the proton's kinetic energy in this plasma, with the energy radiated eventually as photons or perhaps even free pairs. The large cross section for close approach, multiplied by a high dissipation probability, gives a large cross section for capture of the proton [9,10].

Once captured, the proton faces at most finite barriers shielding the monopole; sooner or later it must penetrate within 1 fm and dissolve into quarks. This sets the stage for the approach to the monopole center, and possible catalysis.

3. Catalysis - Can it be? Must it be? In our first, naive approach to catalysis, we neglected Coulomb energies which were much bigger than Fermion energies of interest. This time let us freeze those very high energy degrees of freedom, surrounding the monopole with an imaginary sphere whose radius is much bigger than any length associated with internal monopole structure, yet much smaller than any length associated with known "low energy" physics, e.g. the Compton wavelength of the W Boson or the heaviest third-generation Fermion.

The sphere represents a boundary which is insensitive to monopole dynamics, and also should be unable to distinguish among the different generations of "light" particles. Therefore, we may insist that the constraints imposed at this boundary treat all generations equivalently and that they do not allow to flow in any charge coupled to a known long-range field.

First, consider the case of one generation of Fermions, as proposed by Callan [11]. In the presence of a minimal SU(5) monopole, four types of particle may flow in with no centrifugal barrier, namely electrons and u_1, u_2 or d_3 quarks, where the subscripts stand for color labels. Callan observed that elementary processes which violate baryon number [Δ(B+L) = ± 1] are consistent with the constraint against accumulation by the monople of electric and color charges. Thus these conservation laws do not preclude strong rates for proton decay.

I have interpreted Rubakov as saying something different, and a recent paper appears to confirm my view [12]. There is one more charge which should be conserved, and that is T_{3L} of $SU(2)_{weak}$, which is coupled to the Z Boson. Callan's conditions are compatible with the Z-charge constraint, but for this single-generation case they are actually required by that constraint!

What has happened is that a B+L violation which was built in to the unified SU(5) theory has disappeared from our low energy Lagrangian. However, consistency conditions imposed by the conservation laws of the latter theory compel the violation once again. From the point of view of the low energy theory, there is an anomalous nonconservation of B+L. Without ever writing down a triangle diagram we are led by the monopole to discover anomalies!

While Callan and Rubakov coincide for one generation, they need not coincide for more. Evidently strong violation is still possible with more generations; we may just treat them independently. Furthermore, the original SU(5) soliton boundary conditions, modified as little as possible to obey charge conservation, have the feature that a heavy generation would automatically decouple, leaving the light generations again forced to violate B+L at a strong rate. Thus, it is possible and even plausible that strong violation occurs.

Nevertheless, we may ask if our general ground rules require violation. The answer is that they require some, but it could be weak. Suppose that the boundary conditions preserved the difference in B+L for any pair of generations. In that case, if no other force violated this assumed conservation law, B+L violation would be intrinsically strong but actually unobservable at low energies. The reason is that violation in the lightest generation would require producing antiparticles of the heaviest generation, and sufficient energy would not be available for this. The saving grace is that the charged weak current does mix generations, so that violation for the lightest generation could proceed, albeit at a weak rate. The weak generation-mixing would be the rate-controlling step in catalysis.

I believe that the weak decay rate is the right one, for the following reason. If the true unified theory were left-right symmetric and did not violate B+L, then there should be no anomalous violation in the low energy theory. Such a result would be automatic if we insisted that only anomalies forced by couplings to the low energy gauge fields be included in our boundary conditions. This means that particles whose masses may be larger than we can now observe, but still utterly negligible compared to the monopole mass, should not be assumed to decouple from the very lightest particles. In particular, the heaviest known Fermions should not decouple, hence the intrinsically strong but effectively weak violation seems most natural.

What would all this mean for laboratory experiments? A slow monopole passing through suitable (hydrogen-bearing) material would capture protons with a large cross section. The rate of catalysis could be very low, and yet lead to catalysis in a time shorter than the time between captures. It would be very hard to tell whether the rate was strong or weak.

Of course, for monopoles in a neutron star, the weak rate would imply negligible X-ray emission from the star for any reasonable monopole flux, meaning that the wonderfully ingenious limit on flux times cross section [13,14] should be reinterpreted as a limit on flux times intrinsic catalysis rate, a potentially very different quantity. A proposed limit on flux times cross section from Jupiter [15] is not open to this objection, but does depend on the presence of hydrogen at Jupiter's center.

4. Outlook: If superheavy monopoles occur in nature, and if we already know the essence of the "low energy" dynamics which apply on our side of an imaginary sphere around a pole, then observable catalysis appears inevitable. Failure to see it would be a sign that accessible physics has more particles and interactions than we have yet found, and thereby would also rule out SU(5) as a candidate for a unified theory. If catalysis does occur, but at a weak rate, it may be very hard to predict explicitly the modes of decay for the nucleon-monopole system.

Even if monopoles are not to be seen, the purely theoretical analysis of transmutation has been most instructive already. It is a grand package uniting the notion of explicit baryon destruction characterized by gauge couplings, with 't Hooft's [16] anomaly leading to the same consequences. Only, the monopole makes the exponentially suppressed anomaly into a strong effect. The monopole transmutes baryon transmutation mechanisms!

Recent studies by Kazama [17], Yoneya [18], and Yan [19] have probed the monopole's tendency to break, or at least bend, other conservation laws. They point out that, while charge is conserved globally, there may be short-range charge non-conserving correlations in the vicinity of the monopole. Also, Yan remarks that there is no consistent theory for a purely left-handed doublet interacting with a monopole. I consider this a perfectly acceptable alternative to Witten's much more sophisticated demonstration [20] that SU(2) gauge theory with a left-handed fermion doublet is undefined . The point is simple: A left-handed particle comes in to the pole, but there is no right handed partner to bounce out, and a left handed particle of opposite charge would break charge conservation (required by gauge invariance). Thus, there is no solution short of an anomaly which turns Fermions into anti-Fermions. Knowing that our low energy theory comes from a fundamental interaction which preserves Fermions, we cannot allow such an anomaly. Hence the framework is inconsistent.

It seems safe to guess that monopoles have more surprises in store, not necessarily including personal appearances.

In thinking about this subject, I have gained from discussions with Curtis Callan, Boris Kayser and Robert Shrock. Critical remarks of Edward Witten have sharpened the formulation, but perhaps not to his satisfaction. Most of all, the key ideas were hammered out in exhilarating colloquies with Ashoke Sen, to whom I am deeply indebted. This work was supported in part by the National Science Foundation.

References

[1] C.P. Dokos and T. N. Tomaras, Phys. Rev. D21. 2940 (1980).

[2] G. 't Hooft, Nucl. Phys. B79. 276 (1974).

[3] A. M. Polyakov, JETP Lett. 20, 194 (1974).

[4] A. Abouelsaood, to be published. P. Nelson, Phys. Rev. Lett. 50, 939 (1983). P. Nelson and A. Manohar, ibid, 943 (1983).

[5] C. G. Callan, Jr., Phys. Rev. D26, 2058 (1982).

[6] J. Arafune and M. Fukugita, Phys. Rev. Lett. 50, 1901 (1983).

[7] Stephen J. Parke, SLAC, to be published.

[8] J. S. Trefil, H. P. Kelly, and R. T. Rood, Nature 302, 111 (1983).

[9] A. S. Goldhaber, in Wingspread Workshop on Magnetic Monopoles, R. A. Carrigan, Jr. and W. P. Trower, eds. (Plenum, NY) 1983.

[10] S. Nussinov and L. Stodolosky, unpublished, (1982).

[11] C. G. Callan, Jr., Nucl. Phys. B212, 391 (1983).

[12] V. A Rubakov and M. S. Serebryakov, Nucl. Phys. B128, 240(1983).

[13] E. W. Kolb, S. A. Colgate and J. Harvey, Phys. Rev. Lett 49, 1373 (1982).

[14] S. Dimopoulos, J. Preskill and F. Wilczek, Phys. Lett 119B, 320 (1982).

[15] M. S. Turner, Nature 302, 804 (1983).

[16] G. 't Hooft, Phys. Rev. Lett. 37, 8 (1976).

[17] Y. Kazama, Univ. of Kyoto, to be published.

[18] T. Yoneya, Tokyo Univ., to be published.

[19] T.-M. Yan, Cornell Univ., to be published.

[20] E. Witten, Phys. Lett. 117B, 324 (1982),

THE IBM MONOPOLE EXPERIMENTS

C. D. Tesche +

Abstract

Conventional superconducting magnetometers used for monopole detection must be operated in an extremely stable, low field environment. This places a severe restriction on the cross sectional area of such detectors. In addition, the geometry of these detectors does not easily permit coincidence operation with other particle detectors. Gradiometer monopole detectors developed at IBM permit a significant relaxation of the constraint on the stability of the magnetic field, and on the geometry of the detector. A prototype detector consisting of two 100 cm^2 loops operated in coincidence has been constructed and operated sucessfully in a milligauss environment. No events consistent with a monopole have been observed in 95 days of operation. A larger detector is now being fabricated.

The experiments described in this paper are the result of an ongoing collaboration with C. C. Chi, C. C. Tsuei, P. Chaudhari, and S. Bermon.

+ IBM Research Center, Yorktown Heights, New York 10598

1. Motivation

There are two extremely attractive features of superconducting inductive monopole detectors. First, the detector measures directly the magnetic charge of any incident particle. Second, the response of the detector is independent of all other characteristics of the particle such as mass, charge, etc, and of the detailed nature of the interaction of the particle with matter. In particular, this detection scheme does not require an estimate of the ionization properties of the particle, nor of the probability that the particle may induce proton decay. As a result, superconducting inductive detectors can be used to set a highly reliable, model independent upper limit on the monopole flux.

Commercially available superconducting magnetometers have been used to detect changes in magnetic field intensity for many years. Many of these systems are sensitive enough to detect the passage of a monopole. The most well known superconducting magnetometer which has been used as a monopole detector was constructed and operated by Blas Cabrera. [1] A single event consistent with a monopole was observed in February, 1982. This detector is described in detail in the proceedings of the Third Workshop on Grand Unification. [2] Although systems designed to be operated as conventional magnetometers may have the necessary sensitivity, in general the optimal detection area is relatively small, usually less than a few tens of cm^2. As a result, the total available cross sectional area of detectors of this type is much smaller than that of more conventional high energy particle detectors. Thus, at this time, monopole flux levels set by superconducting magnetometers are several orders of magnitude higher than those set by particle detectors. In addition, the geometry of conventional magnetometers is such that efficient operation in coincidence with other types of particle detectors is all but impossible. This is unfortunate since it would clearly be useful to operate both types of detectors in coincidence in a proton decay experiment.

2. Superconducting Inductive Detectors

Although a conventional magnetometer and an inductive monopole detector may satisfy different design constraints, the two systems have the following common features. Both detectors consist of a continuous loop of superconducting material which is inductively coupled to a SQUID (Superconducting QUantum Interference Device). The loop contains a pick-up coil of inductance L_p in series with a coil of inductance L_c which is coupled through a mutual inductance to the SQUID. The voltage across the output of the SQUID electronics is proportional to the current flowing around the superconducting loop. This current is, in turn, proportional to the change in the ambient magnetic field threading the loop. In a conventional magnetometer, these changes correspond to the signal to be detected. As a result, the pick-up coil usually consists of a single or multiple turn loop. The optimal pick-up coil inductance is equal to the largest inductance, L_c, that can be coupled efficiently to the SQUID loop. Since the resolution of the SQUID increases as the SQUID loop inductance decreases, the coupling and pick-up inductances are usually limited to less than a few microhenries.

For a superconducting magnetometer used as a monopole detector, all changes in the field other than those generated by an incident monopole are considered to be a source of noise. As a result, the device must be operated in an extremely stable magnetic field environment. The passage of a monopole through the loop can be detected if the fluctuations in the ambient field during the measurement time correspond to a change in the flux linking the loop of $\Delta\Phi << 2\Phi_o$, where the flux quantum $\Phi_o = hc/2e = 2.07 \times 10^{-7}$ gauss cm^2 (for a discussion of this point, see reference 2). In a uniform field, the flux $\Delta\Phi = \Delta B_n A$, where B_n is the component of the field normal to the loop and A is the area of the loop. Hence any attempt to increase the area of the pick-up loop, and thus the cross sectional area of the detector, must be accompanied by a corresponding decrease in the influence of the ambient field fluctuations, including those generated by the motion of the detector with respect to the external field. This is usually accomplished by fixing the detector rigidly within some form of magnetic shielding. Room temperature, high permeability shields (μ metal shields) tend to be bulky, with a typical aspect ratio of 3 to 1 for a cylindrical geometry. This effectively prevents close packing of detectors. Superconducting shields are also frequently used. In these shields, the magnetic field lines which penetrate the shield as it is cooled into the superconducting state are trapped in the material, usually at local imperfections called pinning sites. Changes in the ambient field generated by sources exterior to the shield produce screening currents on the outer surface of the shield. As a result, the field inside the shield is extremely stable. Unfortunately, the flux trapped within the shield walls can be activated between pinning sites by a variety of mechanisms. This will produce a step-like change in the flux linking the detector coil. In most cases, the response time of the magnetometer is such that the monopole signal also appears equivalent to a step-like change in the flux. Thus, flux jumps in the shield can generate spurious, monopole-like signals. Furthermore, since the minimum magnetic flux that can be trapped at a single pinning site is Φ_0, the number of flux jumps tends to scale with the cross section of the detector.

The superconducting shields can also affect the characteristics of the monopole signal. The inductance of the superconducting loop containing the pick-up and coupling coils can be screened by the superconducting shields. However, the effective inductance is a function of the geometry of the shields and coils, and not of the monopole trajectory. More importantly, persistent screening currents generated within the shield by the passage of a magnetically charged particle link a flux, Φ_s into the pick-up coil which is some fraction of $2\Phi_0$. As a result, the apparent change in the flux linking the coil as the monopole passes through is decreased below $2\Phi_0$ by an amount which depends not only on the geometry of the detector and shield, but also on the monopole trajectory. In addition, a monopole which nearly misses the detection coil will also generate screening currents which link a fraction of the monopole signal flux into the coil. In both cases, the unique signature of a monopole is lost, making it much more difficult to discriminate against spurious signals or, conversely, much more difficult to determine the magnetic charge of the incident particle.

Three possible solutions to this problem exist. First, extensive simulations can be performed to determine the distribution of the total flux values linked into the coil as a function of the monopole trajectory. These simulations can then be used to estimate the probability that a given event corresponds to a monopole even though the actual trajectory has not been determined, provided that some information about the distribution of the background flux jumps is known. Another solution is to measure directly the location of the penetration of the monopole on the shield using a scanning magnetometer. In this case the stationary pick-up coil becomes redundant, and can be eliminated. All detected events fall into the "near miss" category. Clearly, the resolution of the scanning magnetometer must by sufficient to distinguish between individual pinned flux quanta. In addition, the flux quanta must be pinned tightly enough such that flux jumps which occur spontaneously in the shield between scans can be distinguished from a monopole event. This type of detection scheme, which is under development by Cabrera, requires a low ambient field in addition to a stable magnetic field environment. The third possibility is to develop a detector which is essentially blind, not only to the screening currents on the shield, but to all external sources of magnetic field generated by the motion of charged particles. In this case, not only is the monopole trajectory irrelevant, but the stringent requirements on the stability of the external field may be relaxed. This is the strategy of the IBM monopole experiments.

3. Planar Gradiometer Coils

The pick-up coils for our detectors consist of a set of coplanar superconducting loops connected in series. The passage of a monopole through a pick-up coil of inductance L_p generates a change in the persistent screening current flowing around the entire input coil of

$$\Delta I = (2\Phi_o - \Phi_s)/(L_p + L_c - M_s), \tag{1}$$

where M_s is the mutual inductance between the detection coil and the superconducting shield used to stabilize the field. The pick-up coils are designed so as to be insensitive to field sources external to the loops. As a result, the flux Φ_s and the mutual inductance M_s are reduced. In addition, the detector is also extremely insensitive to the motion of flux which may have been trapped in the shield as it was cooled into the superconducting state and to changes in the orientation of the coil in the ambient field due to stress release within the system. This is an important property of the detector design since it permits the relaxation of the constraint on the ambient magnetic field.

The input coil must be insensitive to external sources for an essentially arbitrary field configuration. This is because, although the geometry of a particular set of superconducting shields or of a particular remnant field inside a set of normal metal shields may be known, the trajectory of an incident monopole may be random. In addition, the distribution of pinning sites within the shield may be quite arbitrary. Thus, we expand the magnetic field in the vicinity of the pick-up loop in a Taylor series, rather than in an expansion determined by the geometry of the shields. All components of the field at the pick-up coil which are generated by localized sources external to the coil fall off inversely with distance. As a result, the contributions of the higher order terms in the Taylor series expansion tend to decrease rapidly. Thus, if the sizes and orientations of the individual loops which comprise the pick-up coil are chosen so as to systematically cancel the lowest order terms in this expansion, the sensitivity of the entire pick-up coil to external sources is reduced. However, since all the loops in the pick-up coil are coplanar, the monopole trajectory will thread only one of the loops. Thus, the apparent signal flux in the pick-up coil is $2\Phi_0$, independent of the detailed geometry of the pick-up coil. As a result, elements in a hierarchy of planar gradiometer coils with these characteristics become progressively less sensitive to external sources, with no loss of sensitivity to an incident monopole.

We have investigated the properties of several classes of coils, both experimentally and numerically. [3] Two examples in which the individual loops in the pick-up coil are assumed to be square or rectangular will be discussed here. Various hierarchies based on an expansion of the field in other than rectangular coordinates appeared to be less effective. In the first example, the loops which are the basic subunits of the pick-up coil will be permitted to be of arbitrary size. In the second case, all the subunits will be assumed to be identical. This is a useful division, since it is possible to fabricate pick-up coils of modest size with high precision, and then assemble a collection of these coils as identical subunits in a larger matrix, all elements of which are connected in series. Furthermore, the analysis of the response of a detector composed of identical subunits is independent of the type of detector, and thus may find application in other fields.

The hierarchy for a pick-up coil of non-identical loops is derived as follows.[4] First, consider a one dimensional chain of closely packed loops (the generalization to a two dimensional array is straightforward). Adjacent loops are assumed to be wound in opposition, with the boundaries of each loop given by x_i, for i = 0, 1, ..., N+1. Take $x_i > x_j$ if $i < j$. In this example, the component of the field normal to the plane of the coil is assumed to depend only on the location x down the length of the coil. In the Taylor series expansion of the field, the Pth derivative of the field is multiplied by a term x^P. This term does not contribute to the flux if its integral over all the loops is zero.

This requirement reduces to

$$\sum_{k=0}^{N} (-1)^k (x_k^m - x_{k+1}^m) = 0, \tag{2}$$

for m = P+1. Eqn. 2 is satisfied for a coil of length 2L for all m = 1, 2, ..., N if

$$x_j = L \cos\left(\frac{j\pi}{N+1}\right). \tag{3}$$

Thus, the lowest order derivative of the field to which the coil is sensitive is N, where the total number of loops is N+1 (i.e. N=0 for a single loop, which is sensitive to the constant component and all higher terms in the series). Note that the number of wire locations needed to cancel the Nth order term is

N+3. Thus the total length of wire in the coil scales with N for large N.

The hierarchy for a coil composed of identical subunits in one dimension is generated as follows.[5] In this case, because the boundaries are not arbitrary, we consider a chain of loops, each of area ds, equally spaced by a distance Δx along the x-axis. The net flux linking the entire array of 2^m loops is

$$\Phi = \sum_{k=1}^{2^m} S(k)\phi(k). \tag{4}$$

The flux $\phi(k) = B_n(x_k)ds$, where the normal component of the field, B_n, is evaluated at $x_k = x_o + k\Delta x$. The orientation function, $S(k) = +1$ or -1, is determined as follows. In the Taylor series expansion of the field, the Pth derivative of B_n is multiplied by $(k\Delta x)^P$. Thus if the S(k) satisfy

$$\sum_{k=1}^{2^m} k^P S(k) = O, \tag{5}$$

then the Pth derivative of the field will not contribute to the net flux linking the detector. Thus, as in the previous example, a detector consisting of a single loop (order N=0) is sensitive to all the terms in the expansion of the field. The first order detector (N=1) is insensitive to the uniform component of B_n provided all the S(k) sum to zero. The lowest value of m which satisfies this constraint is m = 1, with S(1) = 1 and S(2) = -1. The higher order detectors can be determined by induction. The detector of order N+1 is generated by taking the sequence for the detector of order N twice in succession, negating each term in the second block. The pattern for S(k) for the first four detectors are (+) for N=0, (+-) for N=1, (+--+) for N=2, and (+--+-++-) for N=3. Note that the orders N=0,1,2 are identical to the arbitrary area example. However, the number of wire locations in the chain of equal area loops scales like 2^N for large N, rather than N.

A two dimensional array of detectors is generated as follows. If we require the detector to consist of equal area loops located on a symmetric grid in the xy plane, the orientation function is $S_{xy}(j,k) = S(j)S(k)$. An array of order NxN is insensitive to all components with derivatives of order $P<2N$. In a similar fashion, the location of the wires in an array of unequal area loops is determined by Eqn. 3, with x_i and y_i scaled by the detector dimensions, L_x and L_y.

Up to this point, the analysis presented for equal area cells can be applied to detectors other than superconducting magnetometers provided that the individual detectors are identical and can be summed algebraically. For a pick-up loop generated by this scheme, adjacent areas with the same orientation may be unioned. The hierarchy of loops which result is shown in Fig. 1 for N=0,1,2,3, and 4. Adjacent areas are wound in opposition. Note that, although there are some wires which can be deleted, the total length of wire still tends to grow like 2^N, rather than linearly with N as is the case for the unequal area scheme. This is an important point. The inductance of the pick-up coil scales roughly with the total length of wire used, and the current in the input coil scales inversely with the inductance of the coil (Eqn. 1). Thus, the dominant noise source for large N detectors will tend to be the intrinsic noise of the SQUID used to sense the current ΔI, rather than the ambient field fluctuations.

4. Numerical Results

The detectors can be characterized by the response to a point magnetic field source. This source corresponds to an incident monopole, and also approximates the field generated by a flux quantum (or two) trapped in the shield wall. The flux linking a single rectangular loop is

$$\phi_m = \int_{x_1}^{x_2}\int_{y_1}^{y_2} dx'dy'\vec{B}\cdot\hat{z} = (q_m/4\pi)\int_{x_1}^{x_2}\int_{y_1}^{y_2} dx'dy'z\left[(x-x')^2 + (y-y')^2 + z^2\right]^{-3/2}. \quad (6)$$

This integral can be integrated explicitly. The result is

$$\phi_m = g(x_1,y_1) + g(x_2,y_2) - g(x_1,y_2) - g(x_2,y_1)$$

for

$$g(x_i,y_i) = \left(\pi\frac{q_m}{4}\right) \tan^{-1}\left[\frac{(x-x_i)(y-y_i)}{z\sqrt{(x-x_i)^2+(y-y_i)^2+z^2}}\right], \qquad (7)$$

where the source is a monopole of charge $(q_m/4\pi)$ located at (x,y,z). The coordinates of the vertices of the square loop are $(x_1,y_1,0)$ and $(x_2,y_2,0)$. The net flux linking a square detector of order N and total area D^2 is evaluated from Eqn. 7. As an illustration, we plot in Fig. 2 the rms flux as a function of distance z of the monopole above an equal area detector for orders N=0,1,2,3,4. Results for the unequal area case are similar. The flux has been averaged over all monopole locations (x,y,z) such that $(x,y,0)$ lies within the detector. In the limit $z = 0$, the monopole passes through one of the detector loops. Thus, the signal flux approaches Φ_0 for all orders of N. At distances $z > 0$, the response to a point source falls dramatically with increasing N. As a result, higher order coils may be located much closer to a superconducting shield or another detector than conventional single loop detectors. A reasonable separation corresponds roughly to the dimension of the smallest cell in the array.

The response of an equal area detector of order N which is located within a cylindrical superconducting shield to an isotropic flux of monopoles can be approximated by the following Monte Carlo calculation. This calculation is fairly crude but sufficient, because these detectors are designed to be used only to detect direct hit events. The screening currents generated by the passage of the monopole through the shield are approximated by replacing the shield and screening currents by a pair of isolated magnetic charges at the points of penetration of the shield. These sources link a flux, Φ_s, through the detector. In addition, a flux of $2\Phi_0$ or 0 is linked into the detector by the incident monopole, depending on whether or not the trajectory passes directly through one of the detector loops. The net applied flux seen by the detector is thus either reduced below $2\Phi_0$ or increased above 0 by the screening flux Φ_s. The probability that a particular value of net flux will be observed for a

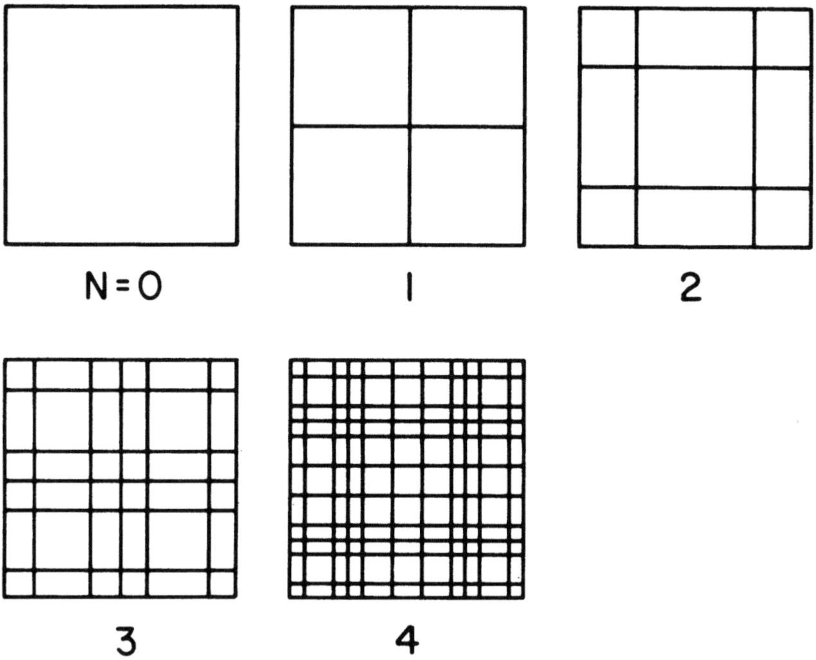

Figure 1. Equal area rectangular planar gradiometer coils.

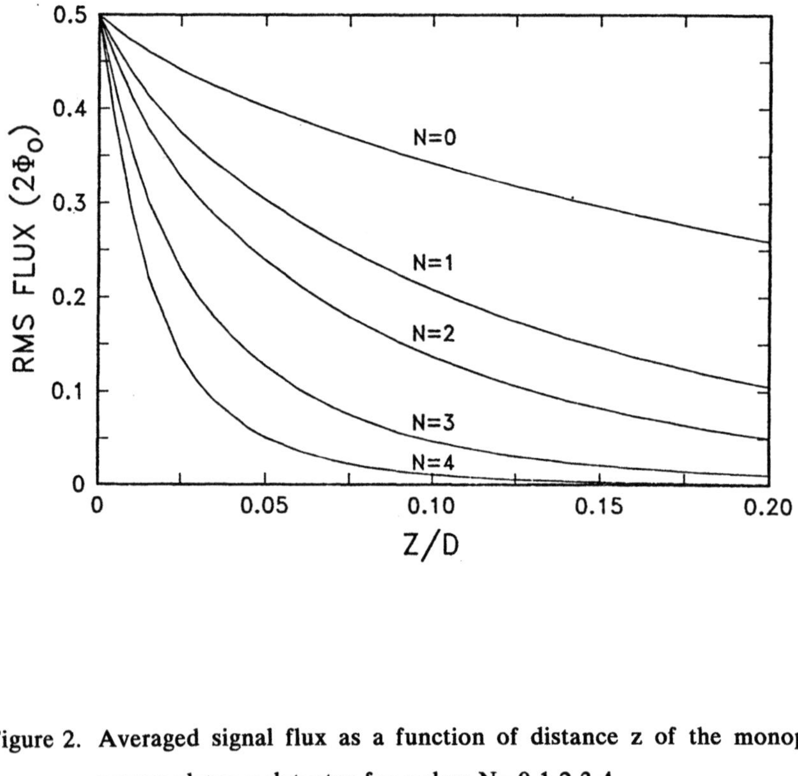

Figure 2. Averaged signal flux as a function of distance z of the monopole source above a detector for orders N=0,1,2,3,4.

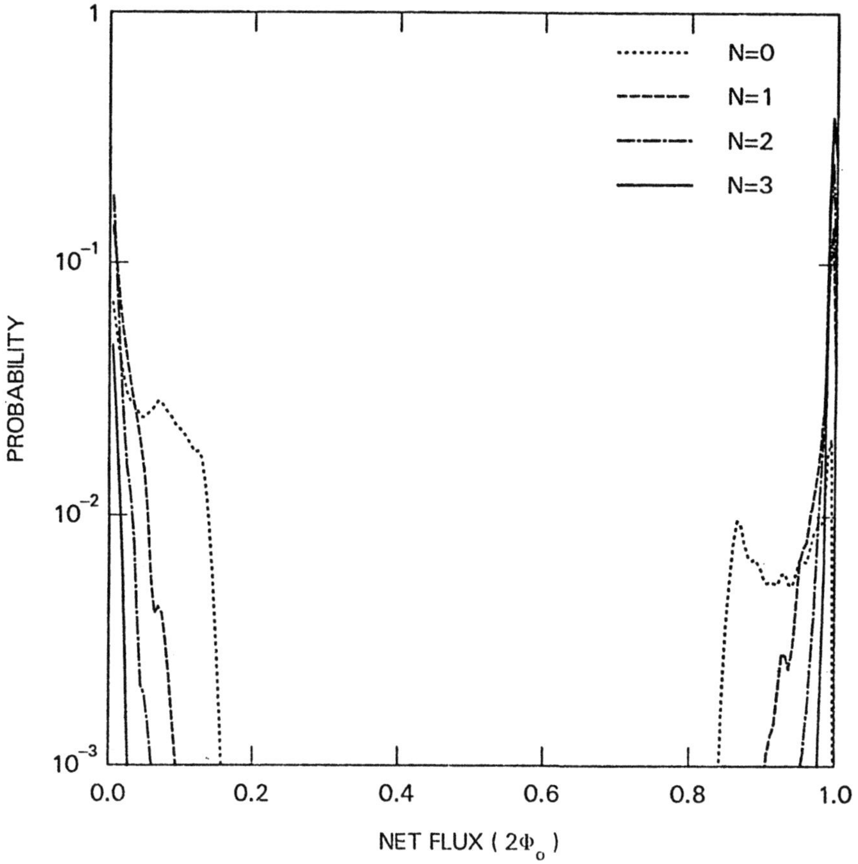

Figure 3. Detector profiles for equal area cells of orders N = 0,1,2,3,4 located within a cylindrical superconducting shield. The probability of observing a net flux is plotted as a function of net flux for an isotropic flux of monopoles. The detector area is 100 cm^2 and the shield diameter is 17.5 cm.

randomly generated trajectory is plotted as a function of net flux in Fig. 3. The shield diameter is 17.5 cm and the area of the detector is 100 cm^2. The detector is located symmetrically within the shield in the plane perpendicular to the axis of the cylinder. Note that, as the order N is increased for constant detector area, the range of values of net flux corresponding to probable monopole events decreases sharply. As a result, spurious flux jumps in the shields may be more easily identified.

5. Coincidence Detection

Because spurious signals cannot be entirely eliminated, coincidence detection between at least two independent detectors is imperative. However, a pair of planar detectors must be separated sufficiently far apart so that the screening current, ΔI, generated in one loop does not link a flux which is a substantial fraction of $2\Phi_0$ into the other detector. Since those particles which intercept only one detector are not recorded as coincident events, the effective area for coincident detection decreases as the separation between the detectors increases. However, if high order derivative detectors are used, the separation between the detectors can be reduced to on the order of the individual loop size, rather than that of the entire detector, without significant flux linkage. Thus the effective detector area can be increased without sacrificing the independence of the detectors. This property is exploited in the prototype detector described in section 6.

In addition, independent detectors can be oriented such that all possible trajectories result in a coincidence detection. For example, if six independent detectors are located on the faces of a cube, an incident monopole will intersect one detector face on entry, and one detector face on exit. Thus, two, and only two, detectors must record an event if the particle is to be a massive monopole. The mutual inductance between adjacent (orthogonal) faces is zero. The derivative configuration reduces the mutual inductance between parallel faces to an insignificant value. The effective planar area for coincidence detection is $3D^2$. This configuration has the advantage of permitting discrimination against disturbances which result in spurious offsets of all detectors.

6. Experimental Results

A prototype gradiometer detector was fabricated as follows. [3] It is designed to be operated in a fairly large field. No μ metal shielding was used. Helmholz coils were used to reduce the ambient field to 10 milligauss over the detector area. This field is more than five orders of magnitude greater than that previously used. [1] A single 2.5 mil thick lead foil shield 17.5 cm in diameter and 60 cm high was used to stabilize the ambient field and to provide a longitudinal shielding factor of 10^5. The shield was mounted over a glass cylinder and was closed at the bottom. The sides of the shield were welded together, and the shield cooled down in direct contact with liquid Helium.

The detector coils were wound out of 5 mil Niobium wire glued into grooves machined into planar substrates. Phenolic was used for the substrate material for convenience. The inductances of the coils were measured with the coils mounted inside the lead foil shield. The outside dimensions of the square coils were 10 cm by 10 cm, yielding an individual detector area of 100 cm^2. Equal area detectors of order N=0,1,2,3,4 and unequal area detectors of order N=7,8,9, and various non-rectangular detectors were tested. The inductance of the pick-up loops was roughly proportional to the total length of wire used in the coils. Thus, the length of wire tended to grow like 2^N in the limit of large N for equal area coils, and like 2N for coils with unequal cell areas. The signal decreased with increasing input coil inductance in agreement with Eqn. 1. The noise was also observed to decrease with increasing order. The best signal-to noise ratio in this particular magnetic field environment was achieved for unequal area detectors with N on the order of 7 or 8. A pair of parallel detectors of order N=7 and N=8 with an effective area for coincidence detection of 50 cm^2 were operated for an extended period with a signal-to-noise ratio of 15:1 in a 1 Hz bandwidth. Mechanical shifts produced by stress release within the phenolic were observed to produce flux jumps coincident with motion of the coils observed on a three axis accelerometer. In addition, deliberately introduced rf pulses and mechanical stresses were seen to produce permanent, and in some cases, coincident, flux jumps. However, no spontaneously occurring signals consistent with a monopole event were observed in 95 days of observation in the prototype detector.

The possibility that a spurious flux jump might be produced by stress release within the detector or shields, or by rf induced flux motion cannot be entirely eliminated, even if mechanically stable detectors are operated in a carefully shielded, low magnetic field environment. However, the use of high-order gradiometer detectors operated in coincidence will increase the reliability of inductive monopole detectors. In addition, the area of such detectors can be increased without demanding a corresponding increase in the stability of the ambient field. A detector incorporating six coils of this type is presently under construction at IBM. In this system, the coils are mounted on the faces of a rectangular parallelopiped and suspended inside a superinsulated cryostat. A set of four, nested high permeability shields are used to reduce the field to the order of a few microgauss over the volume of the coils and SQUID detectors. A reasonably low field is desirable in order to reduce the probability that flux might be trapped in the SQUIDs or in the coils. The volume available within the low field region corresponds to an effective planar detector area of approximately 3000 cm^2 for coincidence detection. Because the volume included within the six detector plates lies within the cryogenic environment, this detector is not particularly appropriate for operation in coincidence with conventional particle detectors. However, gradiometer detectors mounted in a planar shielding and dewar system, are good candidates for coincidence operation with conventional particle detectors.

References

[1] B. Cabrera, Phys. Rev. Lett. 48, 1378 (1982).

[2] B. Cabrera, in Third Workshop on Grand Unification, P.H. Frampton, S.L. Glashow and H. van Dam, eds. (Birkhauser, Boston Mass) pp. 131.

[3] C. D. Tesche, C. C. Chi, C. C. Tsuei, and P. Chaudhari, to appear in Appl. Phys. Lett..

[4] From conversations with M. Gutzwiller, and with M. Gutzwiller in conversations with G. Choodnowski and D. Choodnowski.

[5] From conversations with S. Kirkpatrick.

SEARCHES FOR MAGNETIC MONOPOLES

Peter C. Bosetti

III. Physikalisches Institut der RWTH Aachen, West Germany

Abstract

A review of present and possible future experimental limits on the abundance of magnetic monopoles from experiments and a survey of astrophysical constraints on their flux is given.

I. Introduction

The existence of magnetic monopoles was first hypothezised by P. A. M. Dirac [1] in 1931 to establish symmetry between electricity and magnetism. The value of the magnetic charge g is related to the electric charge e by the relation

$$e * g = n\, h / c \qquad (1)$$

which is called Dirac's quantization condition.

Though the mass of the monopole is an open parameter in Dirac's theory, a large number of experiments have been conducted to search for these objects in cosmic rays, at accelerators, and in bulks of matter. All these experiments, searching mostly for highly ionizing particles with masses up to several GeV/c, failed to see any monopole.

With the advent of unified theories of electromagnetic and weak interactions new interest in the field arose. This is mainly because in 1974 t'Hooft and Polyakov showed [2] that in a variety of these theories there exist solutions for the field equations that represent magnetic monopoles. Furthermore in these theories one can estimate the mass of the monopoles as

$$m(M) \cong \alpha^{-1}\, m(W) \cong 10^4 \text{ GeV} \qquad (2)$$

where m(W) is the mass of the intermediate vector-boson and α is the coupling constant.

Arguments for the existence of magnetic monopoles were strengthened with the "success" of the unification of strong and electro-weak interactions within the Grand Unified Theories (GUTs), as in these theories there exists a (semi-) simple gauge group which is broken down to SU(3) X U(1) and this implies the existence of t'Hooft - Polyakov monopoles. In GUTs their mass is again given by equation (2), however, the mass of the intermediate vector - boson W has to be replaced by the grand unification mass ($\sim 10^{14}$ GeV/c) leading to m(M) $\cong 10^{16}$ GeV/c for the "Grand Unified Monopoles" (GUMs).

It has been known for some time [3] that GUMs in addition to their well-known properties should have baryon number violating interactions. In 1981 V. A. Rubakov [4] and C. Callan [5] proposed that the cross section for these interactions should be comparable in strength to typical strong interaction cross sections.

The analysis showed that a monopole is not an - almost - point-like object, but rather surrounded by a condensate of fermion - antifermion pairs. Thus the size of the monopole is given by 1/m(f) instead of 1/m(M), i. e. it is a typical hadronic size. It is the interaction of hadrons with this fermion cloud that is responsible for the large cross section.

The monopole will essentially remain unchanged by the interaction, it just "supplies" the fermions and can be considered as a catalyzer of the reaction. The form of the interaction depends on the group structure of the group that is broken down to SU(3) X U(1) and it is expected that theories predicting baryon decay at any level also predict monopole catalysis of nucleon decay.

II. Characteristics of magnetic monopoles

As already described above, the mass of the magnetic monopoles in Dirac's original theory is an open parameter. However, in unified and grand unified theories the relation

$$m(M) = \alpha^{-1}\, m(B) \qquad (3)$$

holds, where α is the appropriate coupling constant and m(B) is the mass of the intermediate boson.

As examples, this relation leads to the predictions

$$B = W \longrightarrow m(M) \cong 10^{4}\,\text{GeV} \qquad (4)$$

$$B = X \longrightarrow m(M) \cong 10^{16}\,\text{GeV} \qquad (5)$$

$$B = P \longrightarrow m(M) \cong 10^{21}\,\text{GeV} \qquad (6)$$

where X represents those lepto-quarks in SU(5) that are responsible for nucleon decay, and m(P) is the Planckmass. (Certainly there might be some doubts about the validity of relation (6).)

According to Dirac's quantisation condition the magnetic charge of the monopoles is given by equation (1) and thus the minimum charge is

$$g(\text{min}) = (137/2)\, e = 3.3 * 10^{-8} \text{ cgs units} \quad (7)$$

This leads to an energy extraction in a magnetic field of

$$E(\text{mag}) = g(\text{min})\ B\ 1 = 2 \text{ MeV/Gauss cm} \quad (8)$$

This also leads to an induction of an electric current (in a superconducting coil) of

$$i = 4\pi\ N\ g/1 \quad (9)$$

The binding energy in e.g. iron is 10ev/Å, which is about 100 times the gravitational pull on earth.

Electrically charged monopoles - dyons - are expected to be n * 10^{12} GeV more massive than their electrically neutral counterparts.

The interaction of monopoles with matter due to their magnetic field

$$\underline{B} = g\ (\hat{\underline{r}}/r^2) \quad (10)$$

has been investigated by various authors. Since the minimum

charge g(min) is about 68e, relativistic monopoles would suffer an ionization loss similar to a Z = 68 nucleus. For large $\beta = v/c$ ($\beta > 10^{-2}$) one can estimate

$$dE/dx(GUM) = dE/dx(e^-) * (n*g(min)/e)^2*\beta^2 \quad (11)$$

For lower velocities the question of energy loss of monopoles has been somewhat controversial, however substantial progress has been made recently.

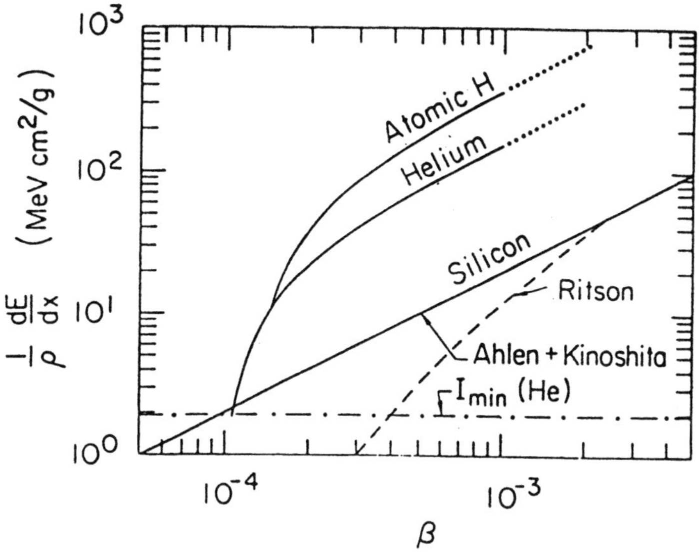

Fig. 1. Energyloss of GUMs as calculated

The energy loss for charged particles is quite well represented for $\beta > 5 * 10^{-3}$ by Lindhard's model [6], developped 30 years ago. Using Maxwell's equations and a degenerate electron sea, this model predicts an energy loss linear in β for $\beta > 10^{-2}$. Ahlen and Kinoshita [7] have applied this model to the case of monopoles and found the same linear β - dependence. Ritson [8] then applied these results to to real materials. The results of these calculations are shown in fig 1. It appears that scintillation counters can be used for monopole detection for monopole velocities $\beta > 3 * 10^{-4}$.

Recently, Drell et al [9] have calculated the cross section for exiting simple atoms by slowly moving monpoles ($10^{-4} < \beta < 10^{-3}$) and find an energy loss much larger then previous studies, namely

$$(1/\rho) * dE/dx =$$
$$15*(\beta/10^{-4})(1-9.3*10^{-5}/\beta^2)^{3/2} \text{ (MeV/g) cm}^2 \quad (12)$$

Their result is about one order of magnitude larger then earlier ones, because they take into account the crossing of energy levels caused by the interaction of the atomic electrons with the magnetic field of the monopole. For comparison, this result is also shown in fig 1.

It has been proposed [10] that monopoles, when passing very close to a nucleon, can capture that nucleon whenever its gyromagnetic ratio is larger than 2. For capture to take place, the nucleon must either suffer a momentum transfer or radiate a photon.

Goebel finds for the interaction cross section

$$\sigma = 8.4 * 10^{-29} * \beta^2 \text{ cm}^2 \qquad (13)$$

and for the mean free path before picking up a nucleon for e.g. $\beta = 5 * 10^{-3}$

$$\lambda = 1.2 \text{ km} \qquad (14)$$

This would imply that monopoles have a larger energy loss when passing through the earth due to the large electric charge of the nucleus, and that a fraction of the monopoles reaching detectors deep underground are accompanied by a nucleus.

We now turn to the possibility of the catalysis of nucleon decay by magnetic monopoles. Parametrizing the mean free path of GUMs between two catalyzed nucleon decays in terms of σ^o ($\sigma^o = 1$ corresponds in strength to a typical hadronic cross section), the interaction length is given by [11]

$$IL = 4300/\rho * \beta/\sigma^o \text{ [cm]} \qquad (15)$$

and the mean time between two interactions is

$$\tau = IL/\beta = .13/\sigma^o \text{ [}\mu\text{sec]} \qquad (16)$$

Estimates for σ^o range from 1 to 10^{-5} and with the expectation

$10^{-5} < \beta < 10^{-3}$ one can deduce the possible ranges for the interaction lengths

$$.043/\rho \text{ [cm]} < IL < 4.3/\rho \text{ [km]} \qquad (17)$$

and correspondingly for the mean times between interactions

$$.13\mu sec < \tau < 13 \text{ msec} \qquad (18)$$

imposing a severe challange to experimenters.

III. Experimental results

In the "early days" as well as today, experiments designed for the detection of monopoles were installed at accelerators, in cosmic ray laboratories, or set up to search for trapped monopoles in bulks of matter.

At accelerators the intention is to detect the monopoles via their large dE/dx directly after their production, usually with plastic detectors or lexan foils.

In fig. 2 the most restrictive upper limits for the production of "classical" monopoles are shown [12]. At fixed target machines the mass range to be explored is limited to $\sim$15 GeV/c and at the ISR to $\sim$30 GeV/c.

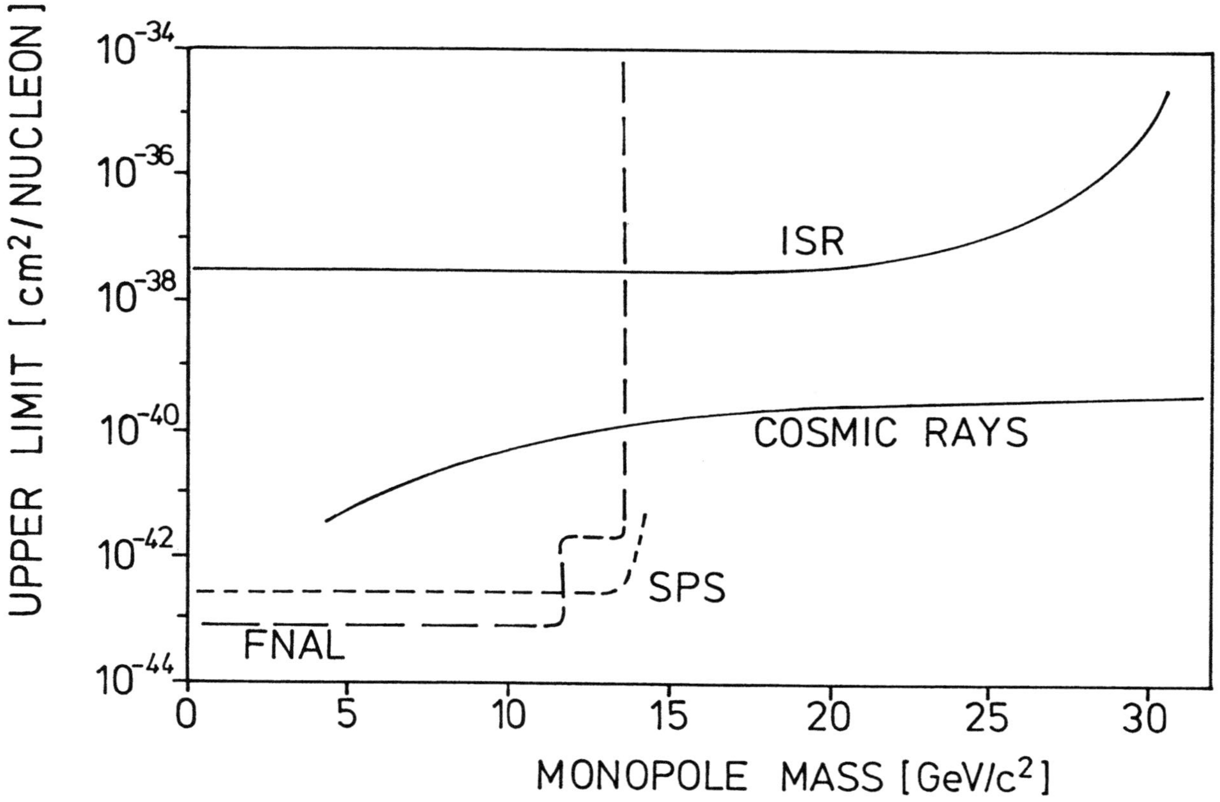

Fig. 2 Results on searches for classical monopoles.

Higher energies are available at cosmic ray experiments. The best upper limits deduced from these experiments is also shown in fig. 2 [12].

Very slow monopoles can be trapped by ferromagnetic materials in the earth or e.g. in lunar material. One can process the material through a superconducting coil and detect an induced current. Using earths or lunar material with this technique has the advantage of sampling the flux over geological timescales.

The pioneering experiment of this type was performed by Ross et. al. [13]. From samples of lunar material they deduced an upper limit of

$$N(\text{Monopoles})/\ N(\text{Nucleons}) < 2 * 10^{-28} \qquad (19)$$

The Wisconsin Group [14] has proposed an experiment to process 10^6 tons of ferromagnetic material, which would improve this limit substantially.

Recently, a quite large number of experiments have obtained results or are beeing set up, which are sensitive to slowly moving GUMs. At Stanford, Cabrera et al. [15] are looking for an induced current by a monopole. This experiment is sensitive to any β. One candidate event has been detected and up to now a flux upper limit is given of 0.03 m^{-2} d^{-1} sr^{-1}. Another experiment of this type is in preparation, with much larger acceptance [16].

There is a large class of experiments searching for the energy loss of monopoles in material using proportional counters, plastic sheets, or scintillation counters, which are sensitive to monopoles with velocities larger than a few times $10^{-4} * c$. They are listed in table 1 together with the limits on the flux obtained so far. Fig. 3 shows these results as a function of velocity β.

Certainly the most exiting signature for monopoles would be a chain of catalyzed nucleon decays inside a detector. From the present lower limit on the nucleon lifetime

$$\tau \text{ (Nucleon)} > 3 * 10^{30} \text{ years} \quad (20)$$

Ellis et al. [11] deduced an upper limit on the GUM flux of

$$F(M) < 2 * 10^{-3} \, \beta/\sigma^{o} \, d^{-1} \, m^{-2} \, sr^{-1} \quad (21)$$

It has to be pointed out, however, that this limit is only valid for very small values of σ^{o}, so that the interaction length of the GUMs is larger than the detector size. Thus applying this limit to real life, one has to multiply this value with σ^{-1}, which makes it worse by a large factor. (Remember that e.g. for $\sigma^{o} = 1$, as proposed by Rubakov, and $\beta = 10^{-4}$, the interactionlength is only 0.4 cm, causing serious problems for at least some of the nucleon decay

Experiment	Detector	β - range	Fluxlimit $[m^{-2}\ sr^{-1}\ d^{-1}]$
Stanford[15]	Superconducting coil	all	.03
BNL[28]	Proportional Counters	$3*10^{-4}$-$1.2*10^{-3}$	$3*10*^{-2}$
Soudan[19]	"	$3*10^{-4}$-$3*10^{-2}$	$3*10^{-4}$
Kolar[30]	"	$4*10^{-3}$-1	$2.5*10^{-5}$
Michigan[71]	Scintillation Counters	$.7*10^{-4}$-$3*10^{-2}$	10^{-1}
Utah-Stanford[32]	"	10^{-4}-10^{-2}	$2*10^{-2}$
Tokyo[33]	"	$3*10^{-4}$	$3.5*10^{-3}$
Bologna[34]	"	$2*10^{-3}$-0.5	$2*10^{-3}$
Baksan[35]	"	$5*10^{-3}$-$3*10^{-1}$	$1.3*10^{-5}$
Aachen-Hawaii Tokyo[17]	Water-Cerenkov	$< 5*10^{-2}$	$7.5*10^{-3}$
IMB[18]	"	10^{-4}-10^{-1}	$4*10^{-6}$
Soudan[19]	Prop.Counters	10^{-4}-10^{-2}	10^{-3}
NUSEX[20]	Streamer tubes	10^{-4}-10^{-2}	$2*10^{-5}$

Table 1. Compilation of the results of GUM search experiments

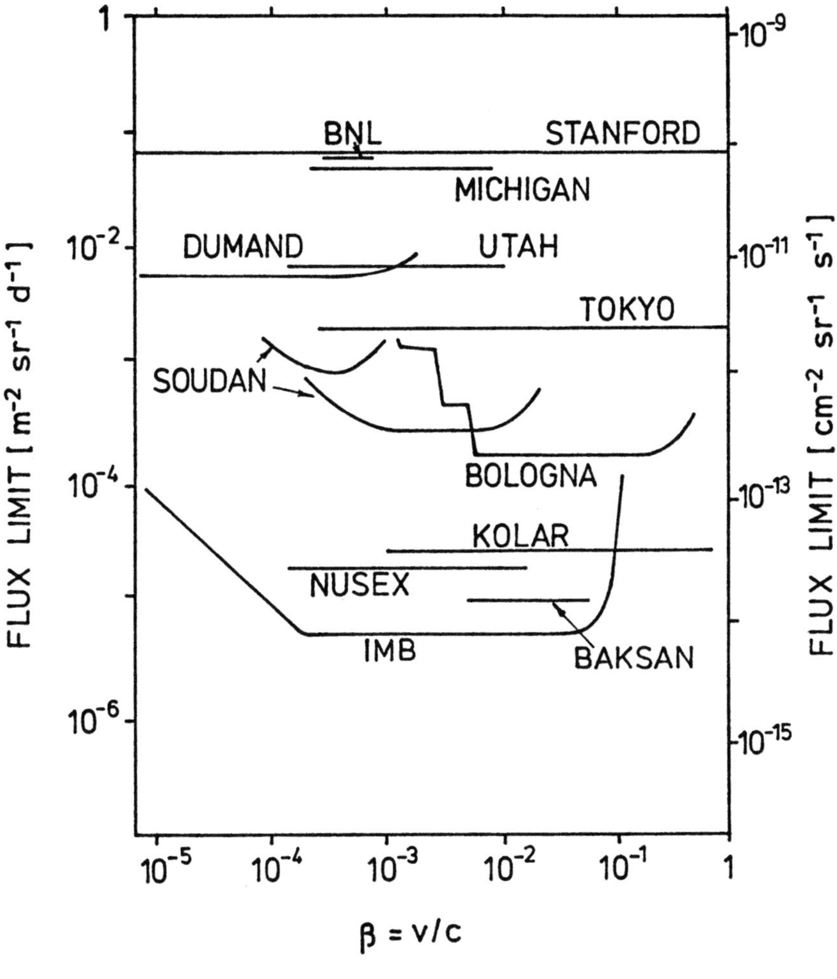

Fig. 3 Results on GUM Searches

detectors to identify such events.)

The first result on a flux limit for GUMs catalyzing nucleon decays has been obtained by the Aachen-Hawaii-Tokyo Collaboration [17], which used a DUMAND test tank at sea level. From the absence of any signal they obtained for $\sigma^0 \cong 1$ and $\beta < 5 * 10^{-2}$

$$F(M) < 7.5 * 10^{-3}\ d^{-1}\ m^{-2}\ sr^{-1} \qquad (22)$$

At this conference, first results from Nucleon-Decay Experiments deep underground were reported. The results for all three collaborations, IMB [18], SOUDAN I [19] , and NUSEX [20] are also summarized in table 1 and displayed in fig 1.

IV. Constraints on GUM fluxes from astrophysics

Grand Unified Monopoles are expected to be created in the early universe shortly after the Big Bang, when the temperature was high enough. If they were produced indeed, there should be some around today, as the lowest monopole state is stable. The only possibility to reduce their number density is by monopole-antimonopole annihilations.

In the standard Big Bang Model, the number of GUMs produced is roughly of the order of the number of baryons, a ridicuously large number, as m(GUM) = 10^{16} m(N) [21]. Many ways have been proposed to solve this so called "Monopole problem". Today, the "New Inflaionary Universe" scenario seems to be the most promising approach. In this scenario the number of monopoles present in our universe could be as small as 0. For the time beeing, however, it seems only safe to say that any number of monopoles between 0 and a detectable number can theoretically be predicted.

The velocities of GUMs today depends on the two main forces acting on them: magnetism and gravitation. The escape velocity from a typical galaxy like ours is ≅ 10^{-3} c (independant of the mass). From the magnetic force they acquire a velocity

$$\beta = 3 * 10^{-3} (L/10^{21}\mathrm{cm})^{1/2} (10^{16}\mathrm{GeV}/M)^{1/2} \qquad (22)$$

where L is the coherence length of the galactic field. For masses $M < 10^{17}$ GeV the magnetic force is larger than the gravitational.

Probably the most reliable limit on the flux of GUMs comes from the requirement that the density of monopoles must not exceed the mass density of the universe, which is $\rho \cong 10^{-29}$ g/cm^3. This leads to

$$N < 10^{-21}\ cm^{-3}\ (10^{16}\ GeV/M) \quad (23)$$

or

$$F = (N\beta/4\pi) < 5 * 10^{-12}\ \beta\ cm^{-2}\ s^{-1}\ sr^{-1} \quad (24)$$

for $M = 10^{16}$ GeV. This limit, however, applies only, if the monopoles are distributed uniformly throughout the universe, i.e. for $\beta > 10^{-3}$.

If the mass is larger than 10^{17} GeV, then the GUMs might be confined to galactic halos, so that their local density could be much larger. Preskill [22] estimates for this case

$$F < 10^{2}\ m^{-2}\ y^{-1}\ (10^{17}\ GeV/M)$$
$$= 2.3 * 10^{-2}\ m^{-2}\ d^{-1}\ sr^{-1} \quad (25)$$

Magnetic Monopoles in a galaxy extract energy from the galactic magnetic field. As this energy extraction must not destroy the fields faster than the regeneration time ($\approx 10^{8}$ years) one can set an upper limit on their abundance inside a galaxy

$$F < 2 * 10^{-4}\ m^{-2}\ y^{-1}$$
$$= 4.6 * 10^{-8}\ m^{-2}\ d^{-1}\ sr^{-1} \quad (26)$$

independent of the mass as long as the mass is smaller than 10^{17} GeV. For $M > 10^{17}$ GeV this limit is

$$F < 4.6 * 10^{-8} * (M/10^{17}\text{GeV})\ m^{-2}\ d^{-1}\ sr^{-1} \quad (27)$$

For velocities $10^{-4} < \beta < 10^{-3}$, Monopoles are confined to galaxies and one can require that their density must not exceed the mass density of the galaxy $\rho = 9 * 10^{-24}$ g/cm^3 which leads to

$$F < 10^{-7} * \beta * (10^{16}\ \text{GeV}/M)\ cm^{-2}\ s^{-1}\ sr^{-1} \quad (28)$$

For $\beta < 10^{-4}$, Monopoles are confined to solar systems and none of the above limits apply. The local GUM abundance in this case can be larger by several orders of magnitude.

Very stringent limits on the GUM fluxes have been set from considerations of GUM - Neutron star interactions. A GUM traversing a star has a mean free interaction length λ of [23]

$$\lambda = 1.5 * 10^{-11} * \beta/\sigma^o\ (.17\ fm^{-3}/n)\ cm \quad (29)$$

wheree n is the density of the star. The catalysis rate is given by

$$\Gamma = 2 * 10^{21}\ (n/.17\ fm^{-3})\ \sigma^o\ s^{-1} \quad (30)$$

For a star with high density like a neutron star ($n \cong .17$ fm^{-3}), this leads to $2 * 10^{21} * \sigma^o$ catalysed nucleon decays per second.

If the GUM traverses the neutron star, no useful limit can be set. If, however, the monopoles are captured, rather strong limits have been calculated. Bais et al [23] estimate that GUMs are stopped inside the neutron star if

$$\beta^2 < 5.3 * 10^{16}\ \sigma^o/M \quad (31)$$

This means that e.g. for $M = 10^{16}$ GeV and $\sigma^o = 1$, all GUMs are stopped, for the same mass and $\sigma^o = 10^{-4}$, all GUMs with $\beta < 10^{-2}$ are stopped.

A not very restrictive limit on the flux of GUMs comes from the requirement that the monopoles should not destroy the neutron star by baryon violating interactions too fast. For the extreme case of $\beta = 10^{-5}$ this leads to

$$F * \sigma^o < 6.4 * 10^{-9}\ cm^{-2}\ s^{-1}\ sr^{-1} \quad (32)$$

However, requiring that the energy output due to catalyzed baryon decays must not exceed the observed energy output leads to more interesting limits. Bais et al [23] have taken the neutron star in the Crab nebula of which we know the age to be about 1000 years. They conclude

$$F * \sigma^o\ (4 + \beta^{-2}) < 6.6 * 10^{-8}\ cm^{-2}\ s^{-1}\ sr^{-1} \quad (33)$$

Colgate, Harvey and Kolb [24] have considered old neutron stars formed 10^{10} years ago. These neutron stars have accumulated over that period GUMs, which then heat the star up , and this in turn leads to X - ray emission. This X - ray emission has to be smaller than the observed upper limit and this leads to

$$F * \sigma o \, (4 + \beta^{-2}) < 10^{-17} \ cm^{-2} \ s^{-1} \ sr^{-1} \quad (34)$$

Bais et al [23] use the same arguments and get a slightly more conservative limit

$$F * \sigma o \, (4 + \beta^{-2}) < 6.6 * 10^{-15} \ cm^{-2} \ s^{-1} \ sr^{-1} \quad (35)$$

Rubakov [25], however, has pointed out recently that the density of monopoles and antimonopoles in old neutron stars might become so large that $M \bar{M}$ annihilations become likely and he concludes that the above quoted limits could be too strong by several orders of magnitude.

V. FUTURE OUTLOOK

When the new accelerators LEP, HERA, and the Tevatron come into operation, new monopole searches will be done along similar lines as the ones performed at present accelerators to search for "classical" monopoles. The next generation of experiments looking for induced currents in superconducting coils is under way [15] or proposed [16].

From continuing and new proton decay experiments, better limits on "Rubakov events" will be obtained. It seems, however, unlikely that new nucleon decay experiments will be built deep underground that are larger than e.g. the IMB detector by an order of magnitude in size. To search for monopoles with such large detectors, a possible approach would be to go underwater. Indeed, as large photomultipliers (PMT's) are available, it is possible, to survey very large volumes for "Rubakov events" with just a few PMT's.

A series of catalyzed nucleon decays could be distinguished from other events deep underwater due to the different characteristics of the Cerenkov light emitted from the decay products [26]. As an example it seems possible to deploy within a rather short time a string of 10 16" PMT's each separated by $\sim$5 meters, as proposed for the Short Prototype String of the DUMAND project [27]. With an expected number of one muon every 16 seconds in 4000 m depth, a signal of at least two nucleon decays is very hard to fake, if one requires a sufficiently large number (>10) of photoelectrons. This also eliminates the K40 background).

One can estimate [26] that within one year of running time of an apparatus like that, a flux upper limit of

$$F < 1.8 * 10^{-15}\ cm^{-2}\ s^{-1}\ sr^{-1} \quad (36)$$

can be obtained for $\sigma^o/\beta > 1$. This detector could detect monopoles with interactionlengths up to 50 m, superior to any present detector.

The full DUMAND array with its 756 modules [27] could set within one year an upper limit of

$$F < 2.4 * 10^{-17}\ cm^{-2}\ s^{-1}\ sr^{-1} \quad (37)$$

for $\sigma^o/\beta > 1$ [26], and is the only proposed device so far that could detect GUMs with interaction lengths up to 500 meters, which for e.g. $\beta = 10^{-4}$ corresponds to $\sigma^o = 10^{-5}$.

The search for magnetic monopoles is a very important and fascinating field in physics. The existence of monopoles would not only be appealing, as this leads to a symmetry between electricity and magnetism, but monopoles are also a necessity in Grand Unified Theories. Furthermore they are particles that could provide us with information on the very first moments of the universe.

If this generation of monopole search experiments fail to detect any of them, new approaches have to be undertaken. The possibility that GUMs might catalyze nucleon decay is one of the most exciting characteristics of them. Under certain circumstances this very fact, however, could cause some problems in interpreting results from nucleon decay experiments. Assume, no "Rubakov event" will be found and nucleon decays would be observed, however at a very small rate. It then will be hard to decide whether one has detected free nucleon decays or decays catalyzed by magnetic monopoles.

Finally, one might remark that setting upper limits on the flux of monopoles is by far not as rewarding as finding them, but almost as important.

REFERENCES

[1] P. A. M.Dirac, Pro. Soc. London, Soc A133, 60 (1931)

[2] G. 't Hooft, Nucl. Phys. 879 (1974) 276
a. Polyakov, ZhETF pis. Red. 20 (1974) 430

[3] C. P. Dokos and T. N. Tomaras, Phys. Rev. D 21(1980) 2940

[4] V. A. Rubakov, Zh. Eksp. Teor. Fiz. Pis'ma Red. 33 (1981) 658
and J.E.T.P. Lett. 33 (1981) 644

[5] C. Callan, Princeton Preprint (1982)

[6] J. Lindhard, Mat. Fig. Med. Dan. Vid. Selsk. 28 (1954)

[7] S. P. Ahlen and K. Kinoshita, Berkeley Preprint

[8] D. M. Ritson, SLAC-PUB-2950 (1982)

[9] S. Drell et. al., SLAC-PUB-3012 (1982)

[10] C. Goebel, private communication

[11] J. Ellis, D. V. Nanopoulos, and K. Olive, CERN-TH 3323 (1982)

[12] See e.g. B.Giacomelli, Talk presented at the Wingspread
Monopole Workshop, Racine, Wisconsin, Oktober 1982

[13] R. Ross et al, Phys. Rev. D8 (1973) 698

[14] D. Cline, Talk presented at the Wingspread Monopole
Workshop, Racine, Wisconsin, Oktober 1982

[15] B. Cabrera, Talk presented at the Wingspread Monopole
Workshop, Racine, Wisconsin, Oktober 1982

[16] C. Tesche, Talk presented at this Workshop

[17] P. C. Bosetti et al, University of Hawaii Preprint HDC-16-82

[18] B. Cortez, Talk presented at this Workshop

[19] E. Peterson, Talk presented at this Workshop

[20] E. Fiorini, Talk presented at this Workshop

[21] See e.g. M. Turner, Talk presented at this Workshop

[22] J. Preskill, Phys. Rev. Lett. 43 (1979) 1365

[23] F. A. Bais et al, CERN-TH 3383

[24] E. W. Kolb, S. A. Colgate, and J. A. Harvey, Los Alamos Preprint, July 1982

[25] V.A. Rubakov and v. Kuzmin, ICTP Preprint IC/83/17

[26] P. C. Bosetti, DUMAND Internal Note

[27] DUMAND Proposal, Hawaii DUMAND Center Report Nov. 1982

[28] J. D. Ullman, Phys.Rev. Lett. 47 (1981) 289

[29] See Ref [19]

[30] See D. Ayres et al, Report prepared for the Division of Particles and Fields, Snow Mass, July 1982

[31] J. Sokolowski and L. Sulak, Univ. of Michigan Preprint June 1982

[32] D. E. Groom, Talk presented at the Wingspread Monopole Workshop, Racine, Wisconsin, October 1982

[33] See Ref [30]

[34] R. Bonarelli et al, Phys. Lett. 112B (1982) 100

[35] See Ref [30]

NEUTRINO MASS AND NEUTRINO OSCILLATIONS

Felix Boehm

1. Introduction

We begin by reviewing the present status of neutrino mass as it emerges from spectroscopic measurements and neutrino oscillation experiments. The following are the principal sources of information on neutrino mass.

1.1 Tritium Decay

The study of the spectral shape near the endpoint of the ^3H electrons by Lubimov *et al.*[1] has led to the conclusion that electron-antineutrinos have finite mass m_ν, in the range $14 \leq m_\nu \leq 46$ eV. A serious problem recently pointed out by Simpson[2], however, is raising concern with the analysis leading to this result. It has to do with the spectrometer resolution function which was derived with the help of an internal conversion line. This line turns out to have a finite width of its own which should have been taken into account in the deconvolution of the ^3H spectrum. We shall discuss this problem in Section 2.

1.2 Double Beta Decay

The other important source for neutrino mass, equally cast in controversy, is neutrinoless double beta decay. From the observation of the rate of double beta decay in ^{128}Te and ^{130}Te by Hennecke *et al.*[3] it follows that the lepton number violating neutrinoless process must be present. A Majorana neutrino mass of about 10 eV would be needed to explain their findings. More recently, a remeasurement of these decay rates by Kirsten *et al.*[4] gives a result in good agreement with the expectation for a lepton number conserving two-neutrino process. From the absence of the neutrinoless process it follows that the neutrino mass must be smaller than about 6 eV. A review of neutrinoless double beta decay will be presented by F. Avignone during this Conference.

1.3 Neutrino Oscillations

Neutrino oscillation experiments, to be discussed in Section 3. offer a more indirect determination of neutrino mass. It is now well known that early results by Reines *et al.*[5] on the relative yields of the charged current reaction, $\bar{\nu}_e d \rightarrow e^+ nn$, and the neutral current reaction, $\bar{\nu}_e d \rightarrow \bar{\nu}_e pn$, are in conflict with recent data by the Caltech-Munich-SIN group[6]. From the latter and from other recent high energy neutrino studies there appears to date no evidence for oscillations.

In 1980 supporting evidence for oscillations was also inferred from CERN beam dump data[7]. This was based on the ν_e/ν_μ ratio found to be 0.5 to 0.6 with an error of 0.2, instead of 0.9 to 1.0 for no oscillations. Recent results from Fermilab[8] however find a ν_e/ν_μ ratio consistent with no oscillations.

1.4 Solar Neutrino Puzzle

One remaining suggestive piece of evidence for oscillations is the solar neutrino puzzle[9]. The discrepancy between the experimentally observed solar ν_e flux and that calculated by the accepted solar model lends support to neutrino oscillations with large mixing. Mass parameters Δm^2 between 10^{-2} and 10^{-11} eV^2 would not be in conflict with present laboratory data. Some of the reaction cross sections responsible for the ^8B neutrinos, however, have come under fire recently[9] and further checks of reaction yields will have to be carried out.

2. The ^3H Experiment

In beta decay the neutrino mass enters in the electron phase space factor through the term[10]

$$N_{e^-}(E,Z) \sim pE[(E_0 - E)^2 - m_\nu^2]^{1/2} (E_0 - E)\, C(E,Z) \quad ,$$

where $N_{e^-}(E)$ is the beta spectrum of electrons with energy E and momentum p, E_0 is the decay energy, and C(E,Z) is the Coulomb factor. By measuring this spectrum in the decay of ^{3}H Lubimov *et al.*[1] have found evidence for a nonzero neutrino mass, bracketed between $14 < m_\nu < 46$ eV. A broad assumption was made for the branching into the 2s and 1s atomic final state of ^{3}He. Although their work is not well documented, it appears that the optical spectrometer resolution was $\Delta p/p = 0.12\%$, or $\Delta E = 45$ eV at $E_0 \approx 18$ keV. To obtain the actual resolution profile the authors have used internal conversion lines from atomic M shells of a 20.74 keV gamma ray, emitted in the decay of ^{169}Yb. It was pointed out by Simpson[2] that the M conversion lines have a Lorentzian width, a consideration overlooked by the ITEP group. Just recently, the width of x ray lines originating from the M levels in Tm were measured by Bennett[11]. The MII level was found to be 9 eV wide. Folding the Gaussian resolution function with this Lorentzian increases the width of the former by an amount $\varepsilon = 4.8$ eV. As a result, by unfolding the measured spectrum with the broadened line an error is introduced which leads to a distortion of the shape of the true spectrum and thus to an apparent neutrino mass. This apparent mass M_ν is related[2] to the true mass m_ν by $M_\nu^2 - m_\nu^2 = 2\varepsilon R$, where R is the instrumental line width.

The relation between M_ν and m_ν is illustrated in Figure 1 for a Lorentzian width of 9 eV. The mean value of $M_\nu = 35$ eV corresponds to $m_\nu = 28$ eV. However, the lower bound of $M_\nu = 14$ eV is consistent with $m_\nu = 0$. In fact the diagram shows that $m_\nu = 0$ is obtained for an apparent mass as large as $M_\nu = 20.8$ eV. Thus we conclude that a neutrino mass $m_\nu = 0$ is well within the 99% c.ℓ. range of the ITEP result.

At this stage the spectroscopic measurements now in progress at Los Alamos[12], Livermore[13], Chalk River[12], Stockholm[14], and Zürich[15] are eagerly awaited. They are designed to have greater sensitivity to small neutrino mass and should also shed light on the problem of the relative population of the final states in ^{3}He.

3. Neutrino Oscillations

In reviewing the status of neutrino oscillations we shall assume that oscillation phenomena[7] can be described by a 2 parameter model with a mass parameter $\Delta m^2 = |m_1{}^2 - m_2{}^2|$ and a mixing amplitude $\sin^2 2\theta$. Thus we consider transitions between two neutrino weak interaction states, $\nu_{\ell 1}$ and $\nu_{\ell 2}$ taking place with a probability,

$$p(\nu_{\ell 1} \rightarrow \nu_{\ell 2}) \sim \frac{\sin^2 2\theta}{2} \left[1 - \cos(2.53\, \Delta m^2 (L/E_\nu))\right] .$$

Here L(m) is the distance between the neutrino source of type $\nu_{\ell 1}$ and detector and E_ν (MeV) is the neutrino energy. Δm^2 is in units $(eV)^2$. The probability of disappearance of a state $\nu_{\ell 1}$ is given by $P(\nu_{\ell 1} \rightarrow X) \sim 1 - P(\nu_{\ell 1} \rightarrow \nu_{\ell 2})$.

We review the experimental results on oscillations in the appearance channels $\nu_\mu \rightarrow \nu_e$ and $\nu_\mu \rightarrow \nu_\tau$. No evidence for oscillations has been reported and the current best upper limits of the parameters Δm^2 and $\sin^2 2\theta$ are shown by the solid lines in Figure 2. The curves which are 90% confidence limit contour lines have been composed from the data

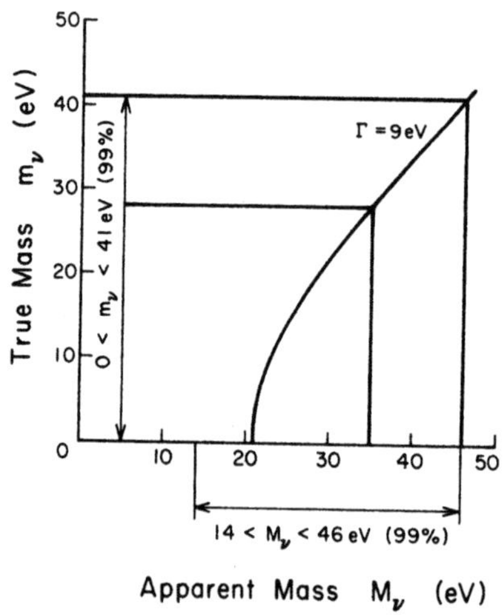

Figure 1 Relation between apparent and true neutrino mass as obtained from the reanalysis of the ITEP data taking into account the finite line width, Γ = 9 eV, of the calibration line. Within the 99% c.ℓ. the true mass can assume the value $m_\nu = 0$.

referred to by the label. The regions to the right of the curves are excluded. A detailed list of references is contained in the review by Baltay[16].

As to forthcoming results, several experiments have been proposed with the aim of exploring the region of smaller Δm^2 and $\sin^2 2\theta$. The sensitivities expected to be reached in these experiments are shown in Figure 2 by dashed lines (the experiments are identified by laboratory and proposal number[17]). It appears that an improvement in sensitivity of more than an order of magnitude should be attainable in the next 2-3 years.

Disappearance experiments have been conducted at high energy accelerators as well as with the help of fission reactors. Their results, in terms of the two parameters Δm^2 and $\sin^2 2\theta$, are summarized in Figure 3. By far the most stringent limits come from a recent $\bar{\nu}_e \rightarrow X$ experiment[6] at the Gösgen power reactor. I shall briefly

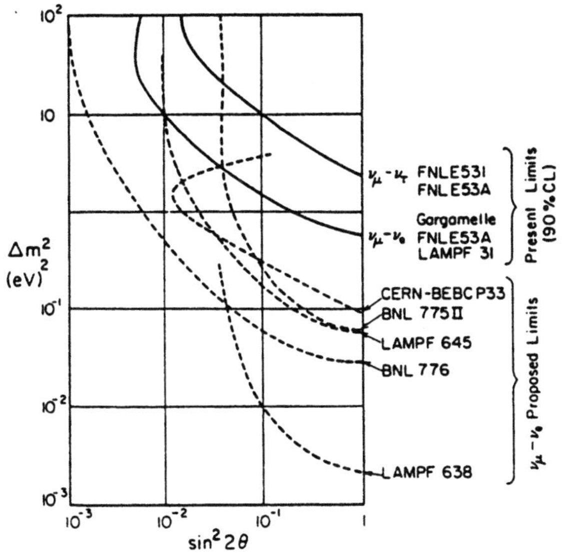

Figure 2 Limits for neutrino oscillations $\nu_\mu \rightarrow \nu_e$ and $\nu_\mu \rightarrow \nu_\tau$. The solid curves represent current experimental limits at 90% c.ℓ. The dashed curves illustrate forthcoming experimental limits.

describe this work here.

The Caltech-Munich-SIN group[6] has recently completed a measurement of the neutrino spectrum at a distance of 38 m from the core of the 2800 MW reactor at Gösgen, Switzerland. The set up of the experiment is sketched in Figure 4. The neutrinos were detected by the reaction $\bar{\nu}_e p - e^+ n$ using a composite liquid scintillation detector and ^3He multiwire proportional chambers. A time correlated e^+,n event constituted a valid signature.

Pulse shape discrimination in the scintillation counter has proved to be a powerful technique to eliminate correlated neutron background events. Cosmic ray induced fast neutrons recoiling on

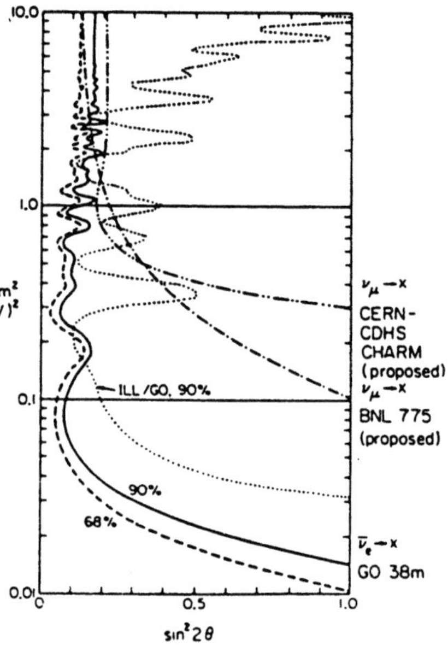

Figure 3 Limits for neutrino oscillations $\nu_\mu \rightarrow X$ and $\nu_e \rightarrow X$. The solid curve represents current experimental limits at 90% c.ℓ. from the Gösgen reactor experiment. Limits obtained from the ratio of the ILL (8.7 m) and Gösgen (L = 38 m) data are shown by the dotted curves labelled ILL/GO. Forthcoming experimental limits are shown by broken lines.

protons in the liquid scintillation counter can give rise to scintillation counter trigger, followed, after a thermalization period, by a neutron capture signal in the ^{3}He counter. The performance of the pulse shape discrimination is shown in Figure 5. The positron peak corresponding to the short pulse-decay time is the neutrino signal. The neutron peaks for reactor on and reactor off have the same height, indicating that there are no reactor associated neutrons.

The signal-to-noise ratio in this experiment was considerably better than in the earlier Grenoble experiment[18]. This was achieved by introducing position correlation cuts for the positron and neutron events.

About 11,000 neutrino induced events were recorded in a six-month reactor-on period. Background was recorded during a one-month reactor-off period. The measured positron spectrum was compared with that expected for no oscillations. The latter was obtained from the on-line beta spectroscopic measurements at the Laue-Langevin reactor by Schreckenbach *et al.*[19] studying ^{235}U and ^{239}Pu fission targets. These two isotopes account for about 89% of the total fission energy at the Gösgen reactor. The remaining 11% are due to fission of ^{238}U and ^{241}Pu. The calculations of Ref. 20 were used to evaluate the contribution to the beta spectra from ^{238}U and ^{241}Pu. The variation in time

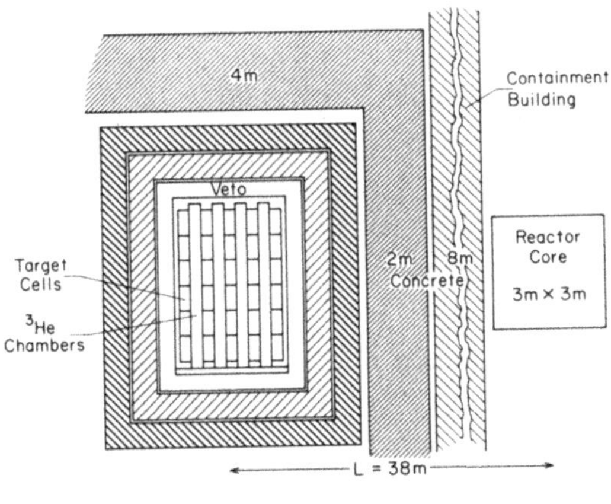

Figure 4 Experimental set up of the Neutrino Detector at the Gösgen Reactor. (The drawing is not to scale).

of the contributions of each fissioning isotope is well known and was taken into account. The calculated spectra to which the experimental spectrum was compared were obtained by multiplying the no-oscillation spectrum with the oscillation functions $P(E_\nu, L, \Delta m^2, \theta)$ and integrating over detector and core dimensions. A χ^2 test to all possible values of Δm^2 and $\sin^2 2\theta$ resulted in the 68% and 90% confidence limits displayed in Figure 3.

The Caltech-Munich-SIN group is now collecting data with the detector at L = 46 m. Comparing the neutrino spectra at two or more positions allows us to eliminate the assumption made for the yield of the no oscillation spectrum. In Figure 6 the observed yields for the 38 m position at Gösgen[6], the 8.7 m position at ILL[18] and some preliminary data for the 46 m position at Gösgen are displayed, in units of the no oscillation yield, as a function of L/E_ν. The exclusion plot of Figure 7 was obtained by considering the ratios of the data at 8.7 m, 38 m and 46 m for each energy bin and fitting it to calculated ratios for various oscillation parameters. This analysis is thus entirely independent on the no oscillation neutrino spectrum. We conclude that there are no neutrino oscillations with parameters larger than those contained to the right of the curve GO in Figure 3, or

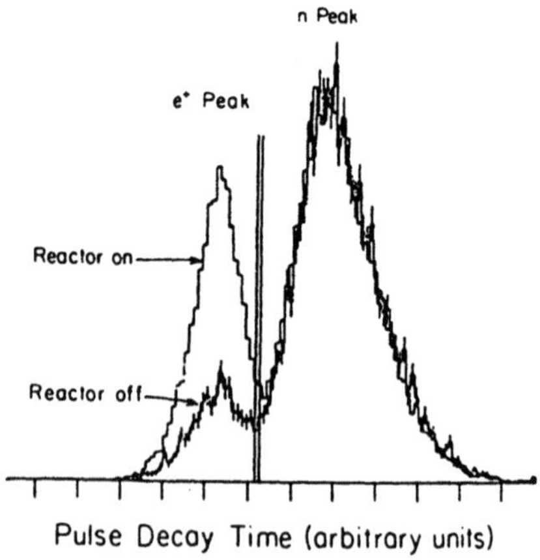

Figure 5 Pulse shape discriminator spectrum in the liquid scintillation counter. The e^+ peak is the neutrino signal.

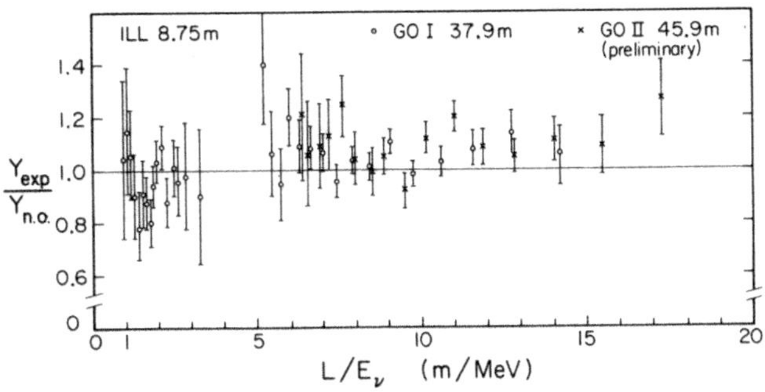

Figure 6 Ratio of experimental to predicted (for no oscillations) positron spectra at 8.7 m, 38 m and 46 m from the core. The errors of the data points shown are statistical.

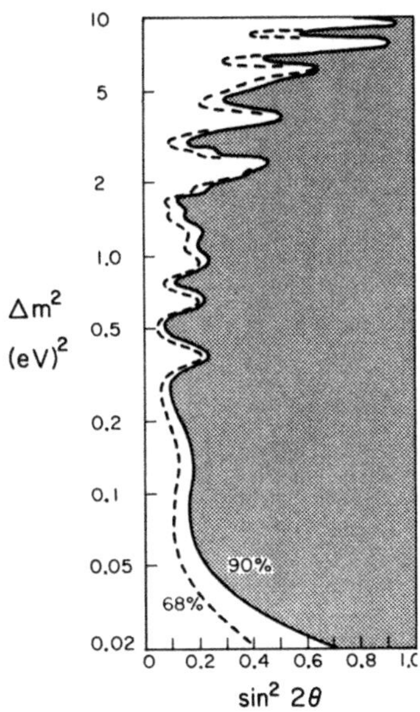

Figure 7 Exclusion plot obtained from the ratios of 8.7 m, 38 m, and 46 m experiments (preliminary result).

Figure 7.

The data also allows us to construct an independent neutrino spectrum and thus to verify the ILL spectrum.

Recent searches[21] for particle-antiparticle transitions $\nu_\mu \to \overline{\nu}_e$ and $\nu_e \to \overline{\nu}_e$ revealed no evidence having oscillation parameters Δm^2 (full mixing) larger than 0.7 eV^2 and 7 eV^2, respectively.

Finally, the important question should be addressed where, in the Δm^2 vs. $\sin^2 2\theta$ plane, should we continue to search for oscillations. Unfortunately there is no guidance from theory whatsoever. As to the mixing angle, we can state that present limits are smaller than the Cabibbo angle. Figure 8 shows these limits, together with other possible dimensional guesses (lepton mass ratios). If the solar experiments are indeed telling us that neutrinos oscillate with large mixing angle, the Δm^2 values must lie between 10^{-2} and 10^{-10} eV^2, a region increasingly more difficult to explore.

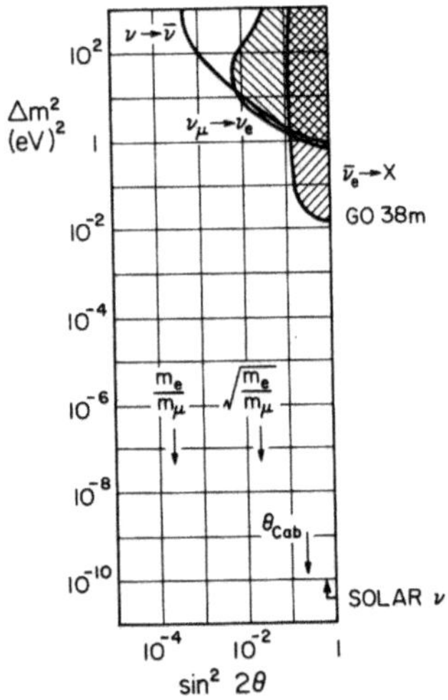

Figure 8 Expanded Δm^2 vs $\sin^2 2\theta$ plane showing current experimental limits and some dimensional guesses.

4. Endnotes

1. V. A. Lubimov *et al.*, Phys. Lett. 94B, 266 (1980).
2. J. J. Simpson, paper presented at ICOMAN 83, Frascati, Italy, January 1983.
3. E. Hennecke *et al.*, Phys. Rev. C 11, 1378 (1975).
4. T. Kirsten *et al.*, Phys. Rev. Lett. 50, 474 (1983).
5. F. Reines *et al.*, Phys. Rev. Lett. 45, 1307 (1980).
6. J.-L. Vuilleumier *et al.*, Phys. Lett. 114B, 298 (1982).
7. See the review by P. H. Frampton and P. Vogel, Phys. Reports 82, 339 (1982), p. 372.
8. R. J. Loveless, Neutrino 82, p. 89, A. Frenkel, editor, Budapest 1982; D. Reeder, Bull. Am. Phys. Soc. 28, No. 4, 643 (1983).
9. See review by J. Bahcall *et al.*, Revs. Mod. Phys. 54, 767 (1982), and recent results on $^{7}Be(p,\gamma)$ by B. W. Fillippone, Phys. Rev. Lett. 50, 412 (1983).
10. See for example: E. J. Konopinski, Beta Radioactivity, Oxford 1966, p. 7.
11. C. Bennett, private communication, to be published.
12. T. Bowles and H. Robertson, private communication.
13. O. Fackler, private communication.
14. K. E. Bergkvist, private communication.
15. W. Kündig *et al.*, Physics Institute, Univ. Zürich, Annual Report 1981/82, p. 48.
16. C. Baltay, Neutrino 81, Vol. II, p. 295, Univ. Hawaii, 1982.
17. For CERN Proposals, see A. L. Grant, Neutrino 81, Vol. II, p. 214; M. Murtagh *et al.*, Brookhaven Proposal E775; A. Pevsner *et al.*, Brookhaven Proposal E776; T. Romanowsky *et al.*, LAMPF Proposal 645; T. Dombeck *et al.*, Los Alamos Proposal 638.
18. H. Kwon *et al.*, Phys. Rev. D 24, 1097 (1981).
19. K. Schreckenbach *et al.*, Phys. Lett. 99B, 251 (1981).
20. P. Vogel *et al.*, Phys. Rev. C 24, 1543 (1981).
21. A. M. Cooper *et al.*, Phys. Lett. B 112, 97 (1982).

Dr. Felix Boehm is affiliated with the California Institute of Technology. This work was supported by the US Department of Energy.

DOUBLE BETA DECAY: RECENT DEVELOPMENTS AND PROJECTIONS

F. T. Avignone, III
University of South Carolina, Columbia, South Carolina 29208
and
R. L. Brodzinski, D. P. Brown, J. C. Evans, Jr., W. K. Hensley, J. H. Reeves and N. A. Wogman, Battelle Pacific Northwest Laboratory, Richland, Washington, 99352

Abstract

A report of recent events in both theoretical and experimental aspects of double beta decay is given. General theoretical considerations, recent developments in nuclear structure theory, geochronological determinations of half lives and ratios as well as laboratory experiments are discussed with emphasis on the past three years. Some projections are given.

Introduction.

The history of double beta decay has been adequately reviewed earlier [1]. Recent discussions of experimental and theoretical topics have been given by Wu [2] and by Rosen [3] respectively. Reviews which place the subject of double beta ($\beta^-\beta^-$) decay in the context of this conference were written by Primakoff and Rosen and by Frampton and Vogel [4]. Here we report recent progress of a number of efforts which address existing controversies, and we also attempt a few important projections. At the current pace of activity of this subject, parts of this paper will be obsolete before the publication of the proceedings of this conference.

Double beta decay is a second order weak process, which is the only possible mode of decay between even-even nuclei whose binding energies are increased by pairing forces sufficiently to render first order beta decay energetically forbidden. Two possible decay modes are:

$$(Z,A) \rightarrow (Z+2,A) + 2\beta^- + 2\bar{\nu}_e \qquad (1)$$

and

$$(Z,A) \rightarrow (Z+2,A) + 2\beta^-. \qquad (2)$$

Decay (1) is an ordinary $\Delta\ell=0$, second-order weak decay, while (2) is lepton nonconserving, $\Delta\ell=2$, and can only occur if the neutrino is a Majorana particle. Studies of the exotic decay mode (2) are designed to yield specific information about the properties of the neutrino under charge conjugation. The most important issues are associated with the $\Delta\ell=2$ decay (2); however, experimental determinations of half-lives of $\Delta\ell=0$ decays are very important for testing our understanding of the nuclear physics involved, and in particular for quantitatively testing the appropriate theoretical nuclear matrix elements.

General Theoretical Considerations.

The $\Delta\ell=2$, $\beta^-\beta^-$-decay depicted in Fig. 1 involves the leptonic current of equation (3). The γ_5-invariance can be broken by

$$J^{\ell}_{\mu} = \psi^{+}_{e}\gamma_4\gamma_{\mu}\{(1+\gamma_5) + \eta(1-\gamma_5)\}\ [\psi_{\bar{\nu}}(x)+\psi_{\nu}(x)] \qquad (3)$$

neutrino mass or explicitly by $\eta(1-\gamma_5)$. Detailed nuclear structure calculations yield an expression for the decay constant λ of the form,

$$\lambda \cong A\ \{a\xi^2 + b\eta^2 + c\eta\xi\}, \qquad (4)$$

where $\xi \equiv m_\nu/m_e$, and where higher order terms in η and ξ are neglected. In principle then, η and ξ can be determined from the same experiment in cases for which the constants A, a, b, and c can be reliably calculated.

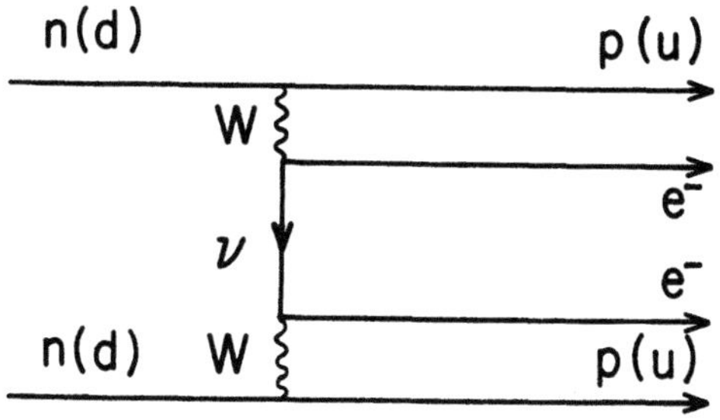

Figure 1. Diagram depicting lepton non-conserving ($\Delta\ell=2$) double beta decay.

The parameter η has interesting implications in the context of left-right symmetric gauge theories in which right-handed currents are mediated by a right-handed weak boson with mass $M(W_R)$. The relationship between the masses $M(W_R)$ and $M(W_L)$ is given by

$$\eta = [M(W_L)/M(W_R)]^2. \quad (5)$$

If we combine the experimental limit, $|\eta| \leq 2.4\times10^{-5}$, with the recent measurement of $M(W_L) = (81\pm5)$ GeV, by Rubbia and his co-workers [5], we see $M(W_R) \geq 1.6 \times 10^4$ GeV.

Recently it was shown by Rosen [6] and by Doi et al. [7], that while decays between $J^\pi = 0^+$ nuclear states can be engendered by an amplitude whose strength depends on the non-zero Majorana neutrino mass and on an explicit right handed current, decays of the type $0^+ \to 1^+, 2^+$, can only be driven by right-handed neutrino currents. This raises the exciting prospect of distinguishing between these mechanisms experimentally in a high energy-resolution measurement by detecting of the coincident γ ray as well as the energy of the two beta particles.

Recent observations by Wolfenstein and by Halprin, Petkov and Rosen [8] indicate that no-neutrino ($\Delta\ell$=2), $\beta^-\beta^-$-decay may also provide a sensitive probe of the neutrino mass matrix to complement ν-oscillation experiments. Wolfenstein points out that two light neutrino mass eigenstates can also be eigenstates of CP with eigenvalues of opposite sign. In this case, the effective mass observed in $\beta^-\beta^-$-decay is of the form,

$$\langle m^{Maj}\rangle_\nu = |m_1 \cos^2\theta - m_2 \sin^2\theta|. \quad (6)$$

In this model, the apparent disagreement between the direct mass measurement of Lubimov et al. (14eV $\leq m_\nu \leq$ 46eV) [9] and the limits from direct searches for $\Delta\ell$=2, $\beta^-\beta^-$-decay by our group and by the Milano group ($m_\nu \leq$ 10eV) and an indirect search by the Heidelberg group ($m_\nu \leq$ 5.6 eV) discussed below, can be reconciled for certain values of m_1, m_2, and θ. Halprin, Petkov and Rosen [8] have extended the argument to one light and one heavy neutrino. The expression (6) then becomes

$$\langle m^{Maj}\rangle_\nu = |m_1 \cos^2\theta - F(m_2,A)\eta_1\eta_2 m_2 \sin^2\theta|, \quad (7)$$

for $m_2 >$ a few MeV. The light neutrino has a photon-like propagator while the heavy neutrino propagator is Yukawa-like. The quantity $F(m_2,A) \equiv \langle\exp(-m_2 r)/r\rangle/\langle r^{-1}\rangle$, r is the distance between the two neutrons while η_1 and η_2 are the phases (±1) resulting from charge conjugation. If one were to assume that recent experimental results discussed below imply $\langle m^{Maj} \rangle_\nu = 0$, observed in the decays of the Te isotopes, values of $m_1 = 15eV$, $m_2 = 150MeV$, $\sin^2\theta = 6.3x10^{-6}$ and $\eta_1\eta_2 = -1$, for example, yield zero for the right hand side of (7). These values on the other hand, yield $\langle m^{Maj} \rangle_\nu = 45,21$ and 18eV for ^{48}Ca, ^{76}Ge and ^{82}Se respectively. This exciting possibility increases the importance of searching for $\Delta\ell=2$, $\beta^-\beta^-$-decay in a variety of nuclei to sensitively investigate the structure of the neutrino mass matrix.

Finally, it is interesting to contemplate possible observable effects of the Δ-mechanism of Primakoff and Rosen [10] when the Majorana neutrino is exchanged between two d-quarks in the same bag as shown in Fig. (2). This process can be significantly enhanced, over that involving two nucleons, because the high energy cut-off of the virtual intermediate neutrinos is set by the mean distance between the particles exchanging the neutrino. In addition, this mechanism involves $\Delta J = 1,2$ operators and, hence, can not connect 0^+ states to 0^+ states. Haxton [11] has estimated the rate of this mode of decay when the exchange is between two neutrons in ^{76}Ge and has found it to be unobservably small. Its observation in one of the ^{76}Ge experiments would certainly give strong support to the picture involving the exchange between two quarks. The Milano group [12] has searched for the $0^+ \to 2^+$, $\Delta\ell=2$, $\beta^-\beta^-$-decay to the first excited 2^+ level in ^{76}Se and set the limit $T_{1/2} \geq 10^{22}y$. A sensitivity of two orders of magnitude better can be projected for several of the ^{76}Ge experiments over the next few years.

The list of exciting consequences of $\Delta\ell=2$, $\beta^-\beta^-$-decay grows continuously and gives ample justification and motivation for the recent increase in experimental activity. Correct interpretation of such experiments will require accurate knowledge of the nuclear matrix elements involved. We now turn our attention to recent developments in nuclear structure theory applied to $\beta^-\beta^-$-decay.

Relevant Developments in Nuclear Structure Theory.

Double beta decay is an effective probe of the fundamental properties of the neutrinos involved only when the nuclear matrix elements are favorable. The failure of earlier efforts [13] to observe $\beta^-\beta^-$-decay of ^{48}Ca was later explained theoretically by a very small Gamow-Teller (GT) matrix element [14]. The most elabo-

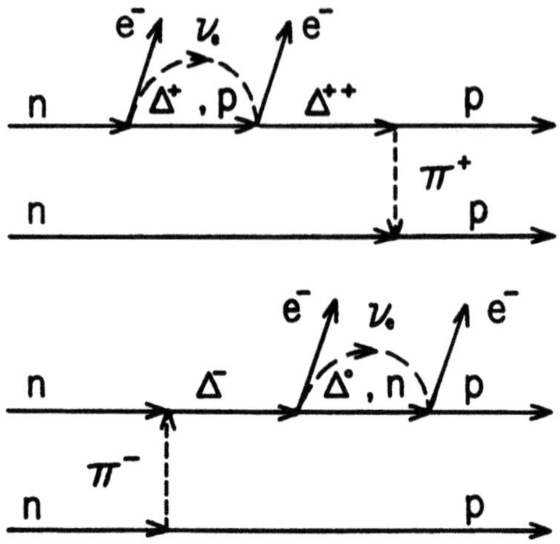

Figure 2. The two-quark process of $\Delta\ell=2$, $\beta^-\nu^-$-decay. There are six diagrams in all because the top diagram above, for example, can have the pion-line coupled to the initial, intermediate or final state in the nucleon line.

rate theoretical treatments of double beta decay are the recent shell model calculations of Haxton, Stephenson and Strottman [15], who treated the cases of ^{76}Ge, ^{82}Se, ^{128}Te and ^{130}Te. The LANL version of the Glasgow code developed by Haxton and Dubach [16] was used with a subsidiary density matrix code, which calculates one- and two-body density matrix elements, between states of good J and T. Neutron and proton valence states were treated separately and then combined in a weak coupling calculation.

In the $^{128,130}Te$ cases, all p and n-hole states in the model space ($2d_{5/2}$, $1g_{7/2}$, $3p_{1/2}$, $2d_{3/2}$, $1h_{11/2}$) were included [16]. The two-body interaction used was derived by Baldrige and Vary from the Kuo bare G-matrix elements [17]. The lowest 50 proton and 50 neutron states were used to form the weak coupling basis. All allowable combinations of these states were used in the p-n interaction matrix, written in the weak coupling basis, which was diagonalized. The two body density matrix was calculated and the $\beta^-\beta^-$-decay matrix elements evaluated. The results exhibited surprising coherence in the addition of the many components of the density matrix, resulting in large GT matrix elements ($M_{GT} \cong 1.48$) for both ^{128}Te and ^{130}Te. These disagree by more than an order of magnitude with those derived from geochronological (geochemical) data ($M_{GT} \cong$ 0.12 for ^{130}Te) [18]. The disagreement in the case of ^{128}Te is at least a factor of 7 ($M_{GT} \cong 0.2$). The calculated matrix elements for the decays of ^{128}Te and ^{130}Te are the same to within 2% and it is tempting to assume that they will cancel in the ratio of decay rates. If the small experimental values are the result of an unpredicted cancellation, however, the assumption that the matrix elements cancel in the ratio may be totally erroneous.

In the calculations of the decays of ^{76}Ge and ^{82}Se [15], the valence space included the $1g_{9/2}$, $1f_{5/2}$, $2p_{3/2}$ and $2p_{1/2}$ levels. The assumed closed core was ^{56}Ni so that the ^{76}Ge ground state consisted of 4-p, 6-n-holes, while the ^{76}Se ground state consisted of 6p, 8n-holes. The wave functions were constructed with all possible combinations of protons and neutron-holes. The Kuo matrix elements appropriate for the ^{56}Ni core, were adjusted to fit 28 observed energy levels in the region with a RMS deviation of 270 keV. The calculated value of the log ft for the decay of ^{81}Kr for example, is 4.7 which agrees well with the experimental value 4.58.

The result for the 2-neutrino ($\Delta\ell=0$), $\beta^-\beta^-$-decay of ^{76}Ge is $T_{1/2} = 2.4 \times 10^{20}$y while for the no-neutrino, ($\Delta\ell=2$) decay, the expression [11] corresponding to eq. (4) is

$$\lambda = 2.22\times10^{-21}\ \mathrm{sec}^{-1}\{\eta^2+1.54\xi^2-0.595\eta\xi\}. \tag{8}$$

This can also be written

$$r^{-2}(\theta) = 10^{-23}[1.27-0.27\cos 2\theta - 0.29\sin 2\theta]T_{1/2}, \tag{9}$$

where $T_{1/2}$ is in years while $r = [\xi^2+\eta^2]^{1/2} \times 10^5$ and θ is measured from the η-axis counterclockwise (see Fig. 3). A similar result given by the Osaka group [19] is

$$\lambda = 1.678\times10^{21}\ \mathrm{sec}^{-1}\{\eta^2 + 2.08\xi^2 + 0.107\ \eta\xi\}, \tag{10}$$

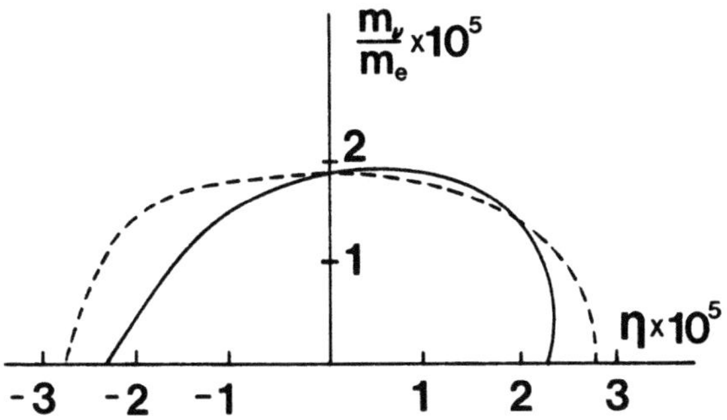

Figure 3. Plots of equations (9) (solid line) and (11) (dashed line) for $T_{1/2} = 1.7 \times 10^{22}$y.

from which we derive

$$r^{-2}(\theta) = 0.77\times10^{-23}\,[1.54-0.54\cos2\theta + 0.054\sin2\theta]. \quad (11)$$

The Osaka curve puts equivalent limits on m_ν but larger limits on η; however, an important matrix element was neglected. To our knowledge, only Haxton and his co-workers have calculated a complete set of matrix elements. From these equations one can show that it is practical to expect that limits of $<m^{Maj}>_\nu$ < 1eV and $\eta \lesssim 10^{-6}$ can be set by future ^{76}Ge, $\beta^-\beta^-$-decay experiments which are already underway and which are discussed later in this paper. The ^{82}Se results are discussed later in the context of a controversy between the calculations, a direct laboratory measurement, and a geochronological determination.

Recently, Zamick and Auerbach [20] considered the $\beta^-\beta^-$-decays of ^{48}Ca and ^{76}Ge in the framework of the Nilsson model with pairing. Their calculations explain the slow decay rate of ^{48}Ca in terms of the Lawson-Nilsson scheme involving the K-selection rule ($\Delta K = 0, \pm 1$). Their result is $M_{GT} = 0.18$, in excellent agreement with the earlier shell model calculation of Khodel [14] and the more recent one by Haxton et al. [11] ($M_{GT} = 0.19$), and also agrees with the experimentally observed suppression [13]. The $\beta^-\beta^-$-decay of ^{76}Ge is described as a transition of two neutrons in the K = 1/2 state at 4.71 $\hbar\omega$ to the K = 3/2 state at 4.443 $\hbar\omega$. A value in good agreement with the shell model prediction was obtained for M_{GT}, however, only for small deformations ($\eta\cong1$). It can be argued that $\eta\cong4$ is much more common in this region of the N-Z plane for which the matrix element M_{GT} is about an order of magnitude smaller than the shell model result ($M_{GT} = 1.28$). This occurs because the two body GT operator, $\vec{\sigma}(1)\cdot\vec{\sigma}(2)\tau_+(1)\tau_+(2)$ has no space dependence while Nilsson states, with different asymptotic quantum numbers, are orthogonal for large deformations. It is shown in ref. (20), however, that when pairing forces are added, the matrix elements of $\vec{\sigma}(1)\cdot\vec{\sigma}(2)$ all have the same sign (-) which results in coherent effects like those found by Haxton et al. [15]. For the Bohr-Mottelson value of the pairing gap, Δ = 1.38 MeV, the result is $M_{GT} = 2.2$. These calculations have been repeated by Haxton and Stephenson [21] using the full Nilsson wave functions with very similar results. Their results are, in fact, in quantitative agreement with their shell model calculations when an equivalent model-space is used. Further, it seems that an increase in the model-space results in a larger value for M_{GT} while truncated model-space calculations seriously underestimate the Gamow-Teller matrix elements for $\beta^-\beta^-$- decay. One way that these results can be interpreted is that in the case of ^{82}Se, one would have to completely neglect pairing effects in order to bring the Nilsson calculations into agreement with the geochronological results. It would be very surprising if pairing effects were not strong in a nucleus in this region.

Geochronological Determinations of Double Beta Decay Lifetimes.

The existence of $\beta^-\beta^-$-decay has definitely been established for ^{130}Te and ^{82}Se [22,23] using geochronological (geochemical) rare gas analyses. This involves mass spectrometry of Xe and Kr from very old ores containing large amounts of Te and Se respectively. Step-by-step heating separates atmospheric contamination from nucleogenic isotopes. These are commonly accepted as well established techniques, applied for decades to both terrestrial and extraterrestrial samples: for example, cosmic ray and solar wind effects in lunar samples and meteorites. Consequently, inherent background effects are generally thought to be well understood; however, the reliability of the results depends heavily on sample quality. An ideal sample is old enough to allow sufficient buildup of the radiogenic daughters. This can be as long as half of the age of the earth. For reliable half-life determination, the formation age of the ore body must be accurately dated. This can be done by potassium-argon dating for example; however, tellurium and selenium ores usually contain little potassium. For selenium ores, spontaneous fission dating can be employed. An alternative is to date the geological formation itself using geochronometers such as rubidium-strontium, lead-lead, etc. Deep burial of the ore for most of its history is required to minimize cosmic ray effects. For the Te-Xe system for example, cosmic ray evaporation processes favor light isotopes which are very underabundant in atmospheric Xe. The absence of excess ^{124}Xe and ^{126}Xe, relative to atmospheric contamination, places stringent constraints on cosmic ray background. Another potential background is spontaneous fission of ^{238}U which favors the heavy Xe isotopes and can also be evaluated by isotopic ratios because the fission yields are known. An ideal sample is also relatively refractory which minimizes gas loss during its history and allows discrimination against atmospheric contamination by using a step-by-step, temperature gas release method.

A very nearly ideal sample of gold telluride was obtained by the Missouri group [24] from Kalgoorlie, Australia. Kalgoorlie telluride has a mineralization age of (2.46 ± 0.08) x 10^9y based on Rb-Sr dating. This is essentially the same as the age of the rock. It contains negligible cosmic ray generated Xe, is very low in uranium content and is suitable for the stepwise heating. Selenium and tellurium are both found in this ore; hence it is also suitable for ^{82}Se half-life determination. The half lives of ^{130}Te and ^{128}Te were both obtained; however, the value obtained for ^{128}Te is not well established [24]. A krypton fraction was isolated for the ^{82}Se half life; however, the results have not been published [25].

The two best determinations of the ^{130}Te half life are those of Hennecke, Manuel and Sabu [24] using a Kalgoorlie gold telluride and those of Kirsten, Richter and Jessburgen [18] using a native tellurium from the Goodhope mine. The half lives are (1.05 ± 0.04) x 10^{21}y and (2.60 ± 0.28) x 10^{21}y respectively. The latter value is longer than the calculated value [16] by a factor $\sim$ 150 and represents one of the interesting current controversies. This disagreement, as well as that between the experimental values themselves, lead us to probe the possible sources of error. The most probable are: the determinations of ^{130}Xe excess, the Te content, background corrections, ore-age determinations and the degree of gas loss. We shall assume that the determination of the ^{130}Xe excess was very probably not a serious candidate because such measurements are routine for both groups. In addition, in the Goodhope sample, 96.5% of the ^{130}Xe was attributable to $\beta^-\beta^-$-decay while in the Kalgoorlie sample ^{130}Xe is enriched by a factor of 700 by $\beta^-\beta^-$- decay. In both cases, background effects were convincingly shown to be small. The Goodhope sample is essentially pure tellurium while the content of the Kalgoorlie sample was accurately measured by neutron activation. The most probable source of error then is the determination of the ore age. The Kalgoorlie ore was dated by the rubidium-strontium method, while the Goodhope sample was dated by the potassium-argon method on the sample itself. The sample was in fact low in potassium hence the measurement rests on a small change in ^{40}Ar to ^{36}Ar ratio relative to that in air. It can be argued that for the Heidelberg ^{130}Te half-life to be correct, the Kalgoorlie sample would have an age greater than that of the earth. If we assume that the Goodhope sample suffered gas loss, then it should have lost a larger fraction of ^{40}Ar. This would result in a shorter half life observed with the Goodhope sample which can not explain the discrepancy with the other measurement nor with theory.

The half life of ^{128}Te is expected to be $\sim 10^3$ times that of ^{130}Te based on the decay energy and the equivalence of the theoretical matrix elements of the two decays [16]. While the half-life would be extremely difficult to measure, the ratio of half lives or decay constants is also sensitive to m_ν and η. This is approximately written as

$$\lambda^{128}/\lambda^{130} \cong 2 + x^2 + y^2 \qquad (12)$$

where $x \equiv (m_\nu/m_e)x10^5$ and $y \equiv (\eta/2)x10^5$ [18]. This ratio is not affected by gas loss nor ore age determinations. Values of $(6.28 \pm 0.70) \times 10^{-4}$ and $(1.03\,^{+1.13}_{-1.03})x10^{-4}$ have been reported by the Missouri [24] and Heidelberg [18] groups respectively. The lower value is consistent with a massless Dirac neutrino, with $<m^{Maj}>_\nu \leq 5.6eV$ and $|\eta| \leq 2.4x10^{-5}$, while the larger value is consistent with $<m^{Maj}>_\nu \cong 10eV$ and $|\eta| \cong 4.5x10^{-5}$. There are important reasons why it would be valuable to analyze the apparently superior Kalgoorlie sample with the Heidelberg mass spectrometry facility. First, Heidelberg has at least one machine which has been maintained in a low background condition and has an extremely modern and elaborate data acquisition system [18]. Second, it appears that the Missouri machine has been exposed to neutron activated iodine samples containing ^{128}Te which can become imbedded in the walls of the spectrometer and released later [25]. This might explain the discrepancy. This experiment could settle this controversy as well as providing data on the half lives of both ^{130}Te and ^{82}Se. It has been pointed out by Kirsten and his coworkers [18] that substantial improvement in sensitivity or on the limits for $<m^{Maj}>_\nu$ and η is not expected because for $<m^{Maj}>_\nu \cong 2$ eV, the Dirac contribution to the decay of ^{128}Te becomes dominant, rendering the ratio insensitive to neutrino mass or to right handed neutrino currents.

An accurate determination of the half life of ^{82}Se would play an important role in the interpretation of the $\beta^-\beta^-$-decay data of ^{76}Ge because these nuclei are very similar in nuclear structure. A difficulty arises because the value calculated by Haxton et al. [15] for the half life of ^{82}Se is 2.35×10^{19}yr while the weighted average of fairly consistent values from geochronological determinations is $(1.45 \pm 0.15) \times 10^{20}$y, a factor of 6.2 longer than the calculated value [18]. This is, of course, far better agreement than in the case of ^{130}Te where there is more than a factor of 150 discrepancy. To further complicate matters, a laboratory measurement made by Moe and Lowenthal [26] yielded a value of $(1.0 \pm 0.4) \times 10^{19}$y, in much closer agreement with the theoretical value but even shorter. An earlier geochronologically determined value was reported by Srinivasan et al. [23] as $(2.76 \pm 0.88) \times 10^{20}$y based on the analysis of selenium and krypton in tellurobismuthite. The half life of ^{130}Te was used to date the source. If, however, the $T_{1/2}(^{130}Te)$ of Hennecke et al. [24] is used, the revised data yield $T_{1/2}(^{82}Se) = (1.07 \pm 0.34) \times 10^{20}$y in better agreement with the Heidelberg results. Experiments, which are now in the final stages of preparation, are expected to shed significant light on this controversy. They concern direct measurements of $T_{1/2}(^{82}Se)$ using a time-projection chamber and are discussed next.

Current Laboratory Searches for Double Beta Decay.

Direct laboratory searches for $\beta^-\beta^-$-decay have a long history and will not be reviewed here. A review of the earlier developments and all but the most recent results was given by Wu [2]. We will, in fact, concentrate on the direct measurements of $T_{1/2}(^{82}Se)$ of Moe and his coworkers as well as on the various ^{76}Ge experiments in preparation or underway at this time. The measurement of the ordinary, $\Delta\ell=0$, $\beta^-\beta^-$-decay half life of ^{82}Se will be extremely valuable in fixing the proper scale of the shell model calculations while the decay of ^{76}Ge has probably the best chance of observing the $\Delta\ell=2$ process as unambiguously as possible.

The sophisticated cloud chamber experiment of Moe and Lowenthal [26] produced results which are in serious disagreement with the geochronological world average reported by Kirsten [18]. The total half life from the cloud chamber experiments is $(1.0 \pm 0.4) \times$

10^{19}y, which is in sharp conflict with the average geochronological value $T_{1/2}(^{82}Se) = (1.45 \pm 0.15) \times 10^{20}$y. The theoretical value calculated by Haxton et al. [15] is 2.6×10^{19}y, where we have included a small correction for the Cabibbo angle factor. It is clearly important to understand the strength of each of these results in determining to what degree the shell model calculations might overestimate, or possibly underestimate the decay rate of $\beta^-\beta^-$-decay isotopes in this region, which includes the important case ^{76}Ge.

The earlier experience of the Columbia group in ^{48}Ca measurement was that the chief contaminant in the source was ^{214}Bi [2]. This can mimic the decay of ^{82}Se because its Q-value of 3.2 MeV is very close to the Q-value of 3.0 MeV for ^{82}Se. In addition, two electrons can result from the single β^-decay of ^{214}Bi to the first excited 2+ state in ^{214}Po when it decays by internal conversion. The decay of the first excited state of ^{214}Po at 0.6094 MeV by internal conversion occurs in approximately 0.5% of the ^{214}Bi decays and would appear to mimic $\beta^-\beta^-$-decay, except for the existence of a delayed α decay of the ground state of ^{214}Po($T_{1/2}$ = 164μs). The ^{82}Se source was evaporated in an atmosphere of argon, forming a particulate suspension which settled in about 24 hours [26]. The compounds of ^{226}Ra, which decays via three α-decays and one β^--decay to ^{214}Bi, were very probably discriminated against in the evaporation process. In addition, if they did carry over, they would be deposited on the surface of the Mylar with the ^{82}Se. A concrete test for the positive identification of ^{214}Bi associated background was accomplished by bleeding ^{222}Rn into the system, which deposits on the 12 Mylar source planes and decays to ^{214}Bi. The background which dies off with the well-known half life of 3.8 days must then be that of ^{214}Bi. A number of stringent tests that allow the positive identification of this background were convincingly discussed in ref. (26). The probability that the α-particle will not get out of the source, and consequently will be missed, was evaluated as 0.23. During the experiment, 72 (β^-+α) events and 16 (β + α + ICe) events were clearly identified. During that same time, 20 clean candidates for $\beta^-\beta^-$ decay were observed. When all corrections were made, the resulting half life is $(1.0 \pm 0.4) \times 10^{19}$y. A careful analysis of the angular distribution of these candidates does reflect the (1-cosθ) suppression of small angles for $\beta^-\beta^-$-decay while not for the known ^{214}Bi events. If these 20 clean candidate events are later shown not to be $\beta^-\beta^-$- decay, and the geochronological result is shown to be correct, a yet unknown background will have to be identified. The disagreement between this extremely well documented laboratory experiment and the standard, reliable, geochronological methods discussed earlier, is most important to resolve. A clear understanding of this case will have important consequences on the case of ^{76}Ge, whose nuclear structure is similar.

A second ^{82}Se $\beta^-\beta^-$-decay experiment, involving a time projection chamber (TPC) is in its final stages of development at UCI. The details of this technique were discussed recently by Moe and Hahn [27]; however, we give a short description here for completeness. The spectrometer is essentially a solenoid with a uniform, axial magnetic induction field and with the source of ^{82}Se at the mid-plane. The ends of the spectrometer are complex crossed-wire counters which have two dimensional sensitivity. The chamber is filled with a gas mixture of 80% He, 20% CH_4 at pressures between 0 and 2 atmospheres. The end counters are biased positive relative to the source so that the electrons formed along helical ionization tracks of the β^--particles are attracted towards the end wire counters. As the electrons drift ($v \sim 0.5$ cm/μs) into the position-sensitive wire chambers, the position information is fed to the computer as a function of time. The helical trajectories can then be reconstructed later during data analysis.

The experiment is scheduled to begin sometime during the summer or fall of 1983. If the earlier cloud chamber result is correct, then a thin, 14 gm source will result in 200, $\beta^-\beta^-$-decay events per month. If, on the other hand, the geochronological result is correct, they will observe 15 events per month. The projected detection limit for no-ν, $\Delta\ell$=2, $\beta^-\beta^-$- decay, using a 38 gm source and two years of counting is 2×10^{23}y which is about one order of magnitude longer than the present limit placed on the $\beta^-\beta^-$-decay of ^{76}Ge. For the same value of $\langle m^{Maj}\rangle_\nu$ and η, ^{82}Se is expected to have a half life somewhat shorter than ^{76}Ge because of the larger energy. In any case, the similarity in nuclear structure of these two cases dictates that they both should be investigated to about the same level of sensitivity. A clear observation of $\Delta\ell=2$, $\beta^-\beta^-$-decay in one isotope should also be observable in the other with a predictable difference in half life. This would be an important check on ^{76}Ge experiments.

There are currently five publicized efforts to observe $\beta^-\beta^-$-decay in ^{76}Ge, using low background Ge-detectors. These efforts are by: the Milano, UC Santa Barbara-Lawrence-Berkeley (UCSB-LBL), Cal Tech, Guelph-APTEC and Battelle-Carolina research groups. The Milano, Guelph-APTEC and Battelle-Carolina approaches differ in planned final volumes of Ge and in the materials of construction, but all involve super-low-background Ge detectors located underground. The Cal-Tech effort is similar but is located above ground with a highly efficient, cosmic ray live shield. The UCSB-LBL effort involves a large volume of Ge with NaI(Tℓ) anticoincidence shielding and is also located above ground. We will discuss the general features of all of these efforts but will concentrate on the Battelle-Carolina experiment.

The present limit is $T_{1/2}(^{76}\text{Ge}, \Delta\ell=2) \geq 1.7 \times 10^{22}$y (90% CL) which is essentially the same as the 2.1×10^{22}y (68% CL) set by the Milano effort. Descriptions of these experiments have been published earlier [12,28]. The Milano experiment has recently been rebuilt and is presently counting under Mont Blanc tunnel. It is a 116 cm^3 intrinsic Ge detector and has a much lower background than that previously published. The Guelph-APTEC detector is also a passively shielded intrinsic Ge detector. It has a volume of 220 cm^3 and is located approximately 1000 ft below the ground in a salt mine in Windsor, Ontario. Preliminary tests have already been run in situ. The Battelle-Carolina experiment presently consists of one intrinsic Ge detector of 130 cm^3 volume located above ground. It has a 4-inch thick plastic scintillator within the lead shield and operated in the veto mode (see Fig. 4). Preliminary results for these improved detectors are given in Table I below. Since all of the detectors are of different volumes, we have normalized the rates by the detector volumes because background rates scale roughly by volume.

Two steps were taken to reduce the level of the background in the Battelle-Carolina experiment. First, the NaI(Tℓ) was eliminated which significantly reduced the background due to ^{214}Bi as well as other sources of radioactivity.

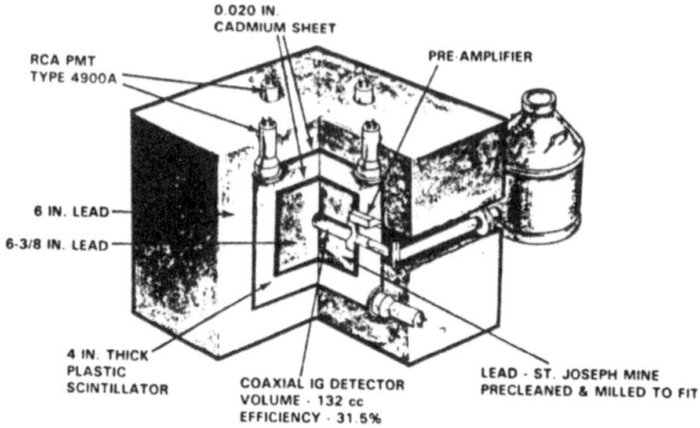

Figure 4. The 1983 Battelle-Carolina $\beta^-\beta^-$-Decay Spectrometer.

TABLE I. Preliminary Background Rates in Second Generation ^{76}Ge $\beta^-\beta^-$-Decay Experiments (given in counts/keV/hr/cm³)

Experiment	Background at 2.041 MeV count/keV/hr/cm³	Location
Milano (1983)	$\sim 1.6\times 10^{-5}$	Mont Blanc Tunnel
Guelph-APTEC (1983)	$\sim 3.2\times10^{-5}$	Windsor Salt Mine
Battelle-Carolina (1983)	$\sim 2 \times 10^{-5}$	Battelle, above ground
Milano (1973)	$\sim 6.2 \times 10^{-5}$	Mont Blanc Tunnel
Battelle-Carolina (1982)	$\sim 6.2 \times 10^{-5}$	Battelle, above ground
Cal. Tech (1983)	$\sim 2 \times 10^{-5}$	Pasadena, above ground

The background was then traced step by step. The aluminum end cap, diode cup, support hardware, and electronic parts close to the detector were considered the primary sources of primordial and man-made radioactivity. The isotopes ^{228}Ac, ^{212}Pb, and ^{208}Tl from the ^{232}Th chain, ^{234m}Pa from the ^{238}U chain, and ^{40}K were the significant primordial contributors remaining, with ^{137}Cs and ^{60}Co being the only discernible man-made radionuclides. Prospective construction materials in quantities of ten to one hundred times that actually used in the detector were assayed for radionuclide contamination using two 30 cm diameter by 20 cm thick NaI(Tℓ) detectors operating in coincidence with each other and in anticoincidence with a large plastic scintillator. Aluminum was found to be the major source of primordial radiation, with capacitors, resistors, field-effect transistors (FET), and rubber o-rings being small contributors. The stainless steel screws were found to contain ^{60}Co.

Samples from three types of copper were analyzed and the one with the least amount of radioactivity (< 0.0003 d/m/g) was selected to replace the aluminum. Brass screws replaced the stainless steel screws, and indium was used for the o-ring vacuum seal. No radioactivity was detected in the resistor or capacitor selected for use in the preamplifier first stage. The FET was modified to exclude the contaminated component, and the cryostat was modified to place the preamplifier outside the shield.

The plastic scintillator reduces the cosmic ray continuum by a factor of about 300. This, combined with radiopurity achieved by materials selection, results in the fact that essentially all of the remaining background in the vicinity of 2 MeV can be attributed to cosmic rays. There is still a very weak line at 2614 keV; however, there is none at 583 keV, indicating that the 2614 keV line is not a result of the β^- decay of ^{208}Tl. This line is due to (n,n') reactions on the lead shield with neutrons produced by cosmic rays. In addition, lines at 1064 and 570 keV appear from neutron capture on ^{206}Pb and (n,n') reactions on ^{207}Pb in the lead shielding. The neutrons are all attributable to cosmic rays. The intensities of well known γ-ray lines indicate that the background due to natural radioactivity has been reduced approximately 2 orders of magnitude, a fact which must be finally documented by either going underground or building an elaborate cosmic ray shield. We have chosen an underground site. The elimination of cosmic rays will reduce our figure of merit given in Table I by a factor of about 25 while the Milano and Guelph-APTEC experiments have already been tested underground and still have natural radioactivity in their materials which must be eliminated.

A small detector ($\sim$ 90 cm^3) at Cal-Tech has been rebuilt in a manner similar to that of the Battelle-Carolina detector but completely independently. Its figure of merit seems to be approximately the same as that of the Battelle-Carolina experiment; however, their cosmic ray active scintillator shield has substantially reduced this source of interference, but they also remain with some small component of radioactivity. The intensity of their 2614 keV line is about an order of magnitude less than that of the Milano experiment [12]. It appears that the Battelle-Carolina experiment has eliminated almost all detectable radioactivity in the region of 2 MeV, but the precise figure of merit, to compare with the other experiments, will have to be obtained in the mine. It is interesting, however, that its figure of merit is excellent even with a significant cosmic ray component remaining.

The projected ultimate sensitivity of ^{76}Ge experiments on neutrino mass and right handed currents will depend on detector volume as well as on background levels. Only the UCSB-LBL and Battelle-Carolina experiments have planned large volumes, 1360 cm^3 and 1600 cm^3 respectively. The UCSB-LBL experiment will consist of 8 intrinsic Ge detectors, 170 cm^3 each, inside of 4 inches of NaI(Tℓ) blocks operated as veto counters. The first two detectors will be operational by the end of the summer of 1983 while the entire detector will be operating, above ground, by the end of the summer of 1984. Considering the experience of the Battelle-Carolina effort that the NaI(Tℓ) and its encasements brought background that it did not eliminate, the UCSB-LBL experiment might well encounter that difficulty. On the other hand, their Ge detector is being constructed of entirely different materials than the Battelle-Carolina detector and can always be operated without the NaI(Tℓ) underground. The cups covering the Ge crystals are quartz, for example, while the cold plates are pure silicon. It would be necessary for the two such different experiments to report the same peak at $\sim$ 2041 keV, with approximately the same intensity per ^{76}Ge atom present, for the result to be accepted as being indicative of no-neutrino, $\Delta\ell=2$, $\beta^-\beta^-$-decay.

If we use the background reduction we have achieved and a volume of 800 cm^3, which we plan to have operating by the summer of 1984, the projected lifetime limit is $\sim 10^{24}$y. This corresponds to a sensitivity of $\langle m^{Maj}\rangle_\nu \leq 1.4$ eV. Combining the results of both experiments, energy-bin by energy-bin, over a period of 4 years, the ultimate sensitivity of about 10^{25} years is possible. This corresponds to $\langle m^{Maj}\rangle_\nu \lesssim$ 0.5 eV and $|\eta| < 10^{-6}$. This is probably the ultimate limit for any direct search for neutrinoless, or $\Delta\ell=2$, $\beta^-\beta^-$-decay because the ^{76}Ge experiment has the best energy resolution and is most sensitive in allowing γ-ray backgrounds to be accurately identified. In addition the nuclear matrix elements appear to be very favorable even though there still exists a controversy concerning the ^{82}Se $\beta^-\beta^-$-decay matrix elements. It is important to point out that even if the shell model overestimates the decay rates by a factor of 6.2, as implied by the geochronological data for ^{82}Se, the mass sensitivity only changes by a factor of 2.5. In this case the ultimate sensitivity would be $\langle m^{Maj}\rangle_\nu \leq 1.2$ eV.

Summary and Conclusions.

The new emphasis on double beta decay experiments and theoretical predictions stems from the important impact this subject has on our understanding of grand unification. Neutrinoless ($\Delta\ell=2$), $\beta^-\beta^-$-decay between 0^+ states can be driven by massive Majorana neutrinos, a small mixture of right-handed neutrino current or a mixture of both. Decays between 0^+ states and J^+ excited states when $J \neq 0$, can only be driven by an explicit γ_5 symmetry breaking term in the neutrino current. The present limits from direct searches [12,28] for the neutrinoless mode are $\langle m^{Maj}\rangle_\nu \leq 10$ eV and $|\eta| \leq 2.4 \times 10^{-5}$ The limit from an indirect search [18] , utilizing the ratio of half lives of ^{130}Te and ^{128}Te is $\langle m^{Maj}\rangle_\nu \leq 5.6$ eV with the same limit on $|\eta|$.

The quantity $\langle m^{Maj}\rangle_\nu$ is an admixture of all of the types of Majorana neutrinos which can be exchanged in the emission and absorption vertices of neutrinoless ($\Delta\ell=2$), $\beta^-\beta^-$-decay as shown in Fig. 1. Cancellation can occur which is dependent on the masses of all of the neutrinos and on the mass of the nucleus [8]. This feature can render this mechanism an extremely sensitive tool for probing the neutrino mass matrix.

Extensive shell model calculations of the $\beta^-\beta^-$-decay matrix elements in ^{128}Te, ^{130}Te, ^{82}Se and ^{76}Ge have been completed over the past few years [15,16]. These calculations result in large GT matrix elements due to coherent effects in the second order sums. This coherence has also been seen in Nilsson-model calculations with pairing forces [20,21]. The Nilsson-model calculations indicate that a further increase in the model-space of the shell model calculations will increase the magnitude of the GT matrix element of $\beta^-\beta^-$-decay of ^{76}Ge and ^{82}Se [21].

Geochronological (usually called geochemical) measurements of the $\beta^-\beta^-$-decay half lives of ^{82}Se and ^{130}Te are in serious disagreement with the shell model calculations. The measured value [18] of $T_{\frac{1}{2}}(^{130}Te)$ is approximately 150 times longer than the calculated value [16] while in the case of ^{82}Se it is about 6 times longer [18,15]. The shell model calculations are in much better agreement with a sophisticated cloud-chamber, direct measurement of $T_{\frac{1}{2}}(^{82}Se)$ which is designed to be capable of distinguishing between $\beta^-\beta^-$-decay and the decay of ^{214}Bi which is the only known background for this particular experiment [26].

A time-projection-chamber (TPC) experiment is in preparation which is expected to shed significant light on the ^{82}Se controversy within a matter of months from the publication of these proceedings [27]. The resolution of this controversy will have significant impact on the accuracy of the interpretation of no-neutrino ($\Delta\ell$=2) $\beta^-\beta^-$-decay experiments involving ^{76}Ge.

There are presently five publicized, serious efforts to sensitively search for neutrinoless $\beta^-\beta^-$-decay in high resolution, ^{76}Ge decay experiments. There have been significant recent breakthroughs in lowering the inherent backgrounds of these experiments by several groups. One can project that within 12 to 18 months the limit on $\langle m^{Maj}\rangle_\nu$ and $|\eta|$ will be about a factor of 4 or 5 smaller while an ultimate sensitivity of $\langle m^{Maj}\rangle_\nu \lesssim 0.5$ eV and $|\eta| \leq 10^{-6}$ is possible within about 4 years. In addition, it is possible that comparable sensitivities will be achieved with the ^{82}Se, TPC experiment [27]. In the case of ^{76}Ge, convincing evidence that neutrinoless, $\Delta\ell$=2, $\beta^-\beta^-$-decay has been observed must consist of independent observations of a spectral peak at $\sim$ 2041 keV, which has the same intensity per ^{76}Ge atom present. This will also have to occur in experiments with vastly different construction materials. Finally, it should be recognized that data from different experiments of comparable quality can be combined to increase the total sensitivity.

We have summarized the current status of the comparison between theory and experiment in Table II below. In addition we have summarized the limits one can place on the quantities $\langle m^{Maj}\rangle_\nu$ and $|\eta|$ from double beta decay experiments in Table III.

TABLE II. Calculated and Experimental Double Gamow-Teller Matrix Elements

Nucleus	$\tau_{1/2}^{exp}$(yr)	M_{GT}^{exp}	M_{GT}^{theory}
^{130}Te	2.55×10^{21} [18]	0.12*	1.48 [15]
^{82}Se	2.76×10^{20} [23]	0.27*	0.94 [15]
	1.06×10^{19} [23]	1.43	
^{76}Ge	$\geq 1.7 \times 10^{22}$ [12,28]		1.28 [15]
^{48}Ca	$\gtrsim 3.6 \times 10^{19}$ [13]	$\lesssim 0.19$	0.22 [15]

*Maximum values determined from geochronological rates.

TABLE III. Present Limits on $\langle m^{Maj}\rangle_\nu$ and $|\eta|$ from double beta decay experiments

Parent Isotope	$\langle m^{Maj}\rangle_\nu$	$\eta \times 10^5$	$\langle m^{Maj}\rangle_\nu^*$	$\eta \times 10^{5*}$	Ref.
^{82}Se	$\leq$ 14 ev	$\leq$ 2	$\leq$ 33 eV	$\leq$ 4.6	[26]$^+$
^{130}Te	$\leq$ 8 eV	$\leq$ 2.3	$\leq$ 100 eV	$\leq$ 15	[18]
^{128}Te	$\leq$ 0.7 eV	$\leq$ 0.3	$\leq$ 8.7 eV	$\leq$ 3.5	[18]
^{48}Ca	$\leq$ 41 eV	$\leq$ 3.9	$\leq$ 44 eV	$\leq$ 4.2	[13]
^{76}Ge	$\leq$ 10 eV	$\leq$ 2.4	$\leq$ 24eV	$\leq$ 4.5	[12,28]

*Values were analyzed with matrix elements renormalized to be in agreement with geochronological results in ^{130}Te and ^{82}Se. There is no compelling reason to do this.

$^+$This value was taken from Cleveland et al. [26].

The authors are grateful for interesting and helpful communications with F. Boehm, I. Campbell, E. Fiorini, W. C. Haxton, T. Kirsten, M. K. Moe, R. H. Pehl, S. P. Rosen, E. Takasugi, and L. Zanotti. This work was supported by the National Science Foundation under grant PHY-8209562 and the U.S. Department of Energy under contract DE-AC06-76RLO 1830.

References

[1] S. P. Rosen and H. Primakoff, in Alpha-beta and gamma-ray spectroscopy, 2nd ed. (ed. K. Siegbahn; North Holland Publ. Co. Amsterdam, 1965), E. Fiorini, Rev. Nuovo Cim. 2 (1971), D. Bryman and C. Picciotto, Rev. Mod. Phys. 50, (1978) 11.

[2] "Science Underground", AIP Conference Proceedings No. 96, ed. M. M. Nieto, W. C. Haxton, C. M. Hoffman, E. W. Kolb, V. D. Sanberg and J. W. Toevs. New York, 1983 (see C. S. Wu, p. 374).

[3] S. P. Rosen, p. 352 of ref. (2).

[4] H. Primakoff and S. P. Rosen, Ann. Rev. Nucl. Part. Sci. 31, (1981) 145.
P. H. Frampton and P. Vogel, Phys. Repts. 82, (1982) 339.

[5] G. Arnison et al., Phys. Lett. 122B, (1983) 103.

[6] S. P. Rosen in "Gauge Theories, Massive Neutrinos and Proton Decay," ed. A. Perlmutter (Plenum Press, New York 1981) p. 333.

[7] M. Doi, T. Kotani, H. Nishimura, K. Okuda and E. Takasugi, Phys. Lett. 103B, (1981) 219.

[8] L. Wolfenstein, Phys. Lett. 107B (1981) 77. See also the discussion by Doi et al., Phys. Lett. 102B, (1981) 323.
A. Halprin, S. T. Petkov and S. P. Rosen, preprint (1983).

[9] V. A. Lubimov, E. G. Novikov, V. Z. Nozik, E. F. Tretyakov and V. S. Kosik, Phys. Lett. 94B, (1980) 266.

[10] H. Primakoff and S. P. Rosen, Phys. Rev. 184 (1969) 1925.

[11] W. C. Haxton (private communication, to be published).

[12] E. Bellotti, E. Fiorini, C. Liguori, A. Pullia, A. Sarracino and L. Zanotti (p. 430 Ref. 2), Phys. Lett. 121B (1983) 72. Also E. Fiorini, These Proceedings.

[13] R. K. Bardin, P. J. Gollon, J. D. Ullman and C. S. Wu, Nucl. Phys. A158 (1970) 337. For ^{82}Se see Cleveland et al. Phys. Rev. Lett. 35, (1975) 737.

[14] V. A. Khodel, Phys. Lett. B32 (1970) 583.

[15] W. C. Haxton, G. J. Stephenson, Jr., and D. Strottman, Phys. Rev. Lett. 47, (1981) 153. (These calculations have been slightly improved by W. C. Haxton by avoiding some of the simplifying approximations). Phys. Rev. D26, (1982) 1805.

[16] W. C. Haxton, G. J. Stephenson, Jr., and D. Strottman, Phys. Rev. D25 (1981) 2360.

[17] W. T. Baldridge and J. P. Vary, Phys. Rev. C14 (1976) 2246; T.T.S. Kuo and G. E. Brown, Nucl. Phys. A114 (1968) 241.

[18] T. Kirsten (in ref. [2]); T. Kirsten, H. Richter and E. Jessberger, Phys. Rev. Lett. 50, (1982) 475.

[19] M. Doi, T. Kotani, H. Nishiura and E. Takasugi, Prog. Theor. Phys. 69, (1983) 602; 66 (1981) 1765; 66 (1981) 1730.

[20] L. Zamick and N. Auerbach, Phys. Rev. C26, (1982) 2185.

[21] W. C. Haxton and G. J. Stephenson, Jr., preprint LA-UR-83-417 submitted to Phys. Rev. C.

[22] T. Kirsten, W. Geutner, and O. A. Schaeffer, Z. Phys. 202 (1967) 273; T. Kirsten, O. A. Schaeffer, E. Norton and R. W. Stoner, Phys. Rev. Lett. 20, (1968) 1300.

[23] B. Srinivasan, E. C. Alexander, Jr., R. D. Beaty, D. Sinclair and O. K. Manuel, Econ. Geol. 68 (1973) 252.

[24] E. Hennecke, O. Manuel, D. Sabu, Phys. Rev. C11, (1975) 1378.

[25] O. Manuel (private communication).

[26] M. K. Moe and D. D. Lowenthal, Phys. Rev. C22 (1980) 2186. Also see Cleveland et al. Phys. Rev. Lett. 35, (1975) 737.

[27] M. K. Moe and A. A. Hahn (in ref. 2).

[28] F. T. Avignone, III, R. L. Brodzinski, D. P. Brown, J. C. Evans, Jr., W. K. Hensley, J. H. Reeves and N. A. Wogman, Phys. Rev. Lett. 50, (1983) 721.

$n\bar{n}$ OSCILLATION EXPERIMENTS

G. Fidecaro
CERN, Geneva, Switzerland

1. Introduction

Phenomenology of neutron-antineutron oscillations has already been discussed several times [1]; in particular, it has been discussed at the previous Workshops on Grand Unification [2-4], so that only a few essential formulae will be recalled in this paper. Theory also will be skipped as several discussions have taken place or will take place in this Workshop. Thus, the main purpose of this paper will be to bring up to date the available indications on the oscillation time of free neutrons, which following Anderson [4] will be called hereinafter mixing time, and to review the status of the existing experiments and proposals, with particular attention to the Grenoble experiment [1,5], which is still the only one producing results. Some considerations will also be reported concerning future experiments to measure $n\bar{n}$ mixing time using free neutrons.

2. $n\bar{n}$ Mixing Time from Proton-Decay Experiments

As is well known, no theory exists which can give indications on the mixing time $\tau_{n\bar{n}}$. The only information useful as a feedback to the theory and for the planning of experiments comes from available information on the stability of matter and, to a lesser extent, from cosmic rays. This information has been summarized in Table 1, which is a revised version of the similar table of Ref. 1, giving lower limits for $\tau_{n\bar{n}}$.

The indications given by Kuz'min [6] and by Glashow [7,8] have now only a historical interest: Kuz'min was the first to point out the possible existence of $n\bar{n}$ oscillations as early as 1970, but the present wave of interest, and in particular the Grenoble experiment, has been

Table 1: Estimates of mixing time

Ref.	Year	Author	Limits τ (y)	Limits $\tau_{n\bar{n}}$ (s)
[6]	1970	Kuz'min		$10^4/2\pi$
[7,8]	1979	Glashow		10^5-10^6
[9,10]	1980	Mohapatra and Marshak	10^{30}	10^5
[11]	1980	Chang and Chang	10^{30}	10^7
[12]	1980	Kazarnovskii et al.	10^{30}	2×10^7
[13]	1981	Chetyrkin et al.	3×10^{30}	3×10^7
[14]	1980	Sandars	10^{30}	1.8×10^7
[15]	1982	Alberico et al.	10^{31}	3×10^7
[16]	1982	Riazzudin	10^{30}	3×10^6
[17]	1983	Dover et al.: for ^{16}O	10^{31}	5.7×10^7
[17]	1983	Dover et al.: for ^{56}Fe	10^{31}	4.7×10^7
[18]	1981	Cowsik and Nussinov (KGF)	10^{31}	5×10^7
[20]	1981	Cherry et al. (Homestake Gold Mine)	1.4×10^{30}	2×10^7
[21]	1983	IMB Collaboration (Morton Salt Mine)	7×10^{30}	2.5-4×10^7
[22]	1983	NUSEX Collaboration (Mont Blanc)	10^{31}	5×10^7
[18]	1981	Cowsik and Nussinov	cos. ray	10^4
[26]	1981	Sawada, Fukugita and Arafune	cos. ray	0.25×10^4
[27]	1982	Arafune and Fukugita	cos. ray	0.7×10^4
[28]	1982	Arafune and Fukugita	cos. ray	0.9×10^4

triggered by the Glashow papers. In both cases the simple one-parameter formula $\delta m \sim (m\hbar/\tau)^{\frac{1}{2}}$ has been used, where δm is the mass splitting between the two eigenstates $n + \bar{n}$ and $n - \bar{n}$ produced by the $\Delta B = 2$ coupling, m is a "typical" hadronic mass, and τ is the decay rate of nuclei under $\Delta B = 2$ transitions. Using this same formula and allowing for a correction factor to take into account the overlap effect of the two nucleons in the nucleus, Mohapatra and Marshak [9,10] got a somewhat shorter mixing time. The result given by L.N. Chang and N.P. Chang [11] does not contain this correcting factor.

The work of Mohapatra and Marshak [9,10] represents perhaps the first attempt to improve the crude model used by Kuz'min and Glashow. In fact, immediately afterwards, Kazarnovskii et al. [12] and Chetyrkin et al. [13], and independently Sandars [14] and later Alberico et al. [15 adopted a non-relativistic optical-model approach with a complex potentia so as to introduce in the theory the annihilation rate, thus obtaining the two-parameter formula $\Gamma = N\delta m^2\Gamma_a/[(\Delta M)^2 + \Gamma_a^2/4]$; where Γ is the width for $\Delta B = 2$ decay of a nucleus with A nucleons and N neutrons, Γ_a the annihilation width, and ΔM the difference between the energies of an n and an $\bar{n}$ in the nucleus. The four slightly different results obtained correspond to different choices of the n-nucleus and $\bar{n}$-nucleus optical potentials obtained either from experiments in the first three cases, or by using a description of the n-nucleus and of the $\bar{n}$-nucleus interaction in terms of explicit meson exchange except for the short distances treated phenomenologically in the last case.

Different lines were followed by Riazzudin [16] and by Dover et al. [17]. The first author assumed that after a neutron is replaced by an antineutron as a result of the $n \to \bar{n}$ transition the subsequent annihilation into pions takes place predominantly through an $I = 1$, $J^P = 1^-$ state.

Dover et al. [17] have solved numerically the coupled Schrödinger equation for n and $\bar{n}$ wave functions in a nucleus, using a shell-model potential which fits the empirical binding energies of the neutron orbits, and a complex $\bar{n}$-nucleus optical potential from fits to $\bar{p}$-atom level shifts. As a result, the mixing times are rather stable when permissible alterations of the parameters of the nuclear model are made.

The above papers aim essentially at the development of a reliable theory which could make it possible to obtain the mixing time from proton-decay experiments. Going down through the table one finds instead the results for the mixing time obtained in the case of specific proton-decay experiments, as they have been reported in the literature.

Cowsik and Nussinov [18], using results obtained by Krishnaswamy et al. [19] in the Kolar Gold Field some time ago, obtain a limit for the mixing time of $\tau_{n\bar{n}} > 5 \times 10^7$ s. No limit has been given by the authors of this experiment.

Cherry et al. [20] have analysed the results of their proton-decay experiment in the Homestake Gold Mine in terms of $n\bar{n}$ oscillations and have reported a limit of $\tau_{n\bar{n}} > 2 \times 10^7$ s.

Two more results have been reported at this Workshop, one by the IMB Collaboration [21] which finds $\tau_{n\bar{n}} > 2.5\text{-}4 \times 10^7$ s, the other one by the NUSEX Collaboration [22] which gives $\tau_{n\bar{n}} > 5 \times 10^7$ s.

The above four results tend to indicate that a lower limit for $\tau_{n\bar{n}}$ beyond 10^7 s should be considered as established. This conclucion seems to be shared by Ellis [23], who has independently examined the results of the same four experiments coming to the conclusion that the oscillation time for neutrons bound in a nucleus should be longer than $(1.0 \pm 0.44) \times$ $\times 10^{31}$ years, corresponding to $\tau_{n\bar{n}} > (2.7 \pm 1.0) \times 10^7$ s. However, Baldo Ceolin [24], quoting the recent results of the Kolar Gold Field experiment [25], concludes that from that experiment one gets $\tau_{n\bar{n}} > 5 \times 10^6$ s only.

Results from cosmic rays are also reported. Comparison of these results with results obtained from experiments using free neutrons or from proton-decay experiments is obviously of interest mainly for astrophysics.

3. Review of experiments

Though experiments using ultracold neutrons have been considered by various authors [1,29], apparently no definite proposal has yet been worked out, so that only experiments of the "beam" type need to be considered here. In experiments of this kind a beam of neutrons is made to travel in vacuum, in a low magnetic-field region, until it reaches a target where the antineutrons annihilate. The number ν of antineutrons, detected with an efficiency of ε through their annihilation products, is given by:

$$\nu = \Phi\, T\, \varepsilon\, (t/\tau_{n\bar{n}})^2 ,$$

where Φ is the neutron flux in n/s, T is the duration of the experiment expressed in seconds, and $t = \ell/v$ is the time it takes a neutron having velocity v, in metres per second, to reach a target placed at a distance of ℓ metres.

The magnetic field has to satisfy the following condition in order not to slow down the $n\bar{n}$ transition rate during the time t:

$$B \ll \hbar/\mu t ,$$

where μ is the magnetic moment of the neutron. For example, in the case $t \sim 10^{-2}$ s, one must have $B \ll 10^{-2}$ G in order to be able to consider the neutrons as free (quasi-free).

It should also be pointed out that the mixing time $\tau_{n\bar{n}} = \hbar/\delta m$ and not its corresponding period $2\pi\tau_{n\bar{n}}$ has been used throughout this paper, contrary to what is sometimes done by other authors.

It should also be recalled that in the case of the experiments considered here cosmic rays are expected to be the main source of background

The status of experiments and proposals has not changed so much since Anderson gave his report at last year's Workshop [4]. For this reason it has been found convenient to reproduce here (Tables 2-4) the three tables presented by Anderson, adding information about two new experiments [30,31]. For the rest, no changes have been introduced except for the correction of a few obvious misprints, since the tables have value more for the purpose of making a comparison between the variou experiments than as a description of the single experiments.

4. The Grenoble experiment

Phase I of the Grenoble experiment, based on a simple Pb-scintillato calorimeter, has been already described [1]. The analysis of the data indicated the need for a track-measuring detector in order to recognize and reject events produced by cosmic rays, and in fact Phase II [32]

Table 2: Neutron sources

Grenoble:	57 MW Reactor: cold neutrons 25°K. Curved beam guide 20 cm × 3 cm low γ, low fast n $\Phi = 10^9$ n/s, $\bar{v} = 160$ m/s, $\lambda = 20$ Å
Pavia:	0.25 MW Reactor: thermal neutrons Area 1.2 m × 1.2 m Bi + paraffin moderator $\Phi = 2 \times 10^{11}$ n/s at target
Oak Ridge:	30 MW Reactor: thermal neutrons Area 1400 cm^2 $\Phi = 5 \times 10^{13}$ n/s at 1 m^2 target at 20 m $\Phi = 7 \times 10^{12}$ n/s at target with Bi + D_2 moderator
Los Alamos LAMPF:	Proton Linac 580 μA heavy metal beam stop, D_2O moderator $\Phi = 2\text{-}4 \times 10^{12}$ thermal neutrons (1 m target at 30 m)
Los Alamos Omega West:	8 MW Reactor: thermal neutrons Channel tangent to core $\Phi = 3 \times 10^{11}$ n/s (1.5 m target at 50 m)
Chalk River:	110 MW Reactor: thermal neutrons Channel tangent to core $\Phi = 1 \times 10^{12}$ n/s (1.5 m target at 50 m)
Grenoble III:	Guide Area 100 cm^2, $\lambda = 10$ Å $\Phi = 1 \times 10^{12}$ n/s (1 m target at 30 m)
Moscow Meson Factory:	Proton Linac 500 μA lead or uranium beam stop, D_2O moderator $\Phi = 6 \times 10^{12}$ n/s (cold neutrons 25°K) (12 targets each 1.6 m diameter at 57 m)
Leningrad:	100 MW Reactor: cold neutron 25°K $\Phi = 10^{14}$ n/s from 12 × 25 cm^2 port (5 × 7 m^2 target at 50 m)

Table 3: Detectors

Grenoble I:	Pb scintillation calorimeter 760 kg 10% active. Acceptance 25% of 4π Thick target ^{6}LiF
Grenoble II:	Limited streamer tubes + Al plates Tracking calorimeter. Acceptance 95% of 4π Thick target ^{6}LiF
Pavia:	Pb-flash chamber tracking calorimeter 65% of 4π acceptance Scintillator hodoscope = resistive plate chambers for trigger Thin target C or Be
Oak Ridge:	Pb-glass Čerenkov counter + scintillation counter hodoscope 90% of 4π acceptance
LAMPF:	Time-of-flight and tracking chambers Temporal + spatial reconstruction 93% of 4π acceptance No calorimetry
Los Alamos Omega West:	Time of flight + calorimetry: using liquid scintillators 95% of 4π acceptance Checks momentum balance
Moscow Meson Factory:	Multiwire limited Geiger planes surrounding each target and forming a single detector Scintillation calorimetry 80-90% acceptance of 4π
Leningrad:	Moscow Meson Factory type detector might be considered

Table 4: $n \to \bar{n}$ experiments

$$\nu = (t/\tau)^2 \, \Phi T \varepsilon$$

$$t = \ell/v \quad T = 1.73 \times 10^7 \text{ s (200 d)} \quad \tau = 10^7 \text{ s}$$

Experiment	Drift length ℓ (m)	Drift time $t = \ell/v$ (s)	Neutron flux Φ (ns^{-1})	Figure of merit Φt^2 (ns)	Detection efficiency ε	No. events for $\tau = 10^7$ s ν	Sensitivity at "zero" background τ_{min} 90% CL (s)
Grenoble II	6	3×10^{-2}	10^9	9×10^5	0.35	0.054	1.5×10^6
Pavia	16	7×10^{-3}	3×10^{11}	1.4×10^7	0.50	1.3	7×10^6
Oak Ridge	20	8×10^{-3}	4×10^{13}	3×10^9	0.50	215	1×10^8
(with Bi-D_2O moderator)			6×10^{12}	4×10^8	0.50	29	4×10^7
LAMPF							
(probable)	30	1.4×10^{-2}	2×10^{12}	4×10^8	0.50	34	4×10^7
(possible)			4×10^{12}	8×10^8	0.50	68	5×10^7
Omega West	50	2.3×10^{-2}	3×10^{11}	1.5×10^8	0.50	13	2×10^7
Grenoble III	35	9×10^{-2}	10^{12}	8×10^9	0.50	700	1.7×10^8
Moscow Meson Factory	57	8×10^{-2}	6×10^{12}	3.8×10^{10}	0.50	3300	5.8×10^8
Leningrad	50	6.7×10^{-2}	10^{14}	4.4×10^{11}	0.50	38000	1.3×10^9

Note: In this table τ *has been used for* $\tau_{n\bar{n}}$.

differs from Phase I essentially because the calorimeter has been replaced by a set of limited streamer tubes of the "Mont Blanc" type [33] surrounding the target on all sides, except for the side through which the beam enters the apparatus (Fig. 1). Eighteen planes were placed in the forward direction and four planes all around the target so as to have a box geometry covering 90% of 4π. All planes were sandwiched between 5 mm thick Al plates, except for the last ten planes of the forward chamber which were sandwiched between 5 mm thick Fe plates.

The detector was completed by two logically separated systems of scintillation counters. The forward system, used for the trigger, was made of a front hodoscope (HX, HY) of 12 + 12 counters, each $48 \times 4 \times 0.6\ cm^3$, arranged in a square matrix outside the first μ-metal shield, and of a set of counters T and a second set of counters B.

The lateral system was composed of a barrel hodoscope (56 counters, each $100 \times 4 \times 0.6\ cm^3$). Both hodoscopes were used for track reconstruction.

The whole apparatus was covered by a system A of 66 anticoincidence counters ($30\ m^2$) providing a veto signal against cosmic rays. For the rest the apparatus was the same as that of Phase I.

The trigger was defined by the coincidence $[(HX \cdot HY) \cdot T \cdot B \cdot \bar{A}]$, thus requiring at least one particle in the forward direction (solid angle for one particle = 31.8% of 4π).

A few events are reproduced in Figure 2 so as to show the resolution of the apparatus. The event reproduced under (d) looks particularly interesting. Assuming that the forward particle stops in the chamber, this event is in fact compatible within the resolution of the apparatus with the elastic or quasi-elastic scattering of a 500 MeV proton generated above the chamber system by a cosmic-ray neutral. This makes possible a more complete discussion of the event, with a view to a better understanding of the behaviour of the detector. In particular, comparison with a Monte Carlo simulated event having nearly the same geometry shows the difficulty of recognizing the nature of events without a measurement of the energy of the particles involved and conversely the importance of more accurate and more complete measurements in the case of higher sensitivity experiments.

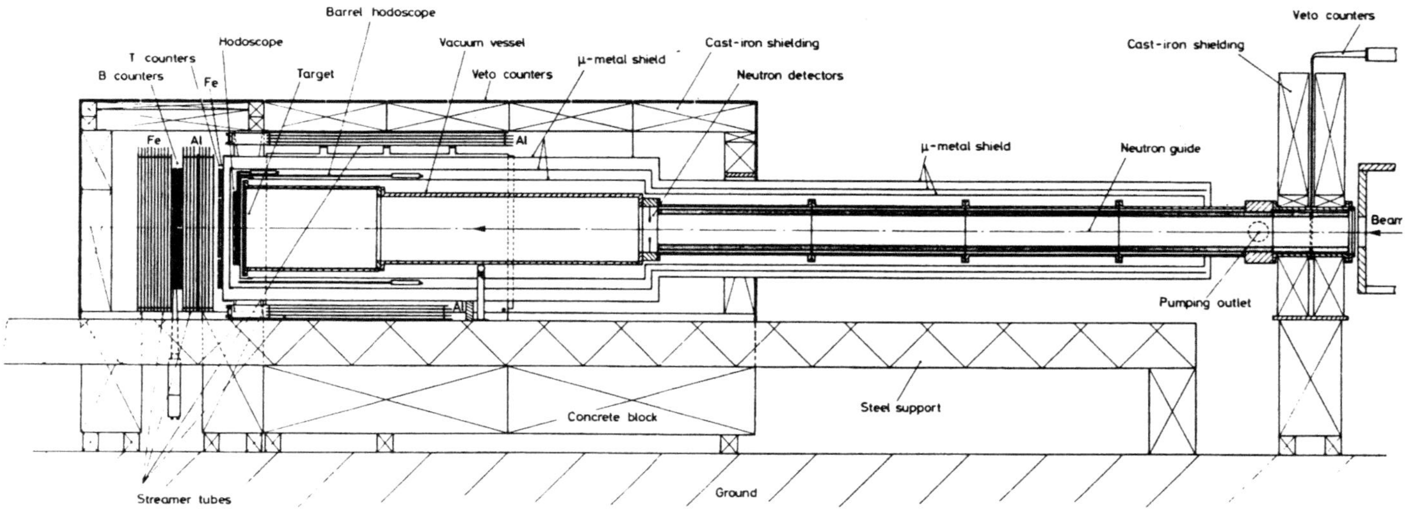

Fig. 1 The Grenoble experiment, phase II

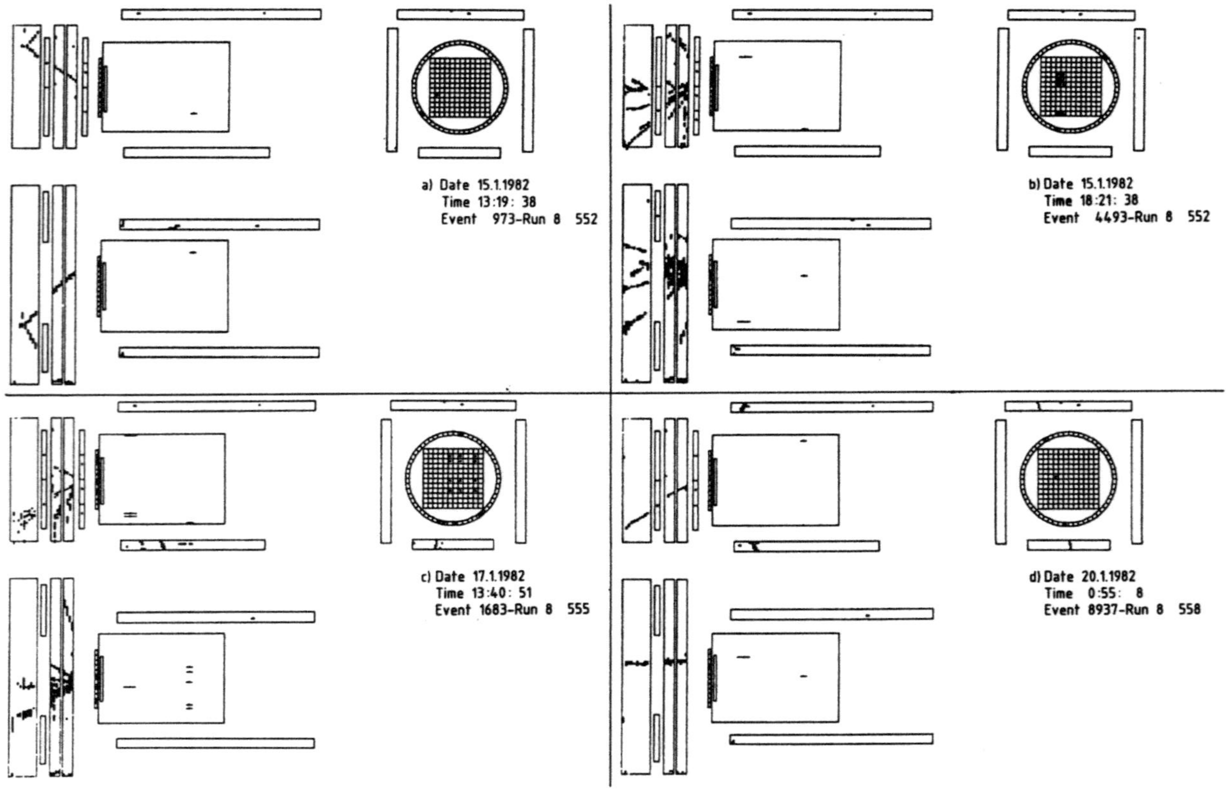

Fig. 2 Events from the Grenoble experiment, phase II

After about 100 days of effective running time (runs were alternated as in Phase I: with the magnetic field of the earth suppressed, with an $n\bar{n}$ oscillation-inhibiting field B superimposed, with reactor off), the apparatus was modified again. In the first place a thin carbon foil 0.125 mm thick was placed inside the vacuum tank, 15 cm before the ^{6}LiF absorber, so as to separate the functions of target and beam dump and to have a very thin target where cosmic rays could produce only negligible background. In the second place a set of scintillation counters was added on the four sides, between the lateral chambers and the Fe shield. These counters, each in coincidence with the corresponding chambers, provided a separate trigger which, working in addition to the forward trigger (also slightly modified as the counters B were removed and a signal from the forward chambers was used in their place), permitted also events with no track in the forward direction to be recorded, thus improving the trigger efficiency.

After 100 more days of effective running data taking was considered as finished and at the end of January 1983 the apparatus was dismantled and taken away from Grenoble.

Data analysis is still going on. However a preliminary result, $\tau_{n\bar{n}} > 10^6$ s, has already been given [5].

5. Considerations on future experiments

As summarized in Section 2, one might tend to conclude that proton-decay experiments already give evidence for a mixing time $\tau_{n\bar{n}}$ which, if not infinite, should at least be longer than a few units $\times\ 10^7$ s. Future results from proton-decay experiments might shift this limit in the direction of a few units $\times\ 10^8$ s, corresponding to about 10^{33} years for the decay time of the proton, but it is difficult to think that with this type of experiment one could easily go well into the region of 10^8 s.

Future $n\bar{n}$ experiments using free neutrons should aim at limits definitely longer than 10^8 s. Table 4 shows that even limits in the region of 10^9 s are not out of reach now, and this without waiting for progress which could still be made in the field of ultracold neutrons.

If one aims at a sensitivity of at least 10^8 s, cosmic-ray backgroun certainly becomes by far the main source of difficulties and the author of this report believes that every effort should be made to increase the selectivity of the apparatus, in particular by improving drastically the measurement of the energy released in the annihilation, so as to be able to reach solid conclusions on the basis of a very few events only, should candidate events show up. With this in mind and also with the intention to be provocative he has done the exercise shown in Figure 3, with reference to the Grenoble Reactor. However, the same ideas could be used alsc at other facilities.

The detector considered here consists of an inner Time Projection Chamber (TPC) working in a magnetic field, surrounded by an electromagnetic (e.m.) calorimeter. The geometric parameters in the charged particles, their momentum and their ionization, are measured in the TPC. If more practical, this TPC could be replaced by other detectors of the same family (drift chambers or multiwire proportional chambers).

The geometric coordinates of the γ-rays from π^0 decay and their energy would be measured by an e.m. calorimeter. No hadron calorimeter is necessary because of the presence of the magnetic field.

Three options have been considered for the e.m. calorimeter:

a) a High-density Projection Chamber (HPC),
b) built-in proportional (or streamer) chambers with charge division read-out of the axial coordinate,
c) streamer tubes equipped with capacitive pads for charge read-out an wire read-out; the axial coordinates could be read either by means of capacitive strips or by charge division.

No trigger system is shown in the figure. Proportional chambers as well as scintillation counters, in particular liquid scintillation counters, could be used. The possibility of using time-of-flight to distinguish outgoing particles from traversing particles depends on the size of the apparatus. Directional Čerenkov counters have also been considered. (They had been considered also for the previous Grenoble experiment.)

It should be recalled that if TPCs are used a continuous on-line selection of the events based on simple criteria could be made. For example, events having their vertex on the target could be recognized

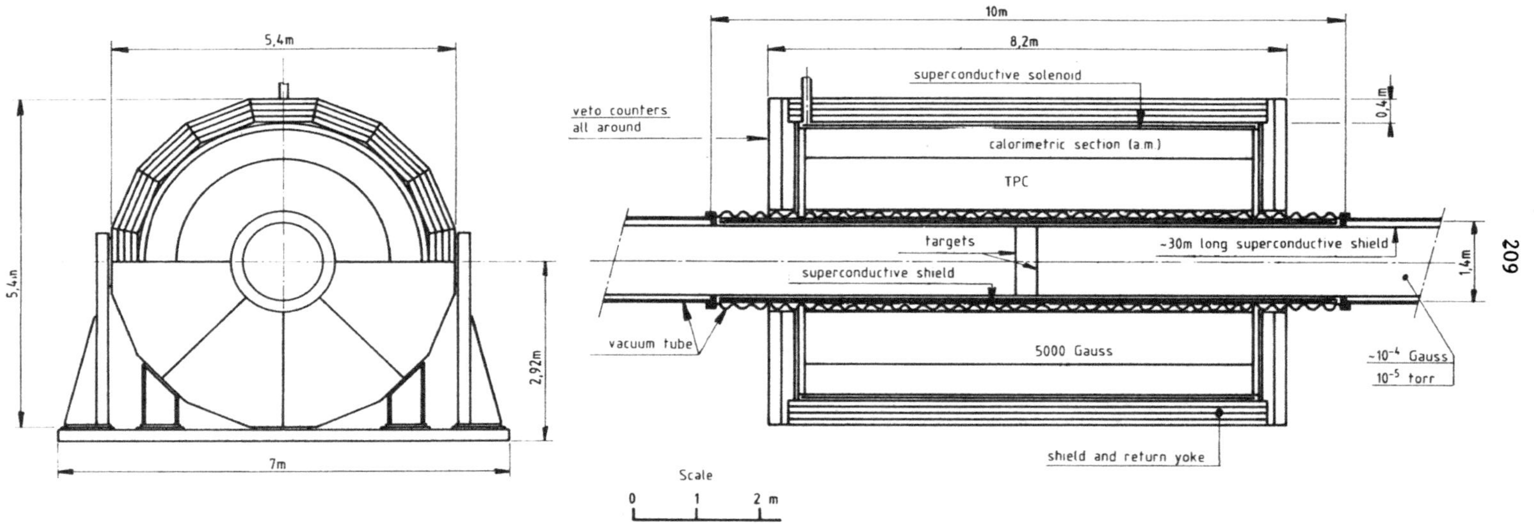

Fig. 3 Tentative design for the Grenoble experiment, phase III

on-line thanks to a relatively simple reconstruction which could take advantage of the fact that the magnetic field is perpendicular to the target. For this reason the apparatus could stand a relatively high trigger rate.

Protection from charged cosmic-ray particles is assumed by an anticoincidence system. The anticoincidence shield should protect also the lower part of the apparatus, so that energetic charged particles would traverse the anticoincidence twice (higher efficicency) and energetic down-going showers produced by neutrals inside the apparatus could also be rejected.

The iron shield needed for protection against self-vetoing of events is used as the end-caps and the return yoke of the magnet. The magnetic field is confined inside the detector by a superconductive shield placed inside the vacuum tank. The intensity of the magnetic field is limited by the magnetic forces acting on this shield: the indicated field of 5 kG results in a 1 atm pressure (pressure increases with the square of the field). The magnet coil could be made of copper because of the small value of the magnetic field. However, the possibility of a later re-use of the same magnet for other experiments (for instance LEP), as well as the fact that a cryogenic system is needed anyhow for the superconductive magnetic-field confining shield makes it advisable to have a superconductive coil.

In the previous Grenoble experiment a triple μ-metal shield was used to reduce the magnetic field of the earth to a few tenths of a milligauss. This could be done also in the case of the new experiment. However, μ-metal shields are somewhat sensitive to transients and have to be demagnetized from time to time. In addition, the reduction of the axial component of the Earth's magnetic field might require an excessively large ratio between the diameters of the successive μ-metal cylinders, and this might make the apparatus too large in diameter and consequently excessively expensive. For all these reasons a superconductive shield has been considered, but the matter has to be studied further. In fact, if on the one hand a superconductive shield is very stable -- and this might be very important in the case of a very large shield -- on the other hand it may turn out to be impractical to use such a sophisticated technology to get fields still as high as 10^{-4} G.

The vacuum chamber inside the detector should be very light in order to reduce multiple scattering. Use of honeycomb structured materials or of a large diameter bellows has been considered. It seems that a thickness equivalent to only a few millimetres of aluminium could be obtained with these techniques in spite of the large diameter. The vacuum tube outside the detector could be made of soft steel so as to use it as part of the magnetic shield, if necessary.

Two carbon targets 1/10 mm thick have been considered. Annihilations would take place in the first one, the second one being used for comparison. To study cosmic-ray background a thick target should be used. However, if superconductive components are used in the experiment it should be possible to have as a target a layer of solid hydrogen deposited on a thin aluminium foil so as to have annihilations without evaporation nucleons. It should be noted that the problem of gas condensation should be considered anyhow if two targets are placed in a low-temperature region.

^{6}LiF diaphragms can be placed inside the vacuum tube so as to have a better definition of the beam on the target. ^{6}LiF can also be used inside the main tank to absorb scattered neutrons.

REFERENCES AND NOTES

[1] See, for example: G. Fidecaro, Neutrino '81, Proc. Int. Conf. on Neutrino Physics and Astrophysics, Maui, Hawaii, July 1981 (Dept. of Phys. and Astrophys., Honolulu, 1981), p. 246, and references therein.

[2] R. Wilson, Proc. First Workshop on Grand Unification, University of New Hampshire, April 1980 (Math. Sci. Press, Brookline, Mass., 1980), p. 215.

[3] K. Green, Proc. Second Workshop on Grand Unification, University of Michigan, Ann Arbor, April 1981 (Birkäuser, Basle, 1981), p. 98.

[4] H.L. Anderson, Proc. Third Workshop on Grand Unification, University of North Carolina, Chapel Hill, April 1982 (Birhäuser, Basle, 1982), Progress in Physics, vol. 41, p. 322.

[5] Preliminary results concerning the second phase of the Grenoble experiment have been reported at various conferences. See for example G. Puglierin, Int. Colloquium on Matter non Conservation (ICOMAN), Frascati, January 1983 and references quoted.

[6] V.A. Kuz'min, JETP Lett. 12, 228 (1970) [Original: Pis'ma Zh. Eksp. Teor. Fiz. 12, 335 (1970)].

[7] S.L. Glashow, Harvard report HUTP-79/A040 (1979).

[8] S.L. Glashow, Harvard report HUTP-79/A059 (1979).

[9] R.N. Mohapatra and R.E. Marshak, Phys. Rev. Lett. 44, 1316 (1980).

[10] R.N. Mohapatra and R.E. Marshak, Phys. Lett. 94B, 183 (1980).

[11] L.N. Chang and N.P. Chang, Phys. Lett. 92B, 103 (1980) and Errata, Phys. Lett. 94B, 551 (1980).

[12] M.V. Kazarnovskii, V.A. Kuz'min, K.G. Chetyrkin and M.E. Shaposhnikov, JETP Lett. 32, 88 (1980) [Original: Pis'ma Zh. Eksp. Teor. Fiz. 32, 88 (1980)].

[13] K.G. Chetyrkin, M.V. Kazarnovskii, V.A. Kuz'min and M.E. Shaposhnikov, Phys. Lett. 99B, 358 (1981).

[14] P.G.H. Sandars, J. Phys. G (Nucl. Phys.) 6, L616 (1980).

[15] Riazzudin, Phys. Rev. D 25, 885 (1982).

[16] W.M. Alberico, A. Bottino and A. Molinari, Phys. Lett. 114B, 266 (1982).

[17] C.B. Dover, A. Gal and J.M. Richard, Phys. Rev. D 27, 1090 (1983).

[18] R. Cowsik and N. Nussinov, Phys. Lett. 101B, 237 (1981).

[19] M.R. Krishnaswamy, M.G.K. Menon, V.S. Narasimham, K. Hinotani, N. Ito, S. Myake, J.L. Osborne, A.J. Parsons and A.W. Wolfendale Proc. Roy. Soc. London A 323, 489 (1971).

[20] M.L. Cherry, K. Lande, C.K. Lee, R.I. Steiberg and B. Cleveland, Phys. Rev. Lett. 50, 1354 (1983).

[21] IMB Collaboration, see these proceedings.

[22] NUSEX Collaboration, see these proceedings.

[23] R. Ellis, CERN-EP/83-26, 15 February 1983.

[24] M. Baldo Ceolin, Proc. Int. Conf. on Unified Theories and their Experimental Tests, Venice, March 1982 (CLEUP, Padova, 1983), p. 127.

[25] M.R. Krishnaswamy, M.G.K. Menon, N.K. Mondal, V.S. Narasimham, B.V. Sreekantan, N. Ito, S. Kawakami, Y. Hayashi and S. Miyake, Phys. Lett. 106B, 339 (1981). See also Neutrino '82, Proc. Int. Neutrino Conf., Balatonfüred, Hungary, June 1982 (Central Research Institute for Physics, Budapest, 1982), p. 256.

[26] O. Sawada, M. Fukugita and J. Arafune, Astrophys. J. 248, 1162 (1981).

[27] J. Arafune and M. Fukugita, Proc. EPS Int. Conf. on High Energy Physics, Lisbon, Portugal, July 1981 (Geneva 1982), p. 989.

[28] J. Arafune and M. Fukugita, Proc. 17th Int. Cosmic Ray Conf., Paris, July 1981 (CEN, Saclay, 1982), Vol. 11, p. 214.

[29] A.G. Petschek, private communication.

[30] A.S. Iljinov et al., Int. Colloquium on Baryon non Conservation (ICOBAN), Bombay, January 1982 (Indian Acad. Sci., Madras, 1982), Pramâna Supplement, p. 179;
N.A. Kuz'min, Int. Colloquiun on Matter non Conservation (ICOMAN), Frascati, January 1983.

[31] A.N. Moskalev and A.P. Serebrov, Leningrad Institute of Nuclear Physics report (1981).

[32] The CERN-ILL-Padova-RAL-Sussex Collaboration includes: G. Fidecaro, M. Fidecaro, L. Lanceri, A. Marchioro (CERN); W. Mampe (ILL); M. Baldo Ceolin, F. Mattioli, G. Puglierin (Padova); C.J. Batty, K. Green, H. Prosper, P. Sharman (RAL); J.M. Pendlebury, K. Smith (Sussex).

[33] G. Battistoni et al. (Frascati-Milano-Torino Collaboration), Proposal for an experiment on nucleon stability with a fine grain detector (1979).

Note added in proof

Arafune and Miyamura [J. Arafune and O. Miyamura, Proc. 17th Int. Cosmic Ray Conf., Paris, July 1981 (CEN, Saclay, 1982) Vol. 5, p. 412] have evaluated the only free parameter of the Glashow one-parameter formula using available equipment information on anti-neutron-nucleus interaction and have obtained $\tau_{n\bar{n}} > 1.2\text{-}2.3 \times 10^7$ s for $\tau > 10^{30}$ years.

INVISIBLE AXIONS?

P. Sikivie
Physics Department
University of Florida
Gainesville, FL 32611

CONTENTS

I. THE STRONG CP PROBLEM AND THE AXION

Consider the lagrangian of QCD

$$\mathcal{L}_{QCD} = \frac{-1}{4} G^a_{\mu\nu} G^{a\mu\nu} + \frac{\theta g^2}{32\pi^2} G^a_{\mu\nu} \tilde{G}^{a\mu\nu}$$
$$+ \sum_{i=1}^{n} \left[\bar{q}_i \gamma^\mu D_\mu q_i - m_i q^\dagger_{Li} q_{Ri} - m^*_i q^\dagger_{Ri} q_{Li} \right] \qquad (1)$$

where $\tilde{G}^a_{\mu\nu} = \frac{1}{2} \varepsilon_{\mu\nu\alpha\beta} G^{a\alpha\beta}$. The second term is a 4-divergence. However, because of instanton [1] effects, that term cannot be neglected [2]. The physics of QCD is θ-dependent. Moreover, it follows from the Adler-Bell-Jackiw anomaly [3] that the physics of QCD is unchanged under the transformation

$$q_j \to e^{i\alpha_j \gamma_5} q_j, \quad m_j \to e^{-2i\alpha_j} m_j$$
$$\theta \to \theta - 2 \sum_{j=1}^{n} \alpha_j . \qquad (2)$$

Both $i\bar{q}\gamma_5 q$ and $G\tilde{G}$ are P and CP odd. One can use (2) to either have all the quark masses real or set $\theta = 0$. But, unless

$$\bar{\theta} = \theta - \arg \det m \qquad (3)$$

vanishes it is impossible to have all the m_j real <u>and</u> $\theta = 0$ at the same time. Thus, if $\bar{\theta} \neq 0$, $\mathcal{L}_{QCD}$ violates P and CP. The present experimental upper limit on the neutron electric dipole moment requires [4] $\bar\theta \lesssim 10^{-8}$. The strong CP problem is the problem of explaining why $\bar\theta$ is so small. Since the quark masses originate in the electroweak interactions which violate P and CP, there is no reason in the standard model for the phase of the quark mass matrix to exactly match θ of $\mathcal{L}_{QCD}$.

The strong CP problem can be solved by following a recipe first given by Peccei and Quinn [5]. Let us replace (1) by

$$\mathcal{L}_{PQ} = \frac{-1}{4} G^a_{\mu\nu} G^{a\mu\nu} + \frac{1}{2} \partial_\mu \phi^\dagger \partial_\mu \phi + \frac{\theta g^2}{32\pi^2} G^a_{\mu\nu} \tilde{G}^{a\mu\nu}$$
$$+ \sum_{i=1}^{n} [\bar{q}_i \gamma_\mu D^\mu q_i - K_i q^\dagger_{Li} q_{Ri} \phi - K^*_i q^\dagger_{Ri} q_{Li} \phi^\dagger] - V(\phi^\dagger \phi) \qquad (4)$$

The scalar potential V has the shape of a Mexican hat. ϕ takes vacuum expectation value $\langle\phi\rangle = ve^{i\alpha}$, and the quarks acquire mass $m_j = K_j \langle\phi\rangle$. Lagrangian (4) thus describes massive QCD, as does (1). However, lagrangian (4) has $\bar\theta = 0 \pmod \pi$ no matter what the values of θ and the complex Yukawa couplings K_i. [Because $\bar\theta$ is cyclic [7] with period 2π and because $\bar\theta$ is odd under P and CP, both $\bar\theta = 0$ and $\bar\theta = \pi$ are P and CP conserving values.]

There are two parts to showing that $\bar\theta = 0 \pmod \pi$ in the theory defined by (4). First, one shows that $\mathcal{L}_{PQ}$ conserves P and CP. This is so because it follows from the Adler-Bell-Jackiw anomaly [3] that the physics of $\mathcal{L}_{PQ}$ is unchanged under the transformations

$$q_j \to e^{i\alpha_j \gamma_5} q_j, \quad \phi \to e^{i\delta}\phi, \quad K_j \to e^{-i(2\alpha_j + \delta)} K_j$$
$$\theta \to \theta - 2 \sum_{j=1}^{n} \alpha_j \qquad (5)$$

and these allow one to have all K_j real and $\theta = 0$ by an appropriate redefinition of the scalar and quark fields. $\mathcal{L}_{PQ}$ is then manifestly P and CP invariant. Second, one must show that P and CP are not violated spontaneously. Here one uses the fact that if a lagrangian possesses a certain symmetry, the effective potential for any vacuum expectation value (V.E.V.) in that theory possesses that symmetry too, and hence the V.E.V.s which are invariant under the symmetry are necessarily extrema of the effective potential. Thus, because $\mathcal{L}_{PQ}$ is invariant under P and CP, we already know that $\bar\theta = 0$ and $\bar\theta = \pi$ correspond to extrema of the effective potential. It only remains to be shown that one or the other corresponds to the absolute minimum. For that, it is enough to calculate the effective potential approximately (but it must be done sufficiently well!). I know of two such calculations. The first one is the dilute instanton gas calculation in Peccei and Quinn's original paper. The second one [8] simply minimizes the Yukawa interaction energy of (4) using the fact that quark-antiquark condensates occur in QCD and that they do not violate P or CP. Both calculations imply that $\bar{\theta} = 0$ or $\bar{\theta} = \pi$ corresponds to the absolute minimum.

The main ingredient to the Peccei-Quinn mechanism to solve the strong CP problem is the $U_{PQ}(1)$ quasi-symmetry. In the case of the toy model (4), the quarks and scalar fields rotate under $U_{PQ}(1)$ as follows

$$q_j \to e^{i\alpha\gamma_5} q_j \text{ (all j)}, \ \phi \to e^{-2i\alpha}\phi. \tag{6}$$

(5) tells us that we need to redefine θ if we do such a transformation, and this means that $U_{PQ}(1)$ is broken by QCD instanton effects. Weinberg and Wilczek pointed out [9] that the spontaneous breaking of $U_{PQ}(1)$ by $\langle\phi\rangle$ implies the existence of a Pseudo-Goldstone boson. This particle, called the axion a, has mass $m_a \simeq \frac{1}{v} f_\pi m_\pi$ and couplings to fermions $\sim i \frac{a}{v} m_f \bar{f}\gamma_5 f$ and to photons $\sim \frac{e^2}{16\pi^2} \frac{a}{v} F_{\mu\nu}\tilde{F}^{\mu\nu}$, all proportional to $1/v$. v can take on any value but, historically speaking, it was first thought that the axion should be associated with the weak interaction scale and $v \simeq 250$ GeV. Such an axion has been looked for in the laboratory but not found [10]. About two years ago, it was pointed out [11,12] that v can be much larger than 250 GeV. For example, the $SU_L(2) \times U_Y(1) \times SU^c(3)$ model of Dine, Fischler and Srednicki [12] has the following Yukawa and scalar self-interactions

$$\begin{aligned} &- \sum_{i,j=1}^{n/2} \Big[K_i^{uj} (u_{Li}^\dagger d_{Li}^\dagger) \begin{pmatrix} \phi_1^0 \\ \phi_1^- \end{pmatrix} u_{Rj} + \text{h.c.} \\ &\quad + K_i^{dj} (u_{Li}^\dagger d_{Li}^\dagger) \begin{pmatrix} -\phi_2^{-*} \\ \phi_2^{0*} \end{pmatrix} d_{Rj} + \text{h.c.}\Big] \\ &\quad - V(\phi_1, \phi_2, \Phi) \end{aligned} \tag{7}$$

where Φ is a color and electroweak singlet scalar field, and V is such that (7) is invariant under the $U_{PQ}(1)$ quasi-symmetry

$$q_i \to e^{i\alpha\gamma_5} q_i \quad (\text{all } i = 1\ldots n),$$

$$\phi_1 \to e^{-2i\alpha}\phi_1, \phi_2 \to e^{2i\alpha}\phi_2, \Phi \to e^{-2i\alpha}\Phi. \tag{8}$$

In this model v is of order the V.E.V. of Φ, which indeed can be as large as one wants. Actually, stellar evolution rules out [13] the range 250 GeV $\lesssim v \lesssim 10^8$ GeV, in which stars lose excessive amounts of energy in the form of axions. One can still have $v \gtrsim 10^8$ GeV. For such large values of v, the axion is so weakly coupled that it was called "invisible". It was thought for awhile that the strong CP problem could be solved without any presently observable consequences whatsoever.

But, during the last year or so, the consideration of the cosmological implications of axion models has gone some way towards making a misnomer out of the expression "invisible axion". In sections II and III, we will discuss the presence of domain walls in axion models and the severe cosmological constraints that can be derived therefrom. Section IV is devoted to the $v \lesssim 10^{12}$ GeV constraint which has been derived from the cosmological abundance of axions. Finally, in section V, the useful role axion models may play in galaxy formation is discussed.

II. DOMAIN WALLS

If a discrete symmetry is spontaneously broken, the vacuum is, of course, discretely degenerate. The number of different vacua equals the number of elements in the coset G/H where G is the discrete symmetry group of the action density (including quantum corrections) and H is the subgroup of G which is not spontaneously broken. Domain walls are the soliton-like boundaries between regions which happen to be in different vacua.

In the case of axion models, the discrete symmetry that gives rise to domain walls is the overlap between the $U_{PQ}(1)$ quasi-symmetry and the anomaly-free part of the flavor transformations on the quarks and on any other color carrying fermions that may be present [8]. For example, for both the original axion model and the Dine-Fischler-Srednicki model one easily shows that [8]

$$[SU_L(n) \times SU_R(n) \times U_V(1)] \cap U_{PQ}(1) = Z(2n) \tag{9}$$

where n is the number of quarks. Z(2n) is an exact symmetry of these models because Z(2n) is a symmetry of their classical action densities by virtue of being a subgroup of $U_{PQ}(1)$ and Z(2n) is anomaly-free by virtue of being a subgroup of $SU_L(n) \times SU_R(n) \times U_V(1)$. The vacuum expectation values of the Higgs fields and the quark-antiquark condensates spontaneously break $Z(2n) \rightarrow Z(2)$. Hence the number N of vacua equals the number n of quarks in these models. This is not a general rule however. For example, Kim's model [11] has the six usual quarks plus one very heavy color neutral electroweak singlet quark. Only this very heavy quark rotates under $U_{PQ}(1)$. The vacuum is unique (N=1) in that case [8]. General formulas for N have been obtained [14,6].

To derive the properties of domain walls in axion models, one considers the effective action for the axion a

$$S_a = \int d^4 x \left[\frac{1}{2} \partial_\mu a \, \partial^\mu a - m_a^2 \frac{v^2}{N^2} f\left(N \frac{a}{v}\right)\right]$$

$$= v^2 \int d^4 x \left[\frac{1}{2} \partial_\mu \alpha \, \partial^\mu \alpha - m_a^2 \frac{1}{N^2} f(N\alpha)\right] \tag{10}$$

where $\alpha = \frac{a}{v}$ denotes collectively all the phases that rotate under a $U_{PQ}(1)$ transformation and f is a periodic function of period 2π, whose Taylor expansion begins with $f(x) = \frac{1}{2} x^2 + \ldots$ A domain wall, in the x-y phane for example, is the static classical solution $\alpha(z)$ obtained by minimizing the energy associated with (10), with the boundary conditions: $\alpha(z) \rightarrow 0$ as $z \rightarrow -\infty$ and $\alpha(z) \rightarrow \frac{2\pi}{N}$ as $z \rightarrow +\infty$. Quantum mechanical fluctuations do not qualitatively change the classical properties of domain walls. The domain walls have thickness of order m_a^{-1}. They have tension equal to their energy per unit surface $\sigma \simeq f_\pi m_\pi v$. The volume energy density inside the domain walls is thus of order $f_\pi^2 m_\pi^2$. These domain walls have formidable gravitational fields which are of a <u>repulsive</u> nature [15]. The interaction of domain walls with the electromagnetic field are studied by considering the effective action (10) augmented by

$$-\frac{1}{4} F_{\mu\nu} F^{\mu\nu} + \frac{e^2 N\alpha}{16\pi^2} F_{\mu\nu} \tilde{F}^{\mu\nu}. \tag{11}$$

One finds that electric and magnetic charges have long range

[$\sim$ (distance)$^{-2}$] interactions with the walls, which behave towards them very much like conductor or dielectric surfaces do [16].

In 1974, Zel'dovich, Kobzarev and Okun [17] pointed out that because of the domain walls the spontaneous breakdown of an exact discrete symmetry is incompatible with standard cosmology. Their argument is very simple but their result very general. The universe starts off at some very high temperature at which the discrete symmetry is unbroken. At some critical temperature, the spontaneous breakdown does occur and the order parameter chooses among several equally probable values (or directions), corresponding to the various vacua of the theory. Different regions of the universe will in general settle into different vacua and hence will be separated by domain walls. In particular, regions which are outside each other's horizon are causally disconnected and thus totally uncorrelated. Consequently, there will be at least on the order of one domain wall per horizon at any given time. The energy density in domain walls today would be

$$\rho_{d.w.}(t_0) \simeq \frac{\sigma}{t_0} = \rho_{crit} \left(\frac{\sigma}{10^{-5}\ \mathrm{GeV}^3}\right) \tag{12}$$

where $t_0 \simeq 10^{10}$ years is the age of the universe today and $\rho_{crit} \simeq 10^{-29}$ gr/cm^3 is its present critical energy density for closure. Since $\sigma \simeq f_\pi m_\pi v \gg 10^{-5}$ GeV^3, it is clear that if axions exist and $N > 1$, our present universe would be domain wall dominated many times over. But this cannot be. If domain wall dominated, our universe would be expanding like $R \sim t^2$ (R is the cosmological scale parameter) [17] and at a much higher rate than what we observe today.

Note that because of the tension σ in the domain walls, they annihilate as fast as allowed by causality, i.e. at speeds $\lesssim$ c. So that one does expect approximately one domain wall per horizon at any given time, i.e. a saturation of the lower limit required by causality.

III. EVADING THE DOMAIN WALL PROBLEM

We will discuss here three possible evasions of the domain wall problem. In the first one, we consider axion models with a unique vacuum (N=1) and their own cosmological consequences. In the second evasion, we entertain the possibility that the Peccei-Quinn quasi-symmetry is weakly broken by someting else besides the QCD gluon anomaly. Finally, we will discuss how the domain wall problem may be solved in a "inflationary universe"-type cosmology.

1. As we mentioned already, it is possible to construct axion models which have a unique vacuum (N=1). Kim's original invisible axion model has that property. Recently, many more examples have been given and the N=1 constraint has been used to say something about the family structure of grand unified theories [14,18]. In addition, Lazarides and Shafi [18] introduced a new way to construct N=1 models by having the Z(N) subgroup of $U_{PQ}(1)$ coincide or lie within the center of the gauge group. The N vacua are then gauge equivalent and therefore not distinct.

Although the argument of the previous section clearly does not apply to axion models with a unique vacuum [19], it is not immediately clear that such models are free of cosmological difficulties. As a matter of fact, the N=1 models have domain walls too. When one traverses these domain walls, $\alpha(x)$ varies from 0 to 2π. The point is that, although the vacuum of N=1 models is unique, the corresponding low energy effective theories (10) have multiply degenerate vacua. The resulting domain walls have many of the properties [thickness $\sim m_a^{-1}$; energy/surface and tension $\sigma \simeq f_\pi m_\pi v$; classical stability] of the domain walls of $N > 1$ models. An important difference is that the domain walls of N=1 models are quantum-mechanically unstable. A hole can be punched into such a wall by pulling $\langle\phi\rangle$ over the top of the "Mexican hat" potential $V(\phi^+\phi)$. The hole will grow classically if its radius $R \gtrsim m_a^{-1}$. The rate of "herniation", i.e. of quantum-mechanical tunneling from a N=1 wall without a hole to one with a hole of radius $R \gtrsim m_a^{-1}$, has been estimated [20]. It is so minuscule however that herniation practically never occurs in domain walls even as old as the universe and correspondingly large.

Nonetheless, it is believed that N=1 axion models do solve the domain wall problem. The reason for this is strings [19]. The string is a topologically stable soliton-like object such that $\alpha(x)$ changes from 0 to 2π when $\vec{x}$ moves around the string once. On the central axis of the string one is on top of the Mexican hat potential $V(\phi^+\phi)$. The strings first appear at the critical temperature $T_{PQ} \simeq v$ when $U_{PQ}(1)$ becomes spontaneously broken by $\langle\phi(x)\rangle = v\, e^{i\alpha(x)}$. They move quite freely through the primordial plasma [21,19,6] and annihilate as fast as allowed by causality, i.e. at speeds $\lesssim$ c. About one string per horizon remains at any given time, till QCD instanton effects switch on at temperatures $\sim$ 1 GeV and the domain walls appear. Each string then becomes the edge of a domain wall since $\alpha(\vec{x})$ varies from 0 to 2π when $\vec{x}$ moves around the string and hence traverses a domain wall. The domain walls bounded by strings have a typical size of 10^{-4} sec, the horizon size at the time QCD has completely switched on (at $\sim$ 100 MeV temperature). The domain walls bounded by strings oscillate, thereby emitting radiation, axions... They disappear before dominating the energy density.

2. The second evasion of the domain wall problem is based on the observation that even a tiny explicit breaking of the Z(N) symmetry is sufficient to make the domain walls disappear before they dominate the energy density[8]. For example, let us softly break the $U_{PQ}(1)$ of the Dine-Fischler-Srednicki model (7) by adding a term of the form $e^{i\delta}\mu^3\Phi$ + h.c. to $V(\phi_1,\phi_2,\Phi)$. This would induce $\bar{\theta} \simeq \mu^3/m_a^2 v$ which is smaller than 10^{-8} provided $\mu^3 \lesssim 10^{-8}\ f_\pi^2 m_\pi^2/v$. The Z(N) symmetry would be explicitly broken down to nothing and shifts in energy density among the vacua would be produced of order $\langle\Delta\mathcal{H}\rangle \simeq \mu^3 v$. These shifts result in pressure $p = \langle\Delta\mathcal{H}\rangle$ on the domain walls in the direction of the domain with the highest vacuum energy density. The corresponding acceleration is c/τ_B with

$$\tau_B = \frac{\sigma}{p} \simeq (5 \times 10^{-5}\ \text{sec})\ \frac{v}{10^{10}\ \text{GeV}} \cdot \frac{f_\pi^2 m_\pi^2 10^{-8}}{v\mu^3} . \tag{13}$$

Once the domain bubbles have average size $c\tau_B$, the differences in volume energy among bubbles is of order their surface energy, the Z(N) breaking effects become important and the true vacuum takes over. Note that in this scenario, density perturbations are produced on the horizon scale at time τ_B, whereas in the N=1 axion models discussed above density perturbations are only produced on the horizon scale at $t_{QCD} \simeq 10^{-4}$ sec. The larger scale perturbations may be useful in explaining galaxy formation (cfr. last section). Finally, we note that an explicit breaking of the Z(N) symmetry is of course very artificial if done by hand. However, this evasion of the domain wall problem would be a natural property of the ultimate theory of the world if it has in its low energy effective theory an automatic [22] $U_{PQ}(1)$ which is then explicitly broken by higher order corrections (gravitational instantons?).

3. Finally we come to the evasion of the domain wall problem afforded by the "inflationary universe" scenario [23]. During inflation, the V.E.V. $\langle\phi(x)\rangle = v\ e^{i\alpha(x)}$ will get aligned over enormous distances. Later when the QCD instanton effects turn on at $\sim$ 1 GeV temperatures, the enormous regions over which α is a constant will entirely fall into the same vacuum and hence will be free of domain walls. For this to work, it is of course necessary that the inflation and subsequent reheating all take place at temperatures below $T_{PQ} \simeq v$. This is a serious constraint in view of the fact that the baryon asymmetry must be produced after inflation and of the $v \lesssim 10^{12}$ GeV constraint derived in the next section.

IV. THE AXION "ENERGY CRISIS"

Our concern here is with a large coherent state of non-relativistic axions which appear when QCD instantons are switched on at $\sim$ 1 GeV temperatures. We will see below that unless $v \lesssim 10^{12}$ GeV, the present day average energy density of these axions is several times ρ_{crit}, i.e too large [24].

After the Peccei-Quinn phase transition at temperature $T_{PQ} \simeq v$, we have

$$\langle\phi\rangle = v\ e^{i\alpha(x)} \quad . \tag{14}$$

The value of α at the bottom of the Mexican hat potential $V(\phi^+\phi)$ is initially randomly chosen, because at those high temperatures QCD instanton effects are too weak to bring α to the CP conserving minimum, say at α=0, for which $\overline{\theta}$=0. When QCD instanton effects have fully turned on at about $t_{QCD} \simeq 10^{-4}$ sec (100 MeV temperature), the misalignment with respect to α=0 implies the presence of a huge axion energy density

$$\begin{aligned}\rho_a(t_{QCD}) &\simeq m_a^2 v^2\ \alpha^2(t_{QCD}) \simeq f_\pi^2 m_\pi^2\ \alpha^2(t_{QCD}) \\ &\simeq \rho_{rad}\ (t_{QCD})\ \alpha^2\ (t_{QCD}),\end{aligned} \tag{15}$$

where ρ_{rad} (t) is the energy density in radiation at time t. These axions are non-relativistic, because $\alpha(x)$ has become uniform over the horizon scale $t_{QCD} \gg m_a^{-1}$ (the wiggles in $\alpha(x)$ start to be red-shifted away as soon as their wavelength falls into the horizon). But non-relativistic energy density decreases with time as $\rho_a \sim R^{-3}$ ($R \sim \sqrt{t}$ is the cosmological scale factor) whereas $\rho_{rad} \sim R^{-4}$. Thus unless $\alpha^2(t_{QCD})$ is very small, the universe will become axion matter dominated too soon. If we require the axion energy density today to be less than 10 times ρ_{crit}, we need

$$\alpha^2(t_{QCD}) \lesssim 10^{-6}. \tag{16}$$

How can $\alpha(t_{QCD})$ be so small? If the switch-on of the axion mass were <u>sudden</u>, we would have $\alpha(t_{QCD}) \sim O(1)$ and all axion models would be ruled out for the simple reason that they predict far too much matter during what is supposed to be the radiation dominated epoch of our universe. "Sudden" means that the switch-on rate $\frac{1}{m_a}\frac{dm_a}{dt}$ is large compared to the frequency m_a at which $\alpha(t)$ oscillates. The opposite of "sudden" is "adiabatic": $\frac{1}{m_a}\frac{dm_a}{dt}$ small compared to m_a. The latter regime is characterized by the adiabatic invariant [25]

$$\oint pdq = \pi A^2(t) m_a(t) \simeq \text{time independent} \tag{17}$$

where $A(t)$ is the amplitude of the oscillation $\alpha(t) = A(t)\cos(m_a t+\delta)$. (17) tells us that in the adiabatic regime the oscillation amplitude decreases while the axion mass is being switched on. Hence, provided the switch-on is sufficiently adiabatic - i.e. provided the axion mass is sufficiently large, or v sufficiently small - the axion "energy crisis" will be avoided. The time dependence of the axion mass follows from its temperature dependence which had already been calculated [26]: $m_a(t) = m_a[T(t)]$. From this the following result was obtained for the axion energy density today [24]

$$\rho_a(t_0) \simeq 10\, \rho_{crit} \left(\frac{v}{10^{12}\ \text{GeV}}\right)^{7/6}. \tag{18}$$

Hence, the constraint $v \lesssim 10^{12}$ GeV which applies to all axion models independently of their vacuum structure and of the history of the universe before the temperature reached 10 - 100 GeV.

It was implicitly assumed in the above discussion that the non-relativistic axionic energy density that appears at ~ 1 GeV temperatures, does not later on convert into radiation. The axion indeed decouples in the large v limit and the $a \to 2\gamma$ decay rate is in fact tiny compared to the age^{-1} of the universe for $v \gtrsim 10^8$ GeV, the constraint from stellar evolution. However, this is somewhat misleading. The axions form a large coherent state and therefore transition amplitudes may be enchanced by factors of $\sqrt{n} = \langle n-1|a|n\rangle$ where n is the relevant state occupation number. An analysis [23] shows that in the presence of the large oscillating axion field, the electromagnetic field exhibits "parametric resonance" [25]: for values of the momentum k near multiples of $\frac{1}{2}m_a$, the electromagnetic field grows exponentially with a growth rate of order $(10^{-3}A)^n m_a$ for $k \simeq \frac{1}{2} n\, m_a$. It turns out however that this parametric resonance is ineffective in pumping the axion energy into photons because the axion mass is for a long time much smaller than the plasma oscillation frequency of the electromagnetic field, and because the growing electromagnetic modes keep being red-shifted out of the instability bands by the universe's expansion. In conclusion, the axion energy does not get converted into radiation and the $v \lesssim 10^{12}$ GeV bound stands.

V. AXIONS AND GALAXY FORMATION

We will now argue that axion models with $v = 10^{10\pm2}$ GeV (the range not yet ruled out by either stellar evolution or the cosmological abundance of axions) may play a useful role in explaining the origin and make-up of galaxies [8,27-29]. More precisely, axion models provide the galaxy maker with both a candidate for the primordial density fluctuations from which the galaxies evolved and a candidate for the dark matter that constitutes the galactic halos [30].

First, if $v \gtrsim 10^{10}$ GeV [cfr. Eq. (18)], axionic matter is abundant enough to make up the galactic halos. Second, axionic halos are automatically dark. Neutrinos [31] are similar to axions in these two respects [32] and they have indeed been a very popular candidate for halo matter. However, neutrino halo models have run into rather serious difficulties because the neutrino phase space density tends to be too small to allow them to cluster into galactic halos [33] and because neutrino free streaming greatly inhibits the growth of all matter density perturbations on all mass scales less than about 10^{15} $M_{\odot}$, the mass scale of rich galactic clusters [34]. If condensations first appear on scales appropriate to rich clusters of galaxies, the neutrinos will pick up random velocities of order 1500 km/s during initial cluster collapse and will have difficulty getting trapped around galaxies whose velocity dispersions are only about 300 km/s. Axions easily circumvent both problems encountered by neutrino halos. Indeed, because they formed a coherent state of non-relativistic bosons since the time of their first appearance at $\sim$ 1 GeV temperature, the axions have enormous phase space density and vanishingly small free streaming distance. They have no difficulty clustering into galactic halos and their density perturbations are preserved and permitted to grow on all mass scales.

Axions are compatible with the "hierarchical clustering" model [35] of galaxy formation, in which condensations first appear on relatively small scales and then build up towards larger and larger scales. It has been estimated that the initial seed mass [36] should be approximately 10^6 $M_{\odot}$ at recombination time. This seed mass may be due to the collapse of density perturbations produced by the presence of domain walls for a limited time period in the early universe. To produce seed masses of that size, the domain walls must remain till and disappear at time $t_B \simeq 10^{-3}$ years, which happens to be just before the domain walls begin to dominate the energy density. This seems to be possible only if the domain wall problem is evaded by a soft weak breaking of the Z(N) symmetry with appropriate strength [see section III].

ACKNOWLEDGEMENT

This work was supported in part by the U.S. Department of Energy under contract No. DSR80136je7.

REFERENCES

[1] A. A. Belavin, A. M. Polyakov, A. S. Schwartz and Y. S. Tyuplin, Phys. Lett. 59B (1975) 85.

[2] G. 't Hooft, Phys. Rev. Lett. 37 (1976) 8, and Phys. Rev. D14 (1976) 3432; R. Jackiw and C. Rebbi, Phys. Rev. Lett. 37 (1976) 172; C. G. Callan, R. F. Dashen and D. J. Gross, Phys. Lett. 63B (1976) 334; for a review see S. Coleman, "The many uses of instantons" in the Proc. of the 1977 International School of Subnuclear Physics, Ettore Majorana [Erice, July 23-Aug. 10].

[3] S. L. Adler, Phys. Rev. 117 (1969) 2426; J. S. Bell and R. Jackiw, Nuovo Cimento A 60 (1969) 47.

[4] V. Baluni, Phys. Rev. D19 (1979) 2227; R. J. Crewther, P. Di Vecchia, G. Veneziano and E. Witten, Phys. Lett. 88B (1979) 123.

[5] R. D. Peccei and H. Quinn, Phys. Rev. Lett. 38 (1977) 1440, and Phys. Rev. D16 (1977) 1791; for a recent review of axion related matters see ref. 6.

[6] P. Sikivie, "Axions in Cosmology", Univ. of Florida preprint UFTP-83-6 (May 1983), to be published in the Proceedings of the Gif-sur-Yvette Summer School on Particle Physics, Sept. 1982.

[7] For example, see ref. 6.

[8] P. Sikivie, Phys. Rev. Lett. 48 (1982) 1156.

[9] S. Weinberg, Phys. Rev. Lett. 40 (1978) 223; F. Wilczek, Phys. Rev. Lett. 40 (1978) 279.

[10] For a recent review, see A. Zehnder, "The Quest of the Axion", SIN preprint PR-83-03 (Jan. 1983), to be published in the Proceedings of the Gif-sur-Yvette Summer School on Particle Physics, Sept. 1982.

[11] J. Kim, Phys. Rev. Lett. 43 (1979) 103; M. Shifman, A. Vainshtein and V. Zakharov, Nucl. Phys. B166 (1980) 493.

[12] M. Dine, W. Fischler and M. Srednicki, Phys. Lett. 104B (1981) 199; H. P. Nilles and S. Raby, SLAC-PhB-2743 (1981).

[13] D. Dicus, E. Kolb, V. Teplitz and R. Wagoner, Phys. Rev. D18 (1978) 1829 and Phys. Rev. D22 (1980) 839; K. Sato and H. Sato, Progr. Theor. Phys. 54 (1975) 1564; M. Fukugita, S. Watamura and M. Yoshimura, Phys. Rev. Lett. 48 (1982) 1522.

[14] H. Georgi and M. Wise, Phys. Lett. 116B (1982) 123.

[15] A. Vilenkin, Phys. Rev. D23 (1981) 852; J. Ipser and P. Sikivie, to be published.

[16] P. Sikivie, to be published.

[17] Y. B. Zel'dovich, I. Y. Kobzarez and L. B. Okun, Zh. Eksp. Teor. Fiz. 67 (1974) 3 [Sov. Phys. JETP 40 (1975) 1].

[18] G. Lazarides and Q. Shafi, Phys. Lett. 115B (1982) 21; B. Holdom, Stanford preprints ITP-712 (1982) and ITP-713 (1982); S. Dimopoulos, P. Frampton, H. Georgi and M. Wise, Phys. Lett. 117B (1982) 185; S. Barr, D. Reiss and A. Zee, Phys. Lett. 116B (1982) 227; S. Barr, X. C. Gao and D. Reiss, Phys. Rev. D26 (1982) 2176; T. Kephart, Phys. Lett. 119B (1982) 92; Y. Fujimoto, K. Shigemoto and Zhao Zhi-yong, ICTP preprint IC-82-140 (1982); B. Grossman, Rockefeller preprint RU82/B/39 (1982); R. Holman, G. Lazarides and Q. Shafi, Phys. Rev. D27 (1982) 995.

[19] A. Vilenkin and A. E. Everett, Phys. Rev. Lett. 48 (1982) 1867; T. W. B. Kibble, G. Lazarides and Q. Shafi, Phys. Rev. D26 (1982) 435.

[20] S. Coleman, unpublished; see ref. 6.

[21] T. W. Kibble, Phys. Rep. 67 (1980) 183.

[22] H. M. Georgi, L. J. Hall and M. B. Wise, Nucl. Phys. B192 (1981) 409.

[23] A. Guth, Phys. Rev. D23 (1981) 347; A. Linde, Phys. Lett. 108B (1982) 389; A. Albrecht and P. Steinhardt, Phys. Rev. Lett. 48 (1982) 1220.

[24] L. Abbott and P. Sikivie, Phys. Lett. 120B (1983) 133; J. Preskill, M. Wise and F. Wilczek, Phys. Lett. 120B (1983) 127; M. Dine and W. Fischler, Phys. Lett. 120B (1983) 137.

[25] L. D. Landau and E. M. Lifshitz, Mechanics (Addison-Wesley, Reading, MA, 1960).

[26] D. J. Gross, R. D. Pisarski and L. G. Yaffe, Rev. Mod. Phys. 53 (1981) 43.

[27] J. Ipser and P. Sikivie, Phys. Rev. Lett. 50 (1983) 925.

[28] F. W. Stecker and Q. Shafi, Phys. Rev. Lett. 50 (1983) 928.

[29] M. S. Turner, F. Wilczek and A. Zee, preprint (March 1982).

[30] J. P. Ostriker, P. J. E. Peebles and A. Yahil, Astrophysical Journal Letters 193 (1974) L1; J. Einasto, A. Kaasik and E. Saar, Nature 250 (1974) 309.

[31] R. Cowsik and J. McClelland, Phys. Rev. Lett. 39 (1972) 669.

[32] M. Gell-Mann, P. Ramond and R. Slansky, unpublished; P. Ramond, Caltech preprint CALT-68-709 (1979); T. Yanagida in Proc. of the Workshop of the Unified Theory and Baryon Number in the Universe, KEK, 1979.

[33] S. Tremaine and J. E. Gunn, Phys. Rev. Lett. 42 (1979) 407.

[34] J. R. Bond, G. Efstathiou and J. Silk, Phys. Rev. Lett. 45 (1980), 1980.

[35] P. J. E. Peebles, Astrophys. J. 142 (1965) 1317; "Physical Cosmology", Princeton Univ. Press, 1971.

[36] W. H. Press and P. Schechter, Astrophys. J. 187 (1974) 425.

INFLATION CIRCA 1983

MICHAEL S. TURNER
Enrico Fermi Institute
The University of Chicago
Chicago, IL 60637

The hot big bang cosmology provides a reliable accounting of the evolution of the Universe from ~ 0.01 s after 'the bang' until the present (10-20 Byr after 'the bang'). This is a truly impressive achievement. There are, however, a handful of very fundamental 'cosmological facts' which the model by itself fails to elucidate. They include: the large-scale homogeneity, the isotropy, the small-scale inhomogeneity, the near critical expansion rate, the predominance of matter, the 'monopole problem', and the extremely tiny value of the present cosmological term. The inflationary Universe scenario proposed by Guth and recently modified by Linde, and Albrecht and Steinhardt (new inflation) offers the possibility of explaining all but the last of these puzzling facts as the result of an early epoch of exponential expansion driven by a large vacuum energy. Inflation makes the present state of the observable Universe virtually insensitive to the initial state of the Universe. Unfortunately, at present there is no model of new inflation which both resolves the cosmological puzzles and leads to sensible particle physics. A general prescription for the cosmologically-desirable Higgs potential does exist.

1. Introduction

The hot big bang model (Friedmann-Robertson-Walker (FRW) cosmology) provides a very simple and accurate description of the evolution of the Universe. It readily accounts for the Universal (Hubble) expansion, the existence of the cosmic microwave background radiation, and through primordial nucleosynthesis, the abundances of D and ^{4}He, and probably ^{3}He and ^{7}Li (see Fig. 1). In short, the model is a reliable framework for understanding the evolution of the Universe from $\simeq 10^{-2}$ s after 'the bang' until today ($\sim 3 \times 10^{17}$ s after 'the bang'). Needless to say this is a remarkable achievement.

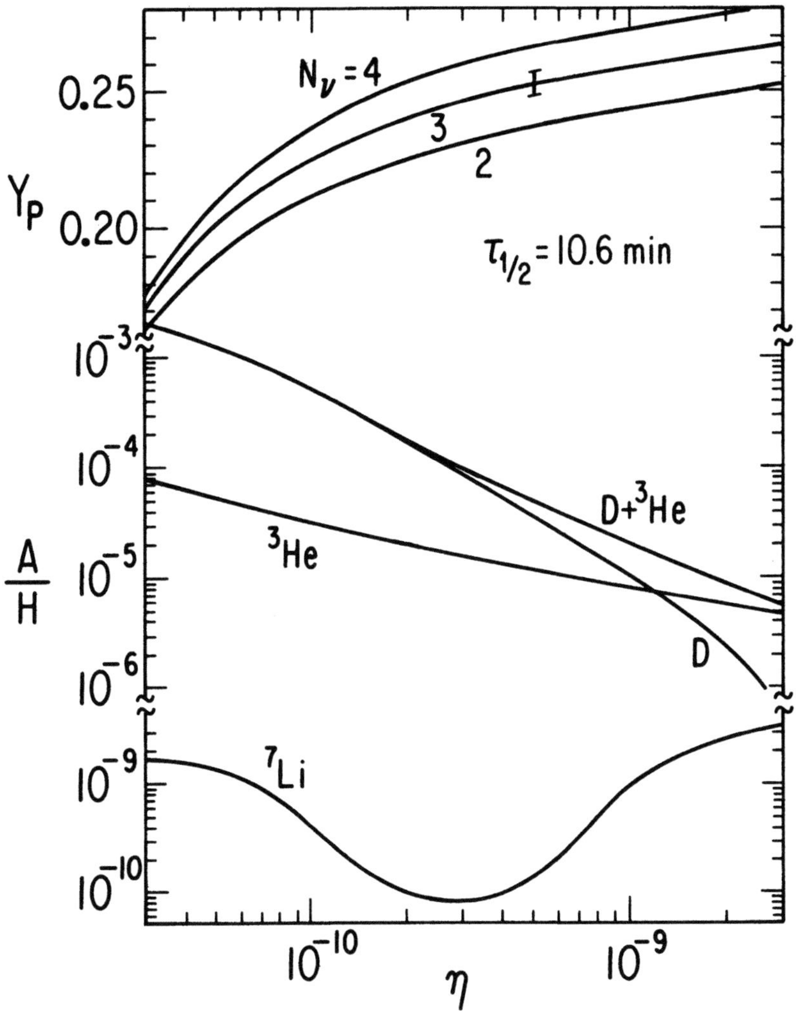

Figure 1 The predicted primordial abundances of D, ^3He, ^4He, and ^7Li ($\tau_{1/2}(n)$ = 10.6 min was used; error bar shows $\Delta\tau_{1/2} = \pm 0.2$ min; Y_P = mass fraction of ^4He). Observed abundances: $Y_P \simeq 0.23 - 0.25$; $(D/H)_P \gtrsim 10^{-5}$; $(D + {}^3He)_P/H \lesssim 10^{-4}$; $({}^7Li/H)_P \simeq (1.1 \pm 0.4) \times 10^{-10}$. Consistency of the predicted abundances with observations can only be achieved for η (= baryon-to-photon ratio) $\simeq (4-7) \times 10^{-10}$, or relaxing the ^7Li constraint, $\simeq (4-10) \times 10^{-10}$, and N_ν (= number of light neutrino species) ≤ 4. For $4 \lesssim \eta/10^{-10} \lesssim 7(10)$, $0.01 \lesssim \Omega_B \lesssim 0.15(0.19)$. See ref. [1] for more details.

Because the FRW cosmology is isotropic and homogeneous the expansion of the Universe can be described by a single quantity, $R(t)$ - 'the cosmic scale factor'. All cosmic distances scale with $R(t)$ -- the separation of two galaxies, the wavelength of a freely propagating photon, etc. The evolution of R is governed by the Friedmann equation,

$$H^2 \equiv (\dot{R}/R)^2 = 8\pi\rho/3m_{pl}^2 - k/R^2, \tag{1}$$

where ρ is the total energy density, $m_{pl} = 1.22 \times 10^{19}$ GeV ($G = m_{pl}^{-2}$), and k is the curvature signature (+1 = closed Universe, -1 or 0 = open Universe). Here and throughout $\hbar = k_B = c = 1$, so that 1 GeV = 1.16×10^{13} K = $(0.197 \text{ fermi})^{-1}$ = $(6.58 \times 10^{-25}\text{ s})^{-1}$. The Hubble parameter H sets the expansion timescale: $R(t)$ e-folds in a time of order H^{-1}. Since all physical lengths e-fold in a time $\simeq H^{-1}$, it also sets the scale for all causal and coherent physical processes, i.e., microphysics can only operate coherently on length scales $\lesssim H^{-1}$. One might call H^{-1} 'the physics horizon'. If the energy density of the Universe is dominated by relativistic and/or non-relativistic particles, then the age of the Universe $t \simeq H^{-1}$ and the particle horizon ($\equiv R(t) \int_0^t dt'/R(t')$) is also about $t \simeq H^{-1}$. [Physically, the particle horizon is the distance that a light signal could have traveled since 'the bang' ($t = 0$).]

Although the energy density contributed by matter (baryons and other non-relativistic particles) is today a factor of about 10^4 greater than that contributed by radiation (cosmic microwave photons and the cosmic neutrino seas), when the scale factor of the Universe was less than about 10^{-4} of its present size, the energy density contributed by radiation dominated the energy density. [The energy density of radiation $\propto R^{-4}$ -- the number density of photons is diluted by R^{-3} and the energy of each photon is redshifted by R^{-1}, while that of matter $\propto R^{-3}$.] Since the curvature term today is at most comparable to the matter energy density term (see next section), for $R \lesssim 10^{-4} R_{today}$ the Universe is radiation-dominated and

$$t \simeq (45/16\pi^3 g_*)^{1/2} m_{pl}/T^2 \simeq 2.4 \times 10^{-6} g_*^{-1/2} T_{GeV}^{-2} \text{ sec.} \tag{2}$$

Here $g_* \equiv \sum g_{Bose} + (7/8) \sum g_{Fermi}$ counts the number of very relativistic degrees of freedom (particles of mass << T). So long as the expansion is adiabatic the entropy density $s \equiv 2\pi^2 g_* T^3/45 \propto R^{-3}$

(total entropy = constant). Whenever $g_* \simeq$ constant this implies that $T \propto R^{-1}$ and $R \propto t^{1/2}$.

Although the standard cosmology has only been rigorously tested back to $t \simeq 10^{-2}$ s after 'the bang' ($T \simeq 10$ MeV) - the beginning of the epoch of primordial nucleosynthesis, the model can be sensibly extrapolated back as early as $t \simeq 10^{-43}$ s ($T \simeq 10^{19}$ GeV) - earlier than this quantum corrections to General Relativity are likely to be significant. [The point-like nature of the quarks and leptons, and the asymptotic freedom exhibited by non-Abelian gauge theories justify the dilute gas approximation embodied in eqn. (2); sensible, of course, does not necessarily mean correct!] Because of the extremely high temperatures reached just after 'the bang', there is a natural relationship between cosmology and high energy physics (see Fig. 2): An understanding of the fundamental particles and their interactions at very high energies is required to extend our understanding of the early evolution of the Universe; the early Universe may provide a 'laboratory' for studying elementary particle physics at energies not accessible in terrestrial laboratories ($E >> 10^3$ GeV). In this article both these aspects of the relationship will manifest themselves.

As great an achievement as the standard cosmology is, it has its shortcomings. By itself, it fails to account for or even elucidate a number of fundamental cosmological facts. They include: the large-scale homogeneity and isotropy, the origin of the small-scale inhomogeneity, the near critical expansion rate today, the predominance of matter in the Universe, 'the monopole problem', and the extreme smallness of the present value of the cosmological term (as measured in any natural system of units). In the past five years or so significant progress has been made toward understanding these cosmological conundrums. This progress has occurred in large measure due to the interplay between cosmology and particle physics, and in particular Guth's inflationary Universe idea. In this article I will review Inflation, Circa 1983. [The standard cosmology is reviewed in greater detail in refs. 2-3.]

2. Some Cosmolgical Conundrums

In this section I briefly review my list of 7 puzzling cosmological facts.

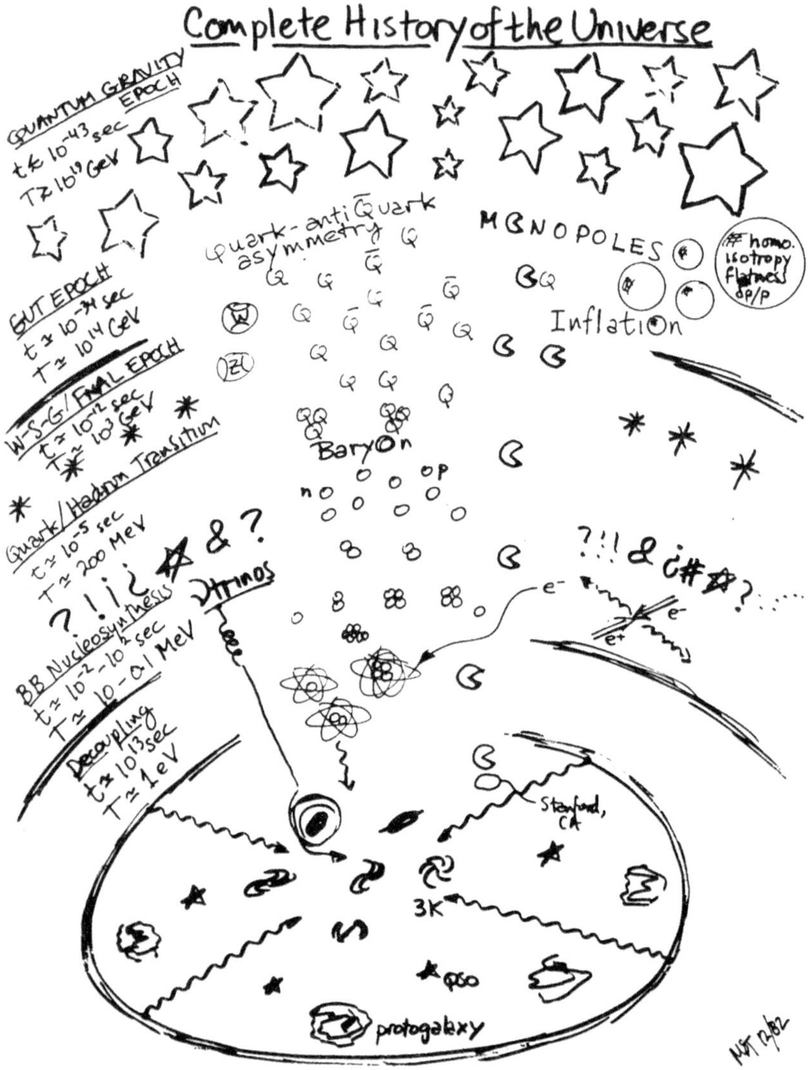

Figure 2 The complete thermal history of the Universe according to the standard big bang cosmology.

2.1 Large-Scale Homogeneity and Isotropy

The observable Universe ($d \simeq H^{-1} \simeq 10^{28}$ cm $\simeq$ 3000 Mpc) is to a high degree of precision isotropic and homogeneous on the largest scales (> 100 Mpc). The best evidence for this is provided by the uniformity of the cosmic background temperature: $\Delta T/T \lesssim 10^{-3}$ (10^{-4} if the dipole anisotropy is interpreted as being due to our peculiar motion through the cosmic rest frame; see Fig. 3 below). Large-scale density inhomogeneities or an anisotropic expansion would result in fluctuations in the microwave background temperature of a comparable size (see, e.g., refs. 4,5). The smoothness of the observable Universe is puzzling if one wishes to understand it as a result of microphysical processes operating in the early Universe. As mentioned in Sec. 1 the standard cosmology has particle horizons, and when matter and radiation last vigorously interacted (decoupling: $t \simeq 10^{13}$ s, $T \simeq 1$ eV) what was to become the present observable Universe was comprised of $\simeq 10^{6}$ causally-distinct regions. Put slightly differently,

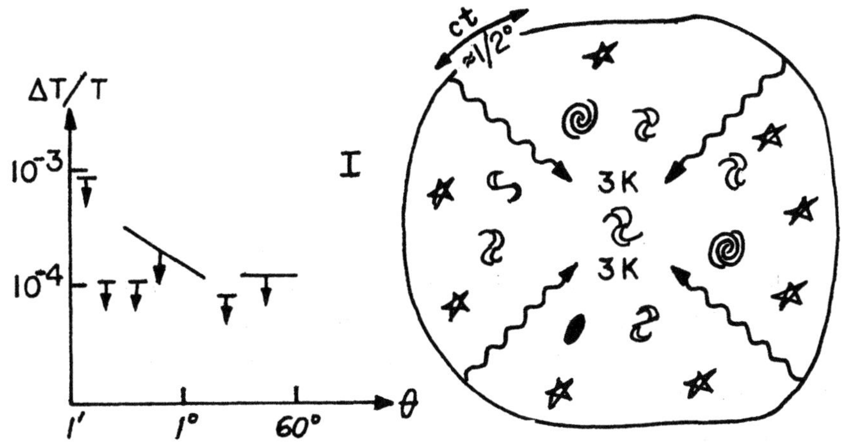

Figure 3 Summary of the measurements of the fluctuations in the microwave background temperature on angular scales $\gtrsim 1'$ (from D. Wilkinson, 1983).

the particle horizon at decoupling only subtends an angle of about $1/2^{\circ}$ on the sky today; how is it that the microwave background temperature is so uniform on angular scales $\gg 1/2^{\circ}$?

2.2 Small-Scale Inhomogeneity

As any astronomer will gladly tell you on small scales ($\lesssim 100$ Mpc) the Universe is very lumpy (stars, galaxies, clusters of galaxies, etc.). The uniformity of the microwave background on very small angular scales ($\ll 1^{\circ}$) indicates that the Universe was smooth, even on these scales at the time of decoupling. [Note, today $\delta\rho/\rho \simeq 10^{5}$ on the scale of a galaxy.] Whence came the structure that is so conspicuous today? Once matter decouples from the radiation and is free of the pressure support provided by the radiation, small inhomogeneities will grow via the Jeans (gravitational) instability: $\delta\rho/\rho \propto t^{2/3} \propto R$ (in the linear regime). [If the mass density of the Universe is dominated by a collisionless particle species, e.g., a light relic neutrino species, or axions, density perturbations in these particles can begin to grow when the Universe becomes matter-dominated, $R \simeq 3 \times 10^{-5}\, R_{today}$ for $\Omega = 1$.] Density perturbations of amplitude $\delta\rho/\rho \simeq 10^{-3}$ or so, on the scale of a galaxy ($\simeq 10^{12}\, M_{\odot}$) at the time of decoupling seem to be required to account for the small-scale structure observed today. Their origin, their spectrum (certainly perturbations should exist on scales other than $10^{12} M_{\odot}$), their nature (adiabatic or isothermal), and the composition of the dark matter (see Primack's contribution to this volume) are all crucial questions for understanding the formation of structure, which to date remain unanswered.

2.3 Flatness

The quantity $\Omega \equiv \rho/\rho_c$ measures the ratio of the energy density of the Universe to the critical energy density ($\rho_c \equiv 3H^2/8\pi G$). Although Ω is not known with great precision, it is safe to say that $0.01 \lesssim \Omega \lesssim$ few (the lower limit provided by the density contributed by luminous matter; the upper limit based on the age of the Universe and the Hubble diagram). Using eqn. (1) Ω can be written as

$$\Omega = 1/(1 - x(t)), \tag{3a}$$
$$x(t) = (k/R^2)/(8\pi G\rho/3). \tag{3b}$$

Note that Ω is not constant, but varies with time since $x(t) \propto R(t)^n$ ($n = 1$ - matter-dominated, or 2 - radiation-dominated). Since

$\Omega \simeq O(1)$ today, x_{today} must be at most $O(1)$. This implies that at the epoch of nucleosynthesis: $x_{BBN} \lesssim 10^{-16}$ and $\Omega_{BBN} = 1 \pm O(\lesssim 10^{-16})$, and that at the Planck epoch: $x_{pl} \lesssim 10^{-60}$ and $\Omega_{pl} = 1 \pm O(\lesssim 10^{-60})$. That is, very early on the ratio of the curvature term to the density term was extremely tiny, or equivalently, the expansion of the Universe proceeded at the critical rate to a very high degree of precision. Since $x(t)$ has apparently always been $\lesssim 1$, our Universe is today and has been in the past closely-described by the $k = 0$ flat model. Were the ratio x not exceedingly small early on, the Universe would have either recollapsed long ago ($k > 0$), or began its coasting phase ($k < 0$) where $R \propto t$. [If $k < 0$ and $x_{BBN} = 1$, then $T = 3K$ for $t \simeq 300$ yr!] The smallness of the ratio x required as an 'initial condition' for our Universe is puzzling. [The flatness puzzle has been emphasized in refs. 6,7.]

2.4 Predominance of Matter Over Antimatter

Although the laws of physics are very nearly matter-antimatter symmetrical (the only violation of this symmetry being the tiny one observed in the $K^0 - \bar{K}^0$ system), the Universe is not. There is no evidence for appreciable quantities of antimatter in the Universe [8]. From primordial nucleosynthesis we can infer that the ratio of baryons to photons $\eta \simeq (4 - 7) \times 10^{-10}$. This means that today the Universe has a small net baryon number; comparing the baryon number to the entropy ($s \simeq 7n_\gamma$) we have $n_B/s \simeq (6 - 10) \times 10^{-11}$ - which remains constant as long as the expansion is adiabatic and baryon number is effectively conserved. The observation that $n_B/s \simeq O(10^{-10})$ suggests that the Universe began with a non-zero, but tiny baryon number (of a rather curious value!). That n_B/s be non-zero when $T \simeq 1$ GeV is essential, otherwise baryons and antibaryons (which at $T >> 1$ GeV were about as abundant as photons) would have annihilated down to an abundance of $n_b/n_\gamma = n_{\bar{b}}/n_\gamma \simeq 10^{-18}$ before they were too scarce to annihilate any further.

Of course, one of the great triumphs of the cosmology/particle physics connection is baryogenesis: interactions predicted by Grand Unified Theories (GUT s) and which do not conserve B, C, or CP allow an initially baryon-symmetric Universe to evolve a small baryon asymmetry of the required magnitude (for a recent review see ref. [9]). However, along with this success come two additional cosmological puzzles.

2.5 The Monopole Problem [10,11]

Most GUTs also predict the existence of stable, superheavy ($m \simeq M_G/\alpha$; $M_G \simeq$ grand scale $\gtrsim 10^{14}$ GeV) 't Hooft-Polyakov magnetic monopoles [12]. These objects are classical configurations of the theory with non-trivial topology. Because of the smallness of the horizon when the spontaneous-symmetry breaking (SSB) of the GUT occurs ($T \simeq M_G$) these monopoles are produced at an abundance of about 1 per horizon volume since the Higgs field cannot be correlated on scales $\gtrsim$ the horizon. This leads to a relic abundance of monopoles of about $(M_G/10^{15}\ \text{GeV})^3$ monopoles per baryon. [The horizon volume at temperature T contains a net baryon number of about $(10^{15}\ \text{GeV}/T)^3$]. For $M_G \simeq 10^{14}$ GeV this results in $\Omega_M \simeq 10^{12}$ today (more precisely, such a model Universe would reach $T \simeq 3$K when $t \simeq 30{,}000$ yr). Within the context of the standard cosmology several solutions have been proposed; however, none of them is particularly compelling or even attractive (for a recent review see ref. [11]).

2.6 The Cosmological Constant

With the possible exception of supersymmetry and supergravity theories, the absolute scale of the effective potential $V(\phi)$ is not determined in gauge theories (ϕ = one or more Higgs field). At low temperatures $V(\phi)$ is equivalent to a cosmological term (i.e., contributes $Vg_{\mu\nu}$ to the stress energy of the Universe). The observed expansion rate of the Universe today ($H \simeq 50 - 100$ km s^{-1} Mpc^{-1}) limits the total energy density of the Universe to be $\lesssim 0(10^{-29}\text{g cm}^{-3})$ $\simeq 10^{-46}\ \text{GeV}^4$. Thus empirically the vacuum energy of our $T \simeq 0$ SU(3) x U(1) vacuum (= $V(\phi)$ at the SSB minimum) must be $\lesssim 10^{-46}\ \text{GeV}^4$. Compare this to the difference in energy density between the false ($\phi = 0$) and true vacua, which is $0(T_c^4)$ ($T_c \simeq$ symmetry restoration temperature): for $T_c \simeq 10^{14}$ GeV, $V_{SSB}/V(\phi = 0) \lesssim 10^{-102}$! At present there is no satisfactory explanation for the vanishingly small value of the $T \simeq 0$ vacuum energy density (equivalently, the cosmological term).

Today, the vacuum energy is apparently negligibly small and seems to play no significant role in the dynamics of the expansion of the Universe. If we accept this empirical determination of the absolute scale of $V(\phi)$, then it follows that the energy of the false ($\phi = 0$) vacuum is enormous ($\simeq T_c^4$), and thus could have played a significant role in determining the dynamics of the expansion of the Universe. Accepting this very non-trivial assumption about the zero of the vacuum energy is the starting point for inflation (see Fig. 4).

3. Generic New Inflation

The basic idea of the inflationary Universe scenario is that there was an epoch when the vacuum energy density dominated the energy density of the Universe. During this epoch $\rho \simeq V \simeq$ constant, and thus R(t) grows exponentially ($\propto \exp(Ht)$), allowing a small, causally-coherent region (initial size $\lesssim H^{-1}$) to grow to a size which encompasses the region which eventually becomes our presently-observable Universe. In Guth's original scenario [7], this epoch occurred while the Universe was trapped in the false ($\phi = 0$) vacuum during a strongly first-order phase transition. Unfortunately, in models which inflated enough (i.e., underwent sufficient exponential expansion) the Universe never made a 'graceful return' to the usual radiation-dominated FRW cosmology [13]. Rather than discussing the original model and its shortcomings in detail, I will instead focus on the variant, dubbed 'new inflation', proposed independently by Linde [14] and Albrecht and Steinhardt [15]. In this scenario, the vacuum-dominated epoch occurs while the region of the Universe in question is slowly, but inevitably, evolving toward the true, SSB vacuum. Rather than considering specific models in this section, I will discuss new inflation for a generic model.

Consider a SSB phase transition which occurs at an energy scale M_G. For $T \gtrsim T_c \simeq M_G$ the symmetric ($\phi = 0$) vacuum is favored, i.e., $\phi = 0$ is the global minimum of the finite temperature effective potential $V_T(\phi)$ (= free energy density). As T approaches T_c a second minimum develops at $\phi \neq 0$, and at $T = T_c$ the two minima are degenerate. [I am assuming that this SSB transition is a first-order phase transition.] At temperatures below T_c the SSB ($\phi = \sigma$) minimum is the global minimum of $V_T(\phi)$ (see Fig. 4). However, the Universe does not instantly make the transition from $\phi = 0$ to $\phi = \sigma$; the details and time required are a question of dynamics.

Assuming a barrier exists between the false and true vacua, thermal fluctuations and/or quantum tunneling must be responsible for taking ϕ across the barrier. The dynamics of this process determine when and how the process occurs (bubble formation, spinodal decomposition, etc.) and the value of ϕ after the barrier is penetrated. For definiteness suppose that the barrier is overcome when the

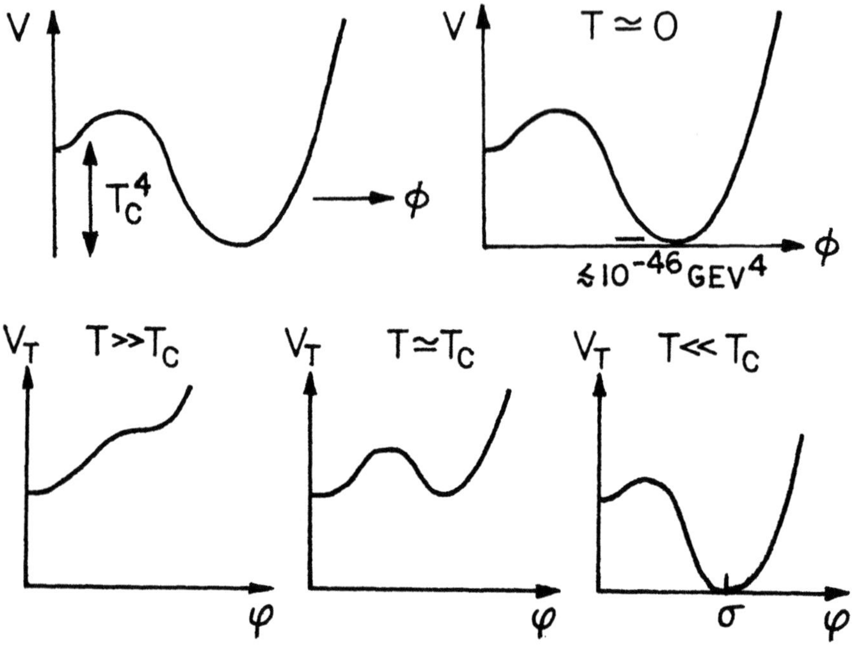

Figure 4 The finite temperature effective potential V_T for $T > T_c$, $T \simeq T_c$, and $T \ll T_c$; here $\phi = \sigma$ is the SSB minimum of V.

temperature is T_{MS} (which could in principle be $\simeq M_G$) and the value of ϕ is ϕ_0. From this point the journey to the true vacuum is downhill (literally) and the evolution of ϕ should be adequately described by the semi-classical equations of motion for ϕ:

$$\ddot{\phi} + 3H\dot{\phi} + \Gamma\dot{\phi} + V' = 0, \qquad (4)$$

where ϕ has been normalized so that its kinetic term in the Lagrangian is $1/2\ \partial_\mu\phi\partial_\mu\phi$, and prime indicates derivative with respect to ϕ. The subscript T on V has been dropped; for $T \ll T_c$ the temperature dependence of V_T can be neglected and the zero temperature potential ($\equiv V$) can be used. The $3H\dot{\phi}$ term acts like a frictional force, and arises because the expansion of the Universe 'redshifts away' the kinetic energy of $\phi (\propto R^{-3})$. The $\Gamma\dot{\phi}$ term accounts for particle creation due to the time-variation of ϕ [16,17]. The quantity Γ is determined by the particles which couple to ϕ and the strength with which they couple ($\Gamma^{-1} \simeq$ lifetime of a ϕ particle). The expansion rate H is determined by the energy density of the Universe [through eqn. (1)]:

$$\rho \simeq 1/2\ \dot{\phi}^2 + V(\phi) + \rho_r , \tag{5}$$

where ρ_r represents the energy density in radiation produced by the time variation of ϕ. For $T_{MS} << T_c$ the original thermal component makes a negligible contribution to ρ. The evolution of ρ_r is given by

$$\rho_r + 4H\rho_r = \Gamma\dot{\phi}^2 , \tag{6}$$

where the $\Gamma\dot{\phi}^2$ term accounts for particle creation by ϕ.

In writing eqns. (4-6) I have implicitly assumed that ϕ is spatially homogeneous. In some small region (inside a bubble or a fluctuation region) this will be a good approximation. The size of this smooth region will be unimportant; take it to be of order the 'physics horizon', H^{-1}. Now follow the evolution of ϕ within the small, smooth patch of size H^{-1}.

If V is sufficiently flat somewhere between $\phi = \phi_0$ and $\phi = \sigma$, then ϕ will evolve very slowly in that region, and the motion of ϕ will be 'friction-dominated' so that $3H\dot{\phi} \simeq -V'$ (in the slow growth phase particle creation is not important). If V is sufficiently flat, then the time required for ϕ to transverse the flat region can be long compared to the expansion timescale H^{-1}, say for definiteness, $\tau_\phi = 100\ H^{-1}$. During this slow growth phase $\rho \simeq V(\phi) \simeq V(\phi = 0)$; both ρ_r and $1/2\ \dot{\phi}^2$ are $<< V(\phi)$. The expansion rate H is then just

$$\begin{aligned} H &\simeq (8\pi V(0)/3m_{pl}^2)^{1/2} \\ &\simeq M_G^2/m_{pl} , \end{aligned} \tag{7}$$

where V(0) is assumed to be of order M_G^4. While $H \simeq$ constant R grows exponentially: $R \propto \exp(Ht)$; for $\tau_\phi = 100\ H^{-1}$ R expands by a factor of e^{100} during the slow rolling period, and the physical size of the smooth region increases to $e^{100}H^{-1}$. This exponential growth phase is called a deSitter phase.

As the potential steepens, the evolution of ϕ quickens. Near $\phi = \sigma$, ϕ oscillates around the SSB minimum with frequency ω: $\omega^2 \simeq V''(\sigma) \simeq M_G^2 >> H^2 \simeq M_G^4/m_{pl}^2$. As ϕ oscillates about $\phi = \sigma$ its motion is damped by particle creation and the expansion of the Universe. If $\Gamma^{-1} << H^{-1}$, the coherent field energy density $(V + 1/2\ \dot{\phi}^2)$ is converted into radiation in less than an expansion time ($\Delta t_{RH} \simeq \Gamma^{-1}$), and the patch is reheated to a temperature $T \simeq O(M_G)$ - the vacuum energy is efficiently converted into radiation ('good reheating'). On the other hand, if $\Gamma^{-1} \gg H^{-1}$, then ϕ continues to oscillate and the coherent field energy redshifts away with the expansion: $(V + 1/2\ \dot{\phi}^2) \propto R^{-3}$. [The coherent field energy behaves like non-relativistic matter; see ref. 18.] Eventually, when $t \simeq \Gamma^{-1}$ the energy in radiation begins to dominate that in coherent field oscillations, and the patch is reheated to a temperature $T \simeq (\Gamma/H)^{1/2} M_G \simeq (\Gamma m_{pl})^{1/2} \ll M_G$ ('poor reheating'). The evolution of ϕ is summarized in Fig. 5.

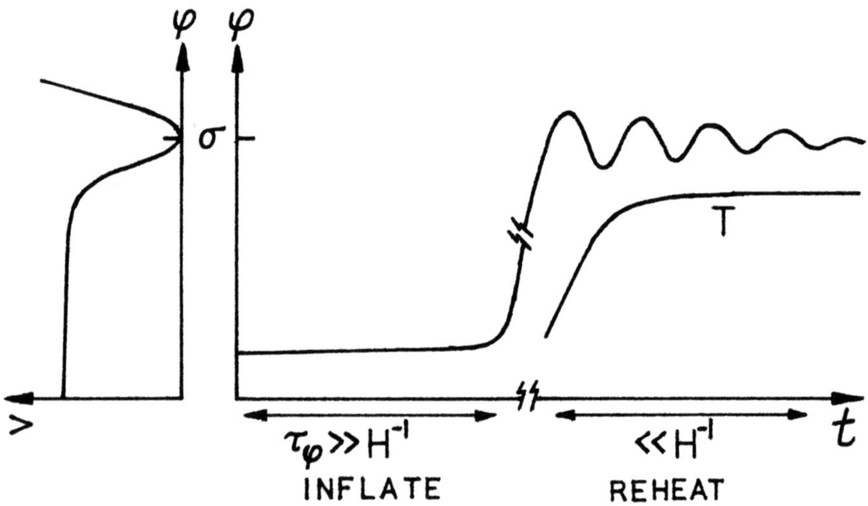

Figure 5 The evolution of $\phi(t)$. During the slow growth phase the time required for ϕ to change appreciably is $\gg H^{-1}$. As the potential steepens ϕ evolves rapidly (timescale $\ll H^{-1}$), eventually oscillating about the SSB minimum. Particle creation damps the oscillations of ϕ in a time $\ll H^{-1}$ (if $\Gamma^{-1} \ll H^{-1}$), reheating the patch to $T \simeq O(M_G)$.

Let us assume 'good reheating' ($\Gamma \gg H$). After reheating the patch has a physical size $e^{100}H^{-1}$ ($\simeq 10^{17}$cm for $M_G \simeq 10^{14}$ GeV), is at a temperature of order M_G, and in the approximation that ϕ was initially constant throughout the patch, the patch is exactly smooth. From this point forward the region evolves like a radiation-dominated FRW model. How have the cosmological conundrums been 'explained'? First, the homogeneity and isotropy; our observable Universe today ($\simeq 10^{28}$cm) had a physical size of about 10 cm ($= 10^{28}$cm x 3K/10^{14} GeV) when T was 10^{14} GeV. Thus it lies well within one of the smooth regions produced by the inflationary epoch (see Fig. 6). At this point the inhomogeneity puzzle has not been solved, since the patch is precisely uniform. Due to the deSitter space quantum fluctuations in ϕ, ϕ is not exactly uniform even in a small patch. In Sec. 4 I will discuss the density inhomogeneities that result from the quantum fluctuations in ϕ. The flatness puzzle involves the smallness of the ratio of the curvature term to the energy density term. This ratio is exponentially smaller after inflation: $x_{after} \approx e^{-200} x_{before}$ since the energy density before and after inflation is $O(M_G^4)$, while k/R^2 has decreased exponentially (by e^{200}). Since the ratio x is reset to an exponentially small value, the inflationary scenario predicts that

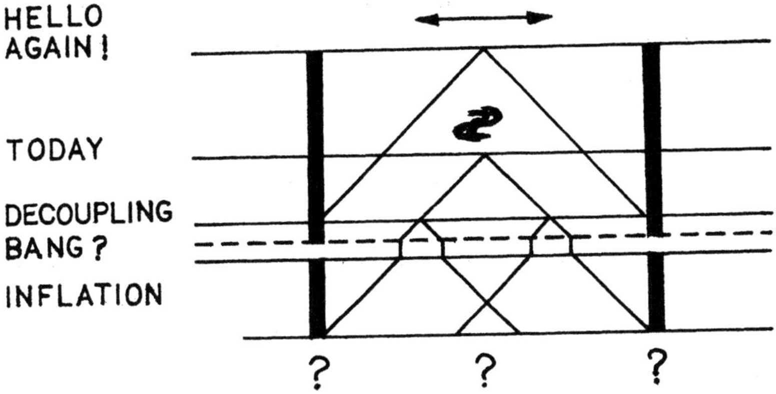

Figure 6 A conformal space-time diagram illustrating the causal structure of the new inflationary Universe (conformal time increases vertically, comoving distance runs horizontally, light rays travel on 45^0 lines). The broken line indicates the singularity in the standard model. The homogeneity puzzle arises because the past light cones of the two events corresponding to the last scattering of microwave photons arriving from opposite directions in the sky never intersect. When the inflationary (deSitter) epoch is added this difficulty is remedied. The solid vertical lines indicate the boundaries of our patch (here assumed to be a fluctuation region and not a bubble). The horizontal arrow at the top indicates the observable Universe today (everything in our past light cone back to decoupling). Eventually we will be able to 'see' all of our patch again (and even outside of it) -- indicated by Hello Again. When this occurs the observable Universe may appear very inhomogeneous.

today Ω should be $1 \pm O(10^{-\text{BIG }\#})$. Assuming the Universe is reheated to a temperature of order M_G, a baryon asymmetry can evolve in the usual way, although the quantitative details may be slightly different [9,19]. Of course, it is absolutely necessary to have baryogenesis occur after reheating since any baryon number (or any other quantum number) present before inflation is diluted by a factor $(M_G/T_{MS})^3 \exp(3H\tau_\phi)$ - the factor by which the total entropy increases.

Since the patch that our observable Universe lies within was once (at the beginning of inflation) causally-coherent, the Higgs field could have been aligned throughout the patch (indeed, this is the lowest energy configuration), and thus there is likely to be $\lesssim 1$ monopole within the entire patch which was produced as a topological defect. The glut of monopoles which occurs in the standard cosmology does not occur. [The production of other topological defects (such as domain walls, etc.) is avoided for similar reasons.] Some monopoles will be produced after reheating in rare, very energetic particle collisions [11,20]. The number produced is exponentially small (and exponentially uncertain) since $T_{RH} \simeq O(M_G) \ll m_M \simeq M_G/\alpha$:

$$\langle F\rangle \simeq 10^{13}(m_M/T_{RH})^3 \exp(-2m_M/T_{RH})\text{cm}^{-2}\text{sr}^{-1}\text{s}^{-1}\,, \tag{8}$$

where T_{RH} is the reheat temperature and $\langle F\rangle$ is the average monopole flux. In minimal SU(5) $2m_M/T_{RH} \simeq 500$, implying less than one thermally-produced monopole in the observable Universe. Note that the ratio m_M/T_{RH} is model-dependent (and smaller in theories where α is larger, e.g., supersymmetric GUT s). Also, the monopole mass $m_M \propto \phi$; when ϕ is $< \sigma$ (e.g., during the slow growth of ϕ phase, or during a downward swing of ϕ during reheating) monopoles are less massive and potentially easier to produce. It has been argued [21] that this effect can significantly reduce the number which goes in the exponent in eqn. (8). The key point is that although monopole production is intrinsically small in inflationary models, the uncertainties in the number of monopoles produced are exponential. It has also been suggested that monopoles might be produced as topological defects in a subsequent phase transition [22].

Finally, the inflationary scenario sheds no light upon the cosmological constant puzzle. Although it can potentially successfully resolve all of the other puzzles in my list, inflation is, in some sense, a house of cards built upon the cosmological constant puzzle.

4. Density Inhomogeneities

4.1 Density Perturbations: 'The Standard Lore'

A density perturbation is described by its wavelength λ or its wavenumber $k(= 2\pi/\lambda)$, and its amplitude $\delta\rho/\rho$ (ρ = average energy density). As the Universe expands the physical (or proper) wavelength of a given perturbation also expands; it is useful to scale out the expansion so that a particular perturbation is always labeled by the same comoving wavelength $\lambda_c \equiv \lambda/R(t)$ or comoving wavenumber $k_c \equiv kR(t)$. [$R(t)$ is often normalized so that $R_{today} = 1$.] Even more common is to label a perturbation by the comoving baryon mass (or total mass in non-relativistic particles if $\Omega_B \neq \Omega_{TOT}$) within a half wavelength $M = \pi\lambda^3 n_B m_N/6$ (n_B = net baryon number density, m_N = nucleon mass).

The relative sizes of λ and H^{-1} (= 'physics horizon' and particle horizon also in the standard cosmology) are crucial for determining the evolution of $\delta\rho/\rho$. When $\lambda \lesssim H^{-1}$ (the perturbation is said to be inside the horizon) microphysics can affect the perturbation. If $\lambda > \lambda_J \simeq v_s H^{-1}$ (physically λ_J, the Jeans length, is the distance a pressure wave can propagate in an expansion time; v_s = sound speed) and the Universe is matter-dominated, then $\delta\rho/\rho$ grows $\propto t^{2/3} \propto R$. Perturbations with $\lambda < \lambda_J$ oscillate as pressure-supported sound waves (and may even damp).

When a perturbation is outside the horizon ($\lambda > H^{-1}$) the situation is a bit more complicated. The quantity $\delta\rho/\rho$ is not gauge-invariant; when $\lambda < H^{-1}$ this fact creates no great difficulties. However when $\lambda > H^{-1}$ the gauge-noninvariance is a bit of a nightmare. Although Bardeen [23] has developed an elegant gauge-invariant formalism to handle density perturbations in a gauge-invariant way, his gauge invariant quantities are not intuitively easy to understand. Physically, only real, honest-to-God wrinkles in the geometry (called curvature fluctuations or adiabatic fluctuations) can 'grow'. In the synchronous gauge ($g_{00} = -1$, $g_{0i} = 0$) $\delta\rho/\rho$ for these perturbations grows $\propto t^n$ ($n = 1$ - radiation dominated, $= 2/3$ - matter dominated). Geometrically, when $\lambda > H^{-1}$ these perturbations are just wrinkles in the space time which are evolving kinematically (since microphysical processes cannot affect their evolution). Adiabatic perturbations are characterized by $\delta\rho/\rho \neq 0$ and $\delta(n_B/s) = 0$; while isothermal perturbations (which do not grow outside the horizon) are characterized by $\delta\rho/\rho = 0$ and $\delta(n_B/s) \neq 0$. [With greater generality $\delta(n_B/s)$ can be replaced by any spatial perturbation in the equation of state $\delta p/p$, where p = p(ρ, . . .).] In the standard cosmology $H^{-1} \propto t$ grows monotonically; a perturbation only crosses the horizon once (see Fig. 8). Thus it should be clear that microphysical processes cannot create adiabatic perturbations (on scales $\gtrsim H^{-1}$) since microphysics only operates on scales $\lesssim H^{-1}$. In the standard cosmology adiabatic (or curvature) perturbations were either there ab initio or they are not present. Microphysical processes can create isothermal (or pressure perturbations) on scales $\gtrsim H^{-1}$ (of course, they cannot grow until $\lambda \lesssim H^{-1}$). Fig. 7 shows the evolution of a galactic mass ($\simeq 10^{12} M_{\odot}$) perturbation: for $t \lesssim 10^8$ s, $\lambda > H^{-1}$ and $\delta\rho/\rho \propto t$; for $10^{13}\text{s} \gtrsim t \gtrsim 10^8\text{s}$, $\lambda < H^{-1}$ and $\delta\rho/\rho$ oscillates as a sound wave since matter and radiation are still coupled ($v_s \simeq c$) and hence $\lambda_J \simeq H^{-1}$; for $t \gtrsim 10^{13}$s, $\lambda < H^{-1}$ and $\delta\rho/\rho \propto t^{2/3}$ since matter and radiation are decoupled ($v_s \ll c$) and $\lambda_J < \lambda_{Galaxy}$. [Note: in an $\Omega = 1$ Universe the mass inside the horizon $\simeq (t/\text{sec})^{3/2} M_{\odot}$.]

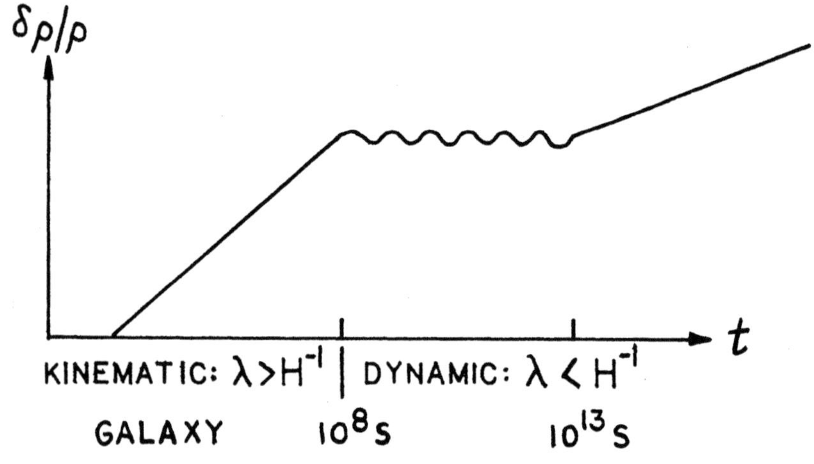

Figure 7 The evolution of a galactic mass adiabatic density perturbation.

Finally, at this point it should be clear that a convenient epoch to specify the amplitude of a density perturbation is when it enters the horizon. It is often supposed (in the absence of knowledge about the origin of perturbations) that the spectrum of fluctuations is a power law (i.e., no preferred scale):

$$(\delta\rho/\rho)_H \propto M^{-n}. \tag{9}$$

If $n > 0$, then on some small scale perturbations will enter the horizon with amplitude $\gtrsim O(1)$ -- this leads to black hole formation; if this scale is $\gtrsim 10^{15}$ g (mass of a black hole evaporating today) there will be too many black holes in the Universe today. On the other hand, if $n < 0$ then the Universe becomes more irregular on larger scales (contrary to observation). In the absence of a high or low mass cutoff, the $n = 0$ so-called Zel'dovich spectrum [24] of density perturbations seems to be the only 'safe' spectrum. It has the attractive feature that all scales cross the horizon with the same amplitude (i.e., it is scale-free). Such a spectrum is not required by the observations; however, such a spectrum with amplitude of $O(10^{-4} - 10^{-3})$ probably leads to an acceptable picture of galaxy formation (i.e., consistent with all present observations - microwave background fluctuations, galaxy correlation function, etc.). [The 'standard lore' of the evolution of density perturbations is discussed in greater detail in refs. 5, 23, 25.]

4.2 Origin of Density Inhomogeneities in the New Inflationary Universe

The basic result is that quantum fluctuations in the scalar field ϕ (due to the deSitter space event horizon which exists during the exponential expansion (inflation) phase) give rise to an almost scale-free (Zel'dovich) spectrum of density perturbations of amplitude

$$\left(\frac{\delta\rho}{\rho}\right)_H = (4 \text{ or } 2/5)\ \frac{H\,\Delta\phi}{\dot{\phi}(t_1)}\ , \tag{10}$$

where 4(2/5) applies if the scale in question reenters the horizon when the Universe is radiation-(matter-) dominated, H is the Hubble parameter during inflation, $\dot{\phi}(t_1)$ is the value of $\dot{\phi}$ when the perturbation left the horizon during the deSitter phase, and $\Delta\phi \simeq H/2\pi$ is the fluctuation in ϕ. This result was independently derived by the authors of refs. [26-29]. Rather than discussing the derivation in detail here, I will attempt to physically motivate the result. This result turns out to be the most stringent constraint on models of new inflation.

The crucial difference between the standard cosmology and the inflationary scenario for the evolution of density perturbations is that H^{-1} (the 'physics horizon') is <u>not</u> strictly monotonic; during the inflationary (deSitter) epoch it is constant. Thus, a perturbation can cross the horizon ($\lambda = H^{-1}$) twice (see Fig. 8)! The evolution of two scales (λ_G = galaxy and λ_H = present observable Universe) is shown in Fig. 8. Earlier than t_1 (time when $\lambda_G \simeq H^{-1}$) $\lambda_G < H^{-1}$ and microphysics (quantum fluctuations, etc.) can operate on this scale. When $t = t_1$ microphysics 'freezes out' on this scale; the density perturbation which exists on this scale, say $(\delta\rho/\rho)_1$, then evolves 'kinematically' until it reenters the horizon at $t = t_H$ during the subsequent radiation-dominated FRW phase with amplitude $(\delta\rho/\rho)_H$.

DeSitter space is exactly time-translationally-invariant; the inflationary epoch is approximately a deSitter phase - ϕ is almost, but not quite constant (see Fig. 8). [In deSitter space $p + \rho = 0$; during inflation $p + \rho = \dot{\phi}^2$.] This time-translation invariance is crucial; as each scale leaves the horizon (at $t = t_1$) $\delta\rho/\rho$ on that scale is fixed by microphysics at some value: $(\delta\rho/\rho)_1$. Because of the (approximate) time-translation invariance of the inflationary phase this value $(\delta\rho/\rho)_1$ is (approximately) the same for all scales. [Recall H, ϕ, $\dot{\phi}$ are all approximately constant during this epoch, and each scale has the same physical size ($= H^{-1}$) when it crosses outside of the horizon.] The precise value of $(\delta\rho/\rho)_1$ is fixed by the amplitude of the quantum fluctuations in ϕ on the scale H^{-1}; for a free scalar field $\Delta\phi = H/2\pi$ (the Hawking temperature). [Recall, during inflation V'' ($\simeq$ the effective mass-squared) is very small.]

While outside the horizon ($t_1 \lesssim t \lesssim t_H$) a perturbation evolves 'kinematically' (as a wrinkle in the geometry); viewed in some gauges the amplitude evolves (e.g., the synchronous gauge), while in others (e.g., the uniform Hubble constant gauge) it remains constant. However, in all gauges the kinematic evolution is independent of scale (intuitively this makes sense since this is the kinematic regime). Given these 'two facts': $(\delta\rho/\rho)_1 \simeq$ scale-independent and the kinematic evolution $\simeq$ scale-independent, it follows that all scales reenter the horizon (at $t = t_H$) with (approximately) the same amplitude, given by eqn. (10). Not only is this a reasonable spectrum (the Zel'dovich spectrum), but this is one of the very few instances that the spectrum of density perturbations has been calculable from first principles. [The fluctuations produced by strings are another such example, see, e.g. ref. 30; however, the initial homogeneity must be assumed.]

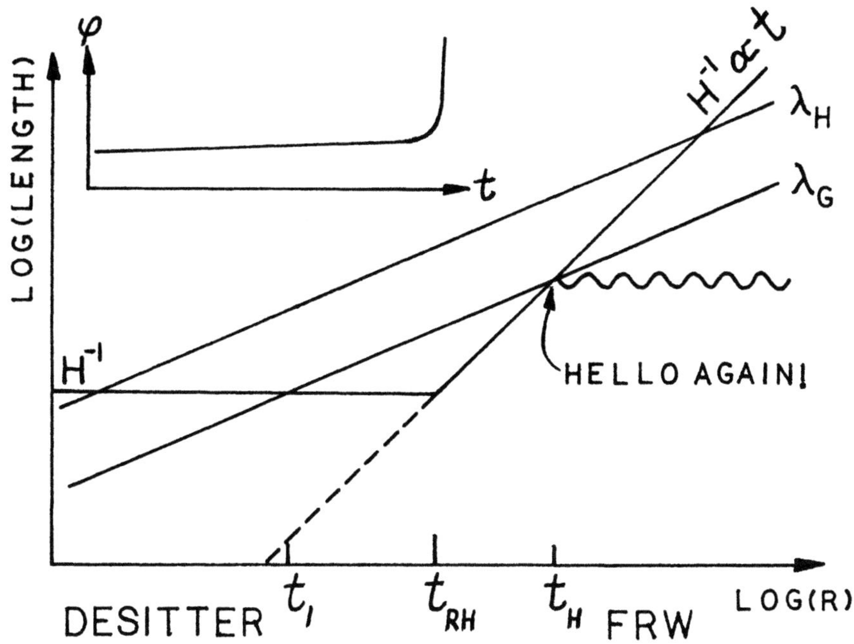

Figure 8 The evolution of the 'physics horizon' ($\simeq H^{-1}$) and the physical sizes of perturbations on the scale of a galaxy (λ_G) and on the scale of the present observable Universe (λ_H) as a function of cosmic scale factor R. Reheating (transition from inflation to a radiation-dominated FRW model) occurs at $t = t_{RH}$. For reference the evolution of ϕ is also shown. The broken line shows the evolution of H^{-1} in the standard cosmology. In the inflationary cosmology a perturbation crosses the horizon twice, which makes it possible for causal microphysics (in this case, quantum fluctuations in ϕ) to produce large-scale density perturbations.

5. Specific Models of New Inflation

5.1 Coleman-Weinberg SU(5) Model

The first model of new inflation [14,15] studied was the Coleman-Weinberg SU(5) model, with T = 0 effective potential

$$\begin{aligned} V(\phi) &= 1/2\ B\sigma^4 + B\phi^4[\ell n(\phi^2/\sigma^2) - 1/2], \\ &\simeq 1/2\ B\sigma^4 - \lambda(\phi)\phi^4 \qquad (\phi \ll \sigma), \end{aligned} \tag{11}$$

where $B = 25g^4/256\pi^2$ (g = gauge coupling constant), $\sigma \simeq 1.2 \times 10^{15}$ GeV, and for $\phi \simeq 10^9$ GeV $\lambda(\phi) \simeq 0.1$. [V may not look familiar; this is because ϕ is normalized so that its kinetic term is $1/2\ \dot{\phi}^2$ rather than the usual $(15/4)\dot{\phi}^2$.] Albrecht and Steinhardt [15] showed that when $T \simeq 10^8 - 10^9$ GeV the metastability limit is reached, and thermal fluctuations drive ϕ over the T-dependent barrier (height $\simeq T^4$) in the finite temperature effective potential. Naïvely, one expects that $\phi_0 \simeq T_{MS}$ since for $\phi \ll \sigma$ there is no other scale in the potential (this is a point to which I will return). The potential is sufficiently flat that the approximation $3H\dot{\phi} \simeq -V'$ is valid for $\phi \ll \sigma$, and it follows that

$$(\phi/H)^2 = (3/2\lambda)[H(\tau_\phi - t)]^{-1}, \tag{12}$$

where $H\tau_\phi \simeq (3/2\lambda)(H/\phi_0)^2$ (recall τ_ϕ= time it takes ϕ to traverse the flat portion of the potential). Physically, $H\tau_\phi$ is the number of e-folds of R which occur during inflation, which to solve the homogeneity-isotropy and flatness puzzles must be $\gtrsim$ 0(60). For this model $H \simeq 7 \times 10^9$ GeV; setting $\phi_0 \simeq 10^8 - 10^9$ GeV results in $H\tau_\phi \simeq$ 0(500-50000) - seemingly more than sufficient inflation.

There is however, a very basic problem here. Equation (12) is derived from the semi-classical equation of motion for ϕ [eqn. (4)], and thus only makes sense when the evolution of ϕ is 'classical', that is when $\phi \gg \Delta\phi_{QM}$ (= quantum fluctuations in ϕ). In deSitter space the scale of quantum fluctuations is set by H: $\Delta\phi_{QM} \simeq H/2\pi$ (on the length scale H^{-1}). Roughly speaking then, eqn. (12) is only valid for $\phi \gg H$. However, sufficient inflation requires $\phi_0 \lesssim H$. Thus the Coleman-Weinberg model seems doomed for the simple reason that all the important physics must occur when $\phi \lesssim \Delta\phi_{QM}$. This is basically the conclusion reached by Linde [31] and Vilenkin and Ford [32] who have analyzed these effects much more carefully. Note that by artificially reducing λ by a factor of 10-100 sufficient inflation can be achieved for $\phi_0 \gg H$ (i.e., the potential becomes sufficiently flat that the classical part of the evolution, $\phi \gg H$, takes a time $\gtrsim 60\ H^{-1}$). In the Coleman-Weinberg model reheating proceeds without a hitch [16].

Let's ignore for the moment these difficulties associated with the need to have $\phi_0 < H$, and examine the question of density fluctuations. Combining eqns. (10 and 12) it follows that

$$\left(\frac{\delta\rho}{\rho}\right)_H \simeq (4 \text{ or } 2/5)100\,\lambda^{1/2}\,[1 + \ell n(M/10^{12}M_\Theta)/171 + \ell n(g\sigma/10^{15}\text{ GeV})/57]^{3/2}\,, \tag{13}$$

where M is the comoving mass within the perturbation. Note that the spectrum is almost, but not quite scale-invariant (varying by less than a factor of 2 from $1M_\Theta$ to $10^{22}M_\Theta$ = present horizon mass). Blindly plugging in $\lambda \simeq 0.1$, results in $(\delta\rho/\rho)_H \simeq O(10^2)$ which is clearly a disaster. [On angular scales $\gg 1^\circ$ the Zel'dovich spectrum results in temperature fluctuations of $\Delta T/T \simeq 1/2(\delta\rho/\rho)_H$ which must be $\lesssim 10^{-4}$ to be consistent with the observed isotropy.] To obtain perturbations of an acceptable amplitude one must artificially set $\lambda \simeq 5 \times 10^{-11}$. [In an SU(5) GUT λ is determined by the value of $\alpha_{GUT} = g^2/4\pi \simeq 1/45$, which implies $\lambda \simeq 0.1$.] As mentioned earlier the density fluctuation constraint is a very severe one; recall that $\lambda \simeq 10^{-2} - 10^{-3}$ would solve the difficulties associated with the quantum fluctuations in ϕ. To say the least, the Coleman-Weinberg SU(5) model seems untenable.

5.2 Reverse-hierarchy Supersymmetric Models

A possible remedy to the problems encountered with the Coleman-Weinberg SU(5) model is to have a potential which is flat for $\phi \gg H$, so that during inflation $\phi \gg H$ and $\dot\phi \gg H^2$ [recall $(\delta\rho/\rho)_H \simeq H^2/\dot\phi$]. Supersymmetric models which employ the Witten [33] reverse hierarchy scheme with O'Raifeartaigh [34] type SSB have just the desired feature. For definiteness consider the geometric hierarchy model of Dimopoulos and Raby [35]. In this model there are 3 scales: $M_I \simeq 10^{12}$ GeV which is the scale of supersymmetry breaking, and two other scales, the grand scale $M_G \simeq m_{pl}$ and the weak scale $M_W \simeq M_I^2/M_G$, both of which are generated radiatively. The scale M_I determines the properties of the phase transition which leads to inflation. For $m_{pl} \gg \phi \gg M_I$ the potential can be written as

$$V(\phi) = M_I^4 [c_1 - c_2 \ln(\phi/m_{pl})]. \tag{14}$$

The shape of the potential near $\phi = 0$ is not of interest here, and the shape near the SSB minimum is not completely determined.

The key feature of this potential is that it becomes very flat for large values of ϕ, $V' \simeq -c_2 M_I^4/\phi$. If $c_1/c_2 \gtrsim O(10^2)$, then the potential is sufficiently flat for $\phi \gtrsim (3c_2/8\pi c_1)^{1/2} m_{pl}$ that ϕ evolves slowly and inflation occurs. During this slow evolution epoch $3H\dot{\phi} \simeq -V'$ and

$$\phi(t) \simeq (3c_2/8\pi c_1)^{1/2} (1 + 2/3\ Ht)^{1/2} m_{pl}, \tag{15}$$

where $H^2 \simeq 8\pi c_1 M_I^4/3m_{pl}^2$ and $t = 0$ marks the start of the slow growth period. The quantity $H\tau_\phi$ (= number of e-folds in R) is $\simeq 4\pi c_1/c_2$. Assuming that $H\tau_\phi \gtrsim 100$, it follows from eqns. (10 and 15) that

$$\left(\frac{\delta\rho}{\rho}\right)_H \simeq 32\left(\frac{2\pi}{3}\right)^{1/2} \frac{c_1}{c_2}^{3/2} \left(\frac{M_I}{m_{pl}}\right)^2 . \tag{16}$$

With $c_1 \simeq O(10)$ and $M_I \simeq 10^{12}$ GeV, c_1/c_2 must be $O(10^7)$ to achieve $(\delta\rho/\rho)_H \simeq 10^{-4}$. In this model c_1/c_2 is not precisely determined; however, c_1/c_2 is naturally $\gtrsim 10^4$.

'The rub' with these models is that in order to have $M_I \gg M_W$ the light fields must be decoupled from ϕ (by many powers of M_I/m_{pl}). Albrecht <u>etal</u>. [36] have studied this model in detail and find that $\Gamma \simeq 10^{-23}$ GeV, so that $T_{RH} \simeq 10$ MeV (sufficiently-high for nucleosynthesis, but not for baryogenesis). One possibility suggested by Albrecht and Steinhardt [37] is to abandon trying to explain the gauge hierarchy with these models, and take $M_I \simeq 10^{16} - 10^{17}$ GeV; in such a model reheating to $10^{10} - 10^{13}$ GeV can be attained.

6. Looking for Ms. Goodbar

The results discussed in Sec. 5 are a little discouraging, but unlike the situation with 'old inflation' a few years ago, not hopeless [13]. From the failures of new inflation thus far, a prescription for the 'right effective potential' has emerged [38]. In addition to predicting sensible particle physics the desired potential should have the following features (see Fig. 9):

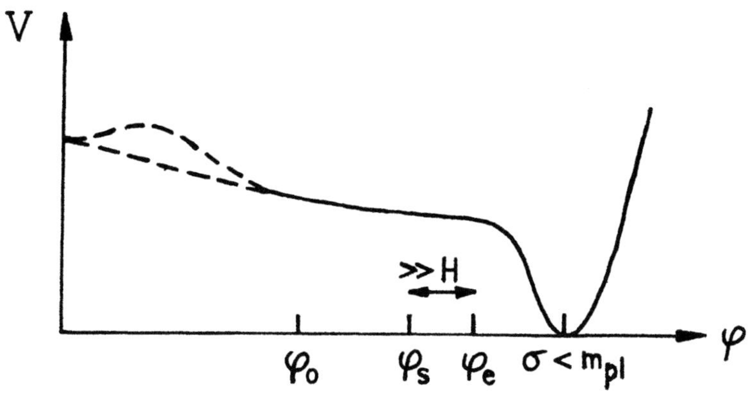

Figure 9 The 'Ms Goodbar Potential'

(1) A flat region where the motion of ϕ is 'friction-dominated', i.e., $\ddot{\phi}$ term negligible so that $3H\dot{\phi} = -V'$. This requires an interval where $V'' \lesssim 9H^2$.

(2) Denote the starting and ending values of ϕ in this interval by ϕ_s and ϕ_e respectively (note: ϕ_s must be $\geq \phi_o$). The length of the interval should be much greater than H (which sets the scale of quantum fluctuations in ϕ): $\phi_e - \phi_s >> H$. This insures that quantum fluctuations will not drive ϕ across the flat region too quickly.

(3) The time required for ϕ to traverse the flat region should be $\gtrsim 60\ H^{-1}$ (to solve the homogeneity-isotropy and flatness problems). This implies that

$$\int_{\phi_s}^{\phi_e} \frac{H^2 d\phi}{-V'} \simeq \frac{H_e^{\ 2}}{V''_m} \ \ln[1 - \Delta\phi\, V''_m / V'_m] \gtrsim 20\,, \qquad (17)$$

where $\Delta\phi = \phi_e - \phi_s$, V''_m and V'_m are evaluated at the point in the interval where $-V'$ attains its minimum value, and $H_e \equiv H(\phi_e)$.

(4) In order to get the correct amplitude for density fluctuations, $(\delta\rho/\rho)_H \simeq H^2/\dot{\phi}(t_1)$, $\dot{\phi}$ must be $\simeq 10^4 H^2$ when a galactic size perturbation crosses outside the horizon. This occurs about 50 Hubble times before the end of inflation; using the value of $\dot{\phi}$ at ϕ_e introduces an error of at most a factor of O(1 - 10). This condition is

then: $-V'(\phi_e) \simeq 3 \times 10^4 H^3$.

(5) The SSB minimum of ϕ must satisfy $\phi_e < \sigma \lesssim m_{pl}$.

(6) To insure sufficient reheating $V''(\sigma)$ must be $\gg H^2$ (so that $\tau_{osc} \ll H^{-1}$) and the reheat temperature must be high enough so that baryogenesis proceeds after reheating. If $\Gamma \gtrsim H$, then $T_{RH} \simeq V(0)^{1/4}$ - which is the maximum reheat temperature (100% conversion of vacuum energy into radiation); if $\Gamma \lesssim H$, then $T_{RH} \simeq (\Gamma m_{pl})^{1/2}$ must be greater than some minimum temperature for baryogenesis, say T_B. This implies $\Gamma \gtrsim T_B^2/m_{pl}$.

Remarkably enough it is possible to find a quartic potential of the form: $V(\phi) = V_0 + \alpha\phi^2 - \beta\phi^3 + \lambda\phi^4$ ($V_0 \simeq 27\beta^4/256\lambda^3$), which satisfies all of these criteria [38]. To do so, the parameters α, β, and λ must be: $\alpha \lesssim 10^{-2} H^2$, $\beta \simeq 10^{-5} H$, $\lambda \lesssim 10^{-9}$, where $H^2 \simeq 8\pi V_0/3m_{pl}^2 \simeq 10^{20}\lambda^3 m_{pl}^2$ and the SSB minimum $\sigma \simeq 10^5\lambda^{1/2} m_{pl}$. Whether or not sensible particle physics can be constructed around such a potential remains to be seen.

In a series of papers [39] the 'CERN group' has studied a large class of supersymmetric and supergravity gravity potentials; however, it does not appear that any of their models satisfy all the criteria necessary for a successful model of new inflation. Whether or not Ms. Goodbar's first name is 'susy' or 'sugar' remains to be seen.

Finally, I have only briefly mentioned the issue of gravitational effects [31,32], and have not discussed at all the topic of homogeneous tunneling solutions in Euclidean space [40,41]. There are still very important issues to be resolved here.

7. Concluding Remarks

New inflation is an extremely attractive cosmological program. It has the potential to 'free' the present state of the Universe (on scales at least as large as 10^{28} cm) from any dependence on the initial state of the Universe, in that the current state of the observable Universe in these models depends only upon microphysical processes which occurred very early on ($t \lesssim 10^{-34}$s). [I should mention that this conjecture of 'Cosmic Baldness' [42] is still just a conjecture; it has not been demonstrated that starting with the most general cosmological solution to Einstein's equations, there exist regions which undergo sufficient inflation. The conjecture however has been addressed perturbatively; pre-inflationary perturbations remain constant in amplitude, but are expanded beyond the present horizon [43], and neither shear nor negative-curvature can prevent inflation from occurring [44].]

At present there exists no successful model of new inflation. However, one should not despair, as I have described in Sec. 6 there does exist a clear-cut and straightforward prescription for the desired ('Ms Goodbar') potential. Whether one can find a potential which fits the prescription and produces sensible particle physics remains to be seen. If such a theory is found, it would truly be a monumental achievement for the inner space/outer space connection (particle physics and cosmology).

Now for some sobering thoughts. The inflationary scenario does not address the issue of the cosmological constant; in fact, the small value of the cosmological constant today is its foundation. If some relaxation mechanism is found to insure that the cosmological constant is always small, the inflationary scenario (in its present form at least) would vanish into the vacuum. It would be fair to point out that inflation is not the only approach to resolving the cosmological puzzles discussed in Sec. 2. The homogeneity, isotropy, and inhomogeneity puzzles all involve the apparent smallness of the horizon. Recall that computing the horizon distance

$$d_H = R(t) \int_0^t dt'/R(t') \tag{18}$$

requires knowledge of $R(t)$ all the way back to $t = 0$. If during an early epoch ($t \lesssim 10^{-43}$s?) R increased as or more rapdily than t (e.g. $t^{1.1}$), then $d_H \to \infty$, eliminating the 'horizon constraint'. The monopole and flatness problems can be solved by producing large amounts of entropy since both problems involve a ratio to the entropy. Dissipating anisotropy and/or inhomogeneity is one possible mechanism for producing entropy. One alternative to inflation is Planck epoch physics. Quantum gravitational effects could both modify the behaviour of R(t) and through quantum particle creation produce large amounts of entropy [see e.g., the recent review in ref. 45].

Two of the key 'predictions' of the inflationary scenario, $\Omega = 1 \pm O(10^{-BIG})$ and scale-invariant density perturbations, are such natural and compelling features of a reasonable cosmological model, that their ultimate verification (my personal bias here!) as cosmological facts will shed little light on whether or not we live in an inflationary Universe. Although the inflationary Universe scenario is not the only game in town, right now it does seem to be the best game in town.

It is a pleasure to acknowledge the many beneficial conversations I have had with Jim Bardeen, Alan Guth, and particularly Paul Steinhardt and the hospitality of the Aspen Center for Physics where this manuscript was written. This work was supported in part by the DOE through contract DE AC02-80ER10773 A003 (at Chicago).

REFERENCES

[1] K. A. Olive, D. N. Schramm, G. Steigman, M. S. Turner, and J. Yang, Astrophys. J. 246, 557 (1981); J. Yang, M. S. Turner, G. Steigman, D. N. Schramm, and K. A. Olive, submitted to Astrophys. J. (1983).

[2] S. Weinberg, Gravitation and Cosmology (Wiley: NY, 1972).

[3] R. V. Wagoner, in Physical Cosmology (Les Houches 1979 Session XXXII), eds. R. Balian etal. (North-Holland: Amsterdam, 1980).

[4] R. Sachs and A. Wolfe, Astrophys. J. 147, 73 (1967).

[5] P. J. E. Peebles, The Large-Scale Structure of the Universe (Princeton Univ. Press: Princeton, 1980).

[6] R. H. Dicke and P. J. E. Peebles, in General Relativity: An Einstein Centenary Survey, eds. S. W. Hawking and W. Israel (Cambridge Univ. Press: Cambridge, 1979).

[7] A. Guth, Phys. Rev. D23, 347 (1981).

[8] G. Steigman, Ann. Rev. Astron. Astrophys. 14, 339 (1976).

[9] E. W. Kolb and M. S. Turner, Ann. Rev. Nucl. Part. Sci. 33, in press (1983).

[10] Ya. B. Zel'dovich and M. Yu. Khlopov, Phys. Lett. 79B, 239 (1978); J. Preskill, Phys. Rev. Lett 43, 1365 (1979).

[11] J. Preskill, in The Very Early Universe, eds. S. W. Hawking, G. B. Gibbons, and S. Siklos (Cambridge Univ. Press: Cambridge, 1983).

[12] G. 't Hooft, Nucl. Phys. B79, 276 (1974); A. M. Polyakov, JETP Lett. 20, 194 (1974).

[13] A. Guth and E. Weinberg, Nucl. Phys. B212, 321 (1983); S. W. Hawking, I. Moss, and J. Stewart, Phys. Rev. D26, 2681 (1982).

[14] A. D. Linde, Phys. Lett. 108B, 389 (1982).

[15] A. Albrecht and P. J. Steinhardt, Phys. Rev. Lett. 48, 1220 (1982).

[16] A. Albrecht, P. J. Steinhardt, M. S. Turner, and F. Wilczek, Phys. Rev. Lett. 48, 1437 (1982).

[17] L. Abbott, E. Farhi, and M. B. Wise, Phys. Lett. 117B, 29 (1982).

[18] M. S. Turner, Phys. Rev. D, in press (1983).

[19] A. D. Dolgov and A. D. Linde, Phys. Lett. 116B, 329 (1982).

[20] M. S. Turner, Phys. Lett. 115B, 95 (1982).

[21] W. Collins and M. S. Turner, "Thermal Production of Superheavy Monopoles in the New Inflationary Universe Scenario", submitted to Phys. Rev. D (1983); A. Goldhaber and A. Guth, in preparation (1983).

[22] G. Lazarides and Q. Shafi, in The Very Early Universe (see ref. 11); I. G. Moss, Newcastle upon the Tyne preprint (1983).

[23] J. M. Bardeen, Phys. Rev. D22, 1882 (1980).

[24] Ya. B. Zel'dovich, Mon. Not. R. Astron. Soc. 160, 1 p (1972); E. R. Harrison, Phys. Rev. D1, 2726 (1970).

[25] W. H. Press and E. T. Vishniac, Astrophys. J. 239, 1 (1980).

[26] S. W. Hawking, Phys. Lett. 115B, 295 (1982).

[27] A. Guth and S. -Y. Pi, Phys. Rev. Lett. 49, 1110 (1982).

[28] A. A. Starobinskii, Phys. Lett. 117B, 175 (1982).

[29] J. M. Bardeen, P. J. Steinhardt, and M. S. Turner, Phys. Rev. D, in press (1983).

[30] Ya. B. Zel'dovich, Mon. Not. R. Astron. Soc. 192, 663 (1980); A. Vilenkin, Phys. Rev. Lett. 46, 1169, 1496(E) (1981).

[31] A. D. Linde, Phys. Lett. 116B, 335 (1982).

[32] A. Vilenkin and L. Ford, Phys. Rev. D26, 1231 (1982).

[33] E. Witten, Phys. Lett. 105B, 267 (1981).

[34] L. O'Raifeartaigh, Nucl. Phys. B96, 331 (1975).

[35] S. Dimopoulos and S. Raby, Nucl. Phys. B, to be published (1983).

[36] A. Albrecht, S. Dimopoulos, W. Fischler, E. W. Kolb, S. Raby, and P. J. Steinhardt, Nucl. Phys. B, to be published (1983).

[37] A. Albrecht and P. J. Steinhardt, Phys. Lett. B, to be published (1983).

[38] P. J. Steinhardt and M. S. Turner, in preparation (1983).

[39] J. Ellis, D. V. Nanopoulos, K. A. Olive, and K. Tamvakis, Nucl. Phys. B, in press (1983); Phys. Lett. 120B, 331 (1983); D. V. Nanopoulos, K. A. Olive, M. Srednicki, and K. Tamvakis, Phys. Lett. 123B, 41 (1983); Phys. Lett. 124B, 171 (1983); D. V. Nanopoulos, K. A. Olive, and M. Srednicki, CERN preprint TH. 3555 (1983).

[40] S. W. Hawking and I. G. Moss, Phys. Lett. 110B, 35 (1982).

[41] E. Mottola and A. Lapedes, Phys. Rev. D27, 2285 (1983); E. Mottola, Phys. Rev. D27, 2294 (1983).

[42] G. W. Gibbons and S. W. Hawking, Phys. Rev. D15, 2738 (1977); W. Boucher and G. W. Gibbons, in The Very Early Universe, [11].

[43] J. Frieman and M. S. Turner, in preparation (1983).

[44] G. Steigman and M. S. Turner, Phys. Lett. B, in press (1983).

[45] J. B. Hartle, in The Very Early Universe, see ref. 11.

DARK MATTER, GALAXIES, SUPERCLUSTERS AND VOIDS

Joel R. Primack
Board of Studies in Physics, University of California
Santa Cruz, CA 95064

and

George R. Blumenthal
Lick Observatory, Board of Studies in Astronomy and Astrophysics
University of California, Santa Cruz, CA 95064

1. Introduction

Two surprising facts of fundamental importance for understanding the large scale structure of the universe have become apparent in the last few years. The first is that most of the mass in the universe is invisible, and the matter it represents may not even be composed of baryons and leptons. The second fact is that on large scales, galaxies are distributed in a network of thick, flattened or filamentary structures, called superclusters, separated by vast voids.

The beautiful idea of cosmic inflation suggests that microphysics on scales $\lesssim 10^{-25}$ cm determines not only the structure of the universe on the scale of the entire horizon (10^{28} cm) and beyond, but also the primordial spectrum of density fluctuations that eventually give rise to all the smaller scale structures we observe.

Since dark matter (DM) dominates gravitationally on all scales larger than galaxy cores, it is likely that the structure of galaxies, clusters of galaxies, and superclusters reflects the properties of the DM as well as the nature of the primordial fluctuations. If the DM is some form of elementary particle, as seems most plausible, then there exists yet another previously unsuspected connection between physical phenomena on very large and very small scales. The dialogue between particle physics and astrophysics is growing increasingly rich!

In this paper we begin by briefly reviewing the observational evidence both for DM dominance and for superclusters and voids. We then discuss three arguments that the dark matter that dominates the present universe is not baryonic - based on excluding specific baryonic models, the deuterium abundance constraint on the ratio of baryon to total density of the universe, and the incompatibility of galaxy formation in a baryon-dominated universe with the absence of small-angle fluctuations in the microwave background radiation.

If the dark matter consists of elementary particles, it may be classified as hot (free streaming erases all but supercluster-scale fluctuations), warm (free streaming erases fluctuations smaller than galaxies), or cold (free streaming is unimportant). We consider scenarios for galaxy formation in all three cases. We discuss several potential problems with the hot (neutrino) case: making galaxies early enough, with enough baryons, and without too much increase in $M_{tot}/M_{\ell um}$ from galaxy to rich cluster scales. The reported existence of dwarf spheroidal galaxies with relatively heavy halos is a serious problem for both hot and warm scenarios. Zeldovich (n = 1) adiabatic initial fluctuations in cold dark matter (axions, or a heavy stable "ino") appear to lead to observed sizes and other properties of galaxies. The big question is whether it also yields large scale structure such as voids and filaments.

2. Observations

2.1 Evidence for Dark Matter

There is abundant observational evidence that dark matter (DM) is responsible for most of the mass in the universe [1]. Dark matter is detected through its gravitational attraction in the massive extended halos of disk galaxies and in groups and clusters of galaxies of all sizes. It is appropriate to call this matter "dark" because it is detected in no other way; it is not observed to emit or absorb electromagnetic radiation of any wavelength. Matter observed in these latter ways we will call "luminous".

The famous astronomer Zwicky pointed out in the 1930s that the velocity dispersion of the galaxies in the nearest rich (i.e. populous) cluster of galaxies, which lies in the direction of the constellation Virgo, are much too high for the cluster to be gravitationally bound unless it contains considerably more mass than is observed in all the galaxies' stars, as determined by the usual mass/luminosity relation for stars. This discrepancy has persisted as the available data has grown. For very rich clusters, of which the nearest is Coma, the mass-to-light ratio is $M/L \approx 650h$. This is enormous compared to $M/L \approx 2.5$ for the solar neighborhood, or $M/L \approx 5$ for the luminous parts of typical spiral galaxies. (Here M is the dynamically measured mass, expressed in units of solar mass $M_\odot = 2.0 \times 10^{33}$g, and L is the total luminosity in units of solar luminosity $L_\odot$. The Hubble parameter $H_0 = 100h$ km sec^{-1} Mpc^{-1} enters because it is the cosmic yardstick, relating distance d to radial velocity, measured via redshift: $d = v/H_0$. H_0 is not yet known precisely because of the uncertainty of distance measurements of distant galaxies. The present consensus is that $\frac{1}{2} \leq h \leq 1$, and in recent years all determinations of h lie in that interval.)

Not only is there a great deal of dark matter in rich clusters, it has become clear in the past decade that there is much dark matter associated with individual galaxies as well. The strongest evidence for this comes from studies of the rotational velocity v versus radius r in disks of spiral galaxies [1,2]. The velocity is determined by the Doppler shift either using radio telescopes (for the 21 cm line of hydrogen) or optically. Both the neutral hydrogen and the stars are found to have the same rotational velocity at the same radius. Rotational velocities have been measured out to ~ 45 kpc (1pc = 3.26 light years) for very large spiral galaxies. If the gravitating mass were concentrated toward the center of the galaxy, as the stars are, then Kepler's Laws imply that the velocity at large distances would fall as $v \propto r^{-\frac{1}{2}}$. This is not seen. Instead, v is constant or increases slightly at large r, which implies that the total mass interior to r is increasing linearly with r, and therefore that $\rho \propto r^{-2}$. See Fig. 1.

How massive are the galaxies? How big are their massive non-luminous halos? Recent studies of the satellites of our galaxy give interesting <u>lower limits</u>. From globular cluster data it is deduced that the massive halo must extend at least to ~ 44 kpc, with a corresponding $M/L = 59 \pm 17$ [3]. The orbital dynamics of the Magellanic clouds suggest [4] that the halo extends at least to ~ 70 kpc; this model will soon be tested by proper motion measurements.

Similar values of M/L are obtained from studies of the dynamics of binaries, groups, and small clusters of galaxies [1,5], although with great uncertainty in the case of binaries because of the sensitivity of the calculated mass to the ellipticity of the orbits [6].

While there is a clear trend for M/L to increase with M [1,5], it does not follow that the more physically relevant quantity $M/M_{\ell um}$

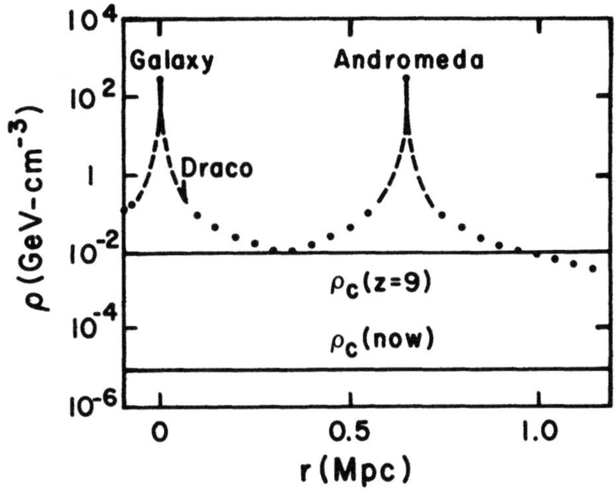

Figure 1. Schematic plot of the density within our Local Group. Luminous mass density is represented by a solid line, observed dark matter by the dashed line, and the dotted line represents an r^{-2} extrapolation of the DM density.

behaves similarly [7,8]. Here $M_{\ell um}$ is the mass deduced from observations of all available electromagnetic radiation, in particular x-rays as well as visible light. For small groups of galaxies, $M/L = 40^{+50}_{-10}$, $M_{\ell um}/L = 2.9 \pm 1.0$, giving $M/M_{\ell um} \simeq 14$. For rich clusters such as Coma, $M_{\ell um}$ (stars)$/L = 6 \pm 1$,[1] $M/L = 325 \pm 50$, so $M/M_{\ell um}$ (stars) $\simeq 54$. But in rich clusters there is considerably more mass in ionized x-ray emitting gas [7,10] than in stars, $M_{\ell um}$ (gas) $\sim 3 M_{\ell um}$ (stars). Thus $M/M_{\ell um}$ is quite comparable for a typical spiral galaxy (including its heavy halo) and a rich cluster, despite the cluster's much larger M/L. (The numbers quoted are from [7] with $h = \frac{1}{2}$; similar conclusions obtain with $h = 1$.)

2.2 Cosmological Density Ω

How does the total mass in galaxies compare with that needed to close the universe? Assuming zero cosmological constant, the universe is closed if $\rho > \rho_c$, where the critical density

$$\rho_c = \frac{3H_0^2}{8\pi G} = 1.9 \times 10^{-29}\, h^2\, \mathrm{g\, cm^{-3}} = 11\, h^2\, \mathrm{keV\, cm^{-3}}. \tag{1}$$

It is convenient to express density in units of ρ_c:

$$\Omega \equiv \rho/\rho_c. \tag{2}$$

Cosmological models in which the universe passes through a very early de Sitter "inflationary" stage predict $\Omega = 1$, the Einstein-de Sitter case.

Bright galaxies ($L^* \simeq 1.0 \times 10^{10}\, h^{-2}\, L_\odot$) have a space number density $n_g \sim 0.02\, h^3\, \mathrm{Mpc^{-3}}$ [11], so taking a typical galaxy rotation velocity to be $v \approx 200\, \mathrm{km\, s^{-1}}$ at the optical radius $r \approx 15\, h^{-1}$ kpc gives [12]

$$\Omega_{\ell um} \approx \frac{n_g v^2 r}{G \rho_c} \sim 10^{-2}. \tag{3}$$

Including massive halos with $M/M_{\ell um} \approx 14$ increases this by an order of magnitude. But we really do not know how far the $\rho \propto r^{-2}$ halos extend. If we extrapolate each galaxy's halo halfway to the next galaxy, then $\Omega \approx 1$ [12]. Equivalently, the luminous mass per bright galaxy is $\sim 10^{11}\, M_\odot$, the mass observed dynamically is $\sim 10^{12}\, M_\odot$, and the mass needed to close the universe is $\sim 10^{13}\, M_\odot$. (More precisely, the characteristic mass needed per bright galaxy is $M^* = 1.5 \times 10^{13}\, \Omega h^{-1} M_\odot$.)

The result of a careful effort [5] to weigh virialized clusters yielded $\Omega \approx 0.07$, and $\Omega \approx 0.15$ if unweighable (relatively isolated) galaxies have the same average mass as those in weighed clusters. It must be borne in mind that this method is sensitive only to mass clustered like luminous matter, and only on scales up to ~ 2 Mpc; thus it can only be used safely to deduce a lower limit: $\Omega \gtrsim 0.15$.

A second method for estimating Ω is based on the peculiar velocity[2] v_V toward the Virgo Cluster of the Local Group (LG) of galaxies (of which our galaxy and M31, the great galaxy in Andromeda, are the prominent members) [13]. This method may represent the best near-term hope of measuring the component of the mass that might be clustered only on scales ~ $10\,h^{-1}$ Mpc. The basic assumption is that v_V(LG) arises from the gravitational acceleration due to the mass concentrated in the Local Supercluster (LS), the flattened or elongated structure of several thousand galaxies surrounding the Virgo cluster. As a result of the agreement between v_V(LG) measured with respect to an ensemble of moderately distant [~ 50 Mpc] galaxies [14] and the value measured from the dipole anisotropy in the microwave background radiation, we can now have some confidence in the result: 400 ± 60 km s^{-1} [13]. A simplified model neglecting flattening and assuming that the mass and galaxy number density enhancements represented by the LS are roughly the same ($\delta M/M \approx \delta N/N \approx 2$) then gives $\Omega = 0.35 \pm 0.15$. Unfortunately, the uncertainties in this result are large. For example, Ω could be larger if the mass density is less concentrated than the galaxy density on SC scales, or if flattening and the effects of possible underdensities outside the LS are accounted for [15]. The present data are certainly not inconsistent with $\Omega = 1$.

A way of determining Ω on very large scales is to measure the deceleration parameter $q_0 = -\ddot{a}\,a\,\dot{a}^{-2}$, where $a = (1+z)^{-1}$ is the scale factor, and the redshift $z = (\lambda - \lambda_0)/\lambda_0$. If the cosmological constant vanishes, as we assume, then $2q_0 = \Omega$. Although q_0 can in principle be measured from the deviation of very distant objects from Hubble's law, the difficulty is in determining their distance (e.g., from their intrinsic luminosity). The traditional approach, based on the assumed constant luminosity of the brightest galaxy in each rich cluster, is frought with uncertainties - in particular, the effects of evolution (time variation in absolute luminosity) and sampling (near and distant samples may not be comparable). Nevertheless, a recent review [16] obtains an upper limit $q_0 \lesssim 1$ ($\Omega \lesssim 2$) from radio galaxies observed in the near-IR having redshifts z in the range ~ 0.5 to ~ 1. Alternative approaches are unfortunately also problematic. Since quasars have by far the highest observed redshifts ($z \lesssim 3.8$), they would provide an ideal sample for determining q_0 if some feature of their spectra could be used to determine their intrinsic luminosity. A recent study, exploiting an observed correlation between the strength of the CIV (triply-ionized carbon) 1550 Å emission line and the luminosity of the underlying continuum in flat radio spectrum quasars, finds $q_0 \approx 1^{+1.5}_{-0.7}$ ($\Omega \approx 2^{+3}_{-1.3}$) assuming no evolution [17]. This result may suffer from possible selection and evolutionary effects [18], and it is based entirely on a correlation whose origin is not well understood.

A final constraint on Ω follows from the age of the universe (see, e.g., [19]): in the standard Friedmann cosmology, a larger value of Ω implies a younger universe. The lower limit on the age of globular clusters from standard stellar evolution theory, $t_0 > 13$ Gy, implies that $\Omega < 2$ (for $h \geq ½$).

To summarize, the accurate measurement of the cosmological density parameter is difficult, but it probably lies in the range $0.1 \lesssim \Omega \lesssim 2$. Large Ω, such as the Einstein-de Sitter value $\Omega = 1$, are excluded unless mass density is distributed considerably more broadly than luminosity density.

2.3 Superclusters and Voids

As a result of the recent dramatic increase in the number of galaxies for which accurate redshifts have been measured, we are at last beginning to see the universe in three dimensions. The most striking conclusion that has emerged is the existence of superclusters and voids [20].

Superclusters are strings of rich clusters joined and surrounded by thousands of galaxies in small groups and clusters. Unlike clusters - which are typically rather well defined and almost certainly gravitationally bound structures representing large density enhancements over the background - superclusters are loose, diffuse, and probably unbound except in their denser regions. Typical sizes are a few tens of Mpc, but some structures may extend over distances ~ 10^2 Mpc.

Voids are large regions devoid, or almost devoid, of visible galaxies. Again, linear dimensions range upward from tens of Mpc. An important question, as yet unsettled, is whether the voids represent an absence of matter or merely of luminous matter.

Other observations that are likely to be critical in determining the origin of this large scale structure include the possible correlations of galaxy and cluster properties with position. Strong hints of such correlations have appeared in the data, the most striking of which is Binggeli's [21] result that there is a strong correlation between the orientation of a cluster's major axis and the direction to the nearest neighboring cluster, for clusters separated by less than ~ 30 Mpc. Already the observations pose tantalyzing problems and challenges to theory.

The thesis explored in this paper is that both large and small scale structure reflect the properties of the gravitationally dominant dark matter. We first explain why the dark matter is probably not just some nonluminous form of ordinary matter, and then consider other possibilities.

3. The DM is Probably Not Baryonic

There are three arguments that the DM is not "baryonic", that is, that it is not made of protons, neutrons, and electrons as all ordinary matter is. As Richard Feynman has said in other contexts, one argument would suffice if it were convincing. All three arguments have loopholes. The arguments that DM $\neq$ baryons are as follows:

3.1 Excluding Baryonic Models [22]

The dark matter in galaxy halos cannot be gas (it would have to be hot to be pressure supported, and would radiate); nor frozen hydrogen "snowballs" (they would sublimate); nor dust grains (their "metals", elements of atomic number ≥ 3, would have prevented formation of the observed low-metallicity Population II stars); nor "jupiters" (how to make so many hydrogen balls too small to initiate nuclear burning without making a few large enough to do so?); nor collapsed stars (where is the matter they must have ejected in collapsing?).

The weakest argument is probably that which attempts to exclude "jupiters": arguments of the form "how could it be that way?" are rarely entirely convincing.

3.2 Deuterium Abundance [23]

In the early universe, almost all the neutrons which "freeze out" are synthesized into ^{4}He. The fraction remaining in D and ^{3}He is a rapidly decreasing function of η, the ratio of baryon to photon number densities. The presently observed D abundance (compared, by number, to H) is $(1 - 4) \times 10^{-5}$. Since D is readily consumed but not produced in stars, 10^{-5} is also a lower limit on the primordial D abundance. This, in turn, implies an <u>upper</u> limit $\eta \leq 10^{-9}$ or

$$\Omega_b \leq 0.035\, h^{-2}\, (T_o/2.7)^3, \tag{4}$$

where Ω_b is the ratio of the present average baryon density ρ_b to the critical density given by Eq. (1).

As discussed in Section 2.2, the observational limits on Ω are $0.1 \leq \Omega \leq 2$. Therefore, in a baryon dominated universe ($\Omega \simeq \Omega_b$), the deuterium bound, Eq. (4), is consistent only with the lower limit on Ω, and then only for the Hubble parameter at its lower limit. An Einstein-de Sitter or inflationary ($\Omega = 1$) or closed ($\Omega > 1$) universe cannot be baryonic.

3.3 Galaxy Formation

In the standard cosmological model, which we will adopt, large scale structure forms when perturbations $\delta \equiv \delta\rho/\rho$ grow to $\delta \gtrsim 1$, after which they cease to expand with the Hubble flow. Let us further assume that perturbations in matter and radiation density are correlated (these are called adiabatic perturbations, since the entropy per baryon is constant; these are the sort of perturbations predicted in grand unified models). Then photon diffusion ("Silk damping") erases perturbations of baryonic mass smaller than [24]

$$M_{Silk,b} \approx 3 \times 10^{13}\, \Omega_b^{-1/2}\, \Omega^{-3/4}\, h^{-5/2}\, M_\odot. \tag{5}$$

Thus galaxies ($M_b \lesssim 10^{11-12} M_\odot$) can form only after the "pancake" collapse of larger-scale perturbations [25]. Perturbations δ in a matter dominated universe grow linearly with the scale factor

$$\delta \propto a = (1+z)^{-1} = T_o/T \tag{6}$$

where $z = (\lambda_o - \lambda)/\lambda$ is the redshift, T is the radiation temperature, and the subscript o denotes the present epoch. In a baryonic universe, δ grows only between the epoch of hydrogen recombination, $z_r \simeq 1300$, and $z \simeq \Omega^{-1}$. It follows that at recombination $\delta T/T \approx \delta\rho/3\rho \gtrsim 3 \times 10^{-3}$ for $M \gtrsim M_{Silk}$, which corresponds to fluctuations on observable angular scales $\theta > 4'$ today. Such temperature fluctuations are an order of magnitude larger than present observational upper limits [26].

The main loophole in this argument is the assumption of adiabatic perturbations. It is true that the orthogonal mode, perturbations in baryonic density which are uncorrelated with radiation (called isothermal perturbations), do not arise naturally in currently fashionable particle physics theories where baryon number is generated in the decay of massive grand unified theory (GUT) bosons, since in such theories $\eta \equiv n_b/n_\gamma$ is determined by the underlying particle physics and should not vary from point to point in space. But galaxies originating as isothermal perturbations do avoid both Silk damping and contradiction with present $\delta T/T$ limits.

A second loophole is the possibility that matter was reionized at some $z \gtrsim 10$, by hypothetical very early sources of uv photons. Then the fluctuations in $\delta T/T$ at recombination associated with baryonic proto-pancakes could be washed out by rescattering.

Despite the loopholes in each argument, we find the three arguments together to be rather persuasive, even if not entirely compelling. If it is indeed true that the bulk of the mass in the universe is not baryonic, that is yet another blow to anthropocentricity: not only is man not the center of the universe physically (Copernicus) or biologically (Darwin), we and all that we see are not even made of the predominant variety of matter in the universe!

4. Three Types of DM Particles: Hot, Warm & Cold

If the dark matter is not baryonic, what _is_ it? We will consider here the physical and astrophysical implications of three classes of elementary particle DM candidates, which we will call hot, warm, and cold. (We are grateful to Dick Bond for proposing this apt terminology.)

Hot DM refers to particles, such as neutrinos, which were still in thermal equilibrium after the most recent phase transition in the hot early universe, the QCD deconfinement transition, which presumably took place at $T_{QCD} \sim 10^2$ MeV. Hot DM particles have a cosmological number density roughly comparable to that of the microwave background photons, which implies an upper bound to their mass of a few tens of eV. As we shall discuss shortly, this implies that free streaming destroys any perturbations smaller than supercluster size, $\sim 10^{15} M_\odot$.

Warm DM particles interact much more weakly than neutrinos. They decouple (i.e., their mean free path first exceeds the horizon size) at $T > T_{QCD}$, and consequently their number density is roughly an order of magnitude lower, and their mass an order of magnitude higher, than hot DM particles. Perturbations as small as large galaxy halos, $\sim 10^{12} M_\odot$, could then survive free streaming. It was initially suggested that, in theories of local supersymmetry broken at $\sim 10^6$ GeV, gravitinos could be DM of the warm variety [27]. Other candidates are also possible, as we will discuss.

Cold DM consists of particles for which free streaming is of no cosmological importance. Two different sorts have been proposed, a cold Bose condensate such as axions, and heavy remnants of annililation or decay such as heavy stable neutrinos. As we will see, a universe dominated by cold DM looks remarkably like the one astronomers actually observe.

It is of course also possible that the dark matter is NOTA - none of the above! A perennial candidate, primordial black holes, is becoming increasingly implausible [28-30]. Another possibility which, for simplicity, we will not discuss, is that the dark matter is a mixture, for example "jupiters" in galaxy halos plus neutrinos on large scales [23].

5. Galaxy Formation with Hot DM

The standard hot DM candidate is massive neutrinos [23-25], although other, more exotic, theoretical possibilities have been suggested, such as a "majoron" of nonzero mass which is lighter than the lightest neutrino species, and into which all neutrinos decay [31]. For definiteness, we will discuss neutrinos.

5.1 Mass Constraints

Left-handed neutrinos of mass $\lesssim 1$ MeV remain in thermal equilibrium until the temperature drops to $T_{\nu d}$, at which point their mean free path first exceeds the horizon size and they essentially cease interacting thereafter, except gravitationally [32]. Their mean free path is, in natural units ($\hbar = c = 1$), $\lambda_\nu \sim [\sigma_\nu n_{e\pm}]^{-1} \sim [(G_{wk}^2 T^2)(T^3)]^{-1}$, and the horizon size is $\lambda_h \sim (G\rho)^{-\frac{1}{2}} \sim M_{P\ell} T^{-2}$, where the Planck mass $M_{P\ell} \equiv G^{-\frac{1}{2}} = 1.22 \times 10^{19}$ GeV $= 2.18 \times 10^{-5}$ g. Thus $\lambda_h/\lambda_\nu \sim (T/T_{\nu d})^3$, with the neutrino decoupling temperature

$$T_{\nu d} \sim M_{P\ell}^{-1/3} G_{wk}^{-2/3} \sim 1 \text{ MeV} \tag{7}$$

After T drops below 1 MeV, e^+e^- annihilation ceases to be balanced by pair creation, and the entropy of the e^+e^- pairs heats the photons. Above 1 MeV, the number density $n_{\nu i}$ of each left-handed neutrino species (counting both ν_i and $\bar{\nu}_i$) is equal to that of the photons, n_γ, times the factor 3/4 from Fermi vs. Bose statistics; but e^+e^- annihilation increases the photon number density relative to that of the neutrinos by a factor of 11/4.[3] Thus today, for each species,

$$n_\nu^o = \frac{3}{4} \cdot \frac{4}{11} n_\gamma^o = 109 \left(\frac{T_\gamma}{2.7K} \right)^3 cm^{-3}. \tag{8}$$

Since the present cosmological density is

$$\rho = \Omega\rho_c = 11 \ \Omega h^2 \, keV \, cm^{-3}, \tag{9}$$

it follows that

$$\sum_i m_{\nu i} < \rho/n_\nu^o \le 100 \, \Omega h^2 \, eV, \tag{10}$$

where the sum runs over all neutrino species with $m_{\nu i} \lesssim 1$ MeV.[4] Observational data imply that Ωh^2 is less than unity [23]. Thus if one species of neutrino is substantially more massive than the others and dominates the cosmological mass density, as for definiteness we will assume for the rest of this section, then a reasonable estimate for its mass is $m_\nu \sim 30$ eV.

At present there is apparently no reliable experimental evidence for nonzero neutrino mass. Although one group reported [35] that $14 \text{ eV} < m_{\nu_e} < 40$ eV from tritium β end point data, according to Boehm [36] their data are consistent with $m_{\nu_e} = 0$ with the resolution corrections pointed out by Simpson. The so far unsuccessful attempts to detect neutrino oscillations also give only upper limits on neutrino masses times mixing parameters [36].

5.2 Free Streaming

The most salient feature of hot DM is the erasure of small fluctuations by free streaming. It is easy to see that the minimum mass of a surviving fluctuation is of order $M_{P\ell}^3/m_\nu^2$ [37,24].

Let us suppose that some process in the very early universe - for example, thermal fluctuations subsequently vastly inflated, in the inflationary scenario [38] - gave rise to adiabatic fluctuations on all scales. Neutrinos of nonzero mass m_ν stream relativistically from decoupling until the temperature drops to m_ν, during which time they will traverse a distance $d_\nu \approx \lambda_h(T = m_\nu) \sim M_{P\ell}\, m_\nu^{-2}$. In order to survive this free streaming, a neutrino fluctuation must be larger in linear dimension than d_ν. Correspondingly, the minimum mass in neutrinos of a surviving fluctuation is $M_{J,\nu} \sim d_\nu^3 m_\nu n_\nu(T = m_\nu) \sim d_\nu^3 m_\nu^4 \sim M_{P\ell}^3\, m_\nu^{-2}$. By analogy with Jeans' calculation of the minimum mass of an ordinary fluid perturbation for which gravity can overcome pressure, this is referred to as the (free-streaming) Jeans mass. (See Fig. 2). A more careful calculation [24,39] gives

$$d_\nu \approx 41\,(m_\nu/30\text{ eV})^{-1}(1+z)^{-1}\,\text{Mpc}, \tag{11}$$

and

$$M_{J,\nu} \approx 1.77\, M_{P\ell}^3\, m_\nu^{-2} = 3.2 \times 10^{15}\,(m_\nu/30\text{ eV})^{-2}\, M_\odot, \tag{12}$$

which is the mass scale of superclusters. Objects of this size are the first to form in a ν-dominated universe, and smaller scale structures such as galaxies can form only after the initial collapse of supercluster-size fluctuations.

5.3 Growth of Fluctuations

The absence of small angle $\delta T/T$ fluctuations is compatible with this picture. When a fluctuation of total mass $\sim 10^{15}\, M_\odot$ enters the horizon at $z \sim 10^4$, the density contrast of the radiation plus baryons δ_{RB} ceases growing and instead starts oscillating as an acoustic wave, while that of the neutrinos δ_ν continues to grow linearly with the scale factor $a = (1+z)^{-1}$. Thus by recombination, at $z_r \approx 1300$, $\delta_{RB}/\delta_\nu < 10^{-1}$, with possible additional suppression of δ_{RB} by Silk damping (depending on the parameters in Eq. (5)). This picture, as well as the warm and cold DM schemes, predicts small angle fluctuations in the microwave background radiation just slightly below current observational upper limits [26].

In numerical simulations of dissipationless gravitational clustering starting with a fluctuation spectrum appropriately peaked at $\lambda \approx d_\nu$, the regions of high density form a network of filaments, with the highest densities occurring at the intersections and with voids in between [25,40-42]. The similarity of these features to those seen in observations [43,20] is certainly evidence in favor of this model.

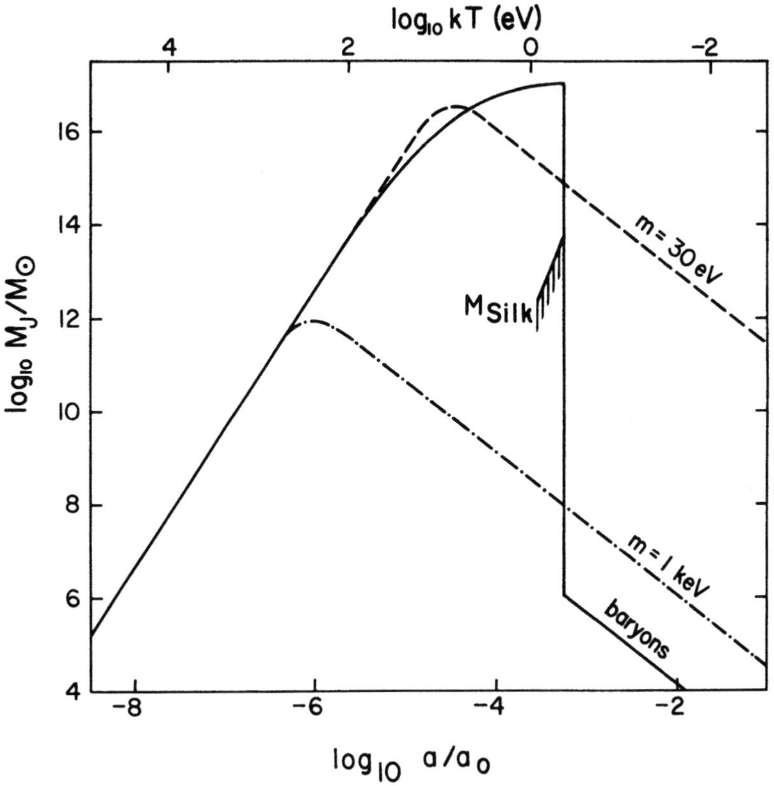

Figure 2. The Jeans mass versus scale factor for a baryon, hot DM(m = 30 ev), and warm DM (m = 1 keV) dominated universe with $\Omega h^2 = 1$. Hot and warm DM perturbations with $M < M_J$ at any time are dissipated by free-streaming.

5.4 Potential Problems with ν DM

A number of potential problems with the neutrino dominated universe have emerged in recent studies, however. (1) From studies both of nonlinear [42] clustering ($\lambda \lesssim 10$ Mpc) and of streaming velocities [44] in the linear regime ($\lambda > 10$ Mpc), it follows that supercluster collapse must have occurred recently: $z_{sc} \lesssim 0.5$ is indicated [44], and in any case $z_{sc} < 2$ [42]. But then, if QSOs are associated with galaxies, their abundance at $z > 2$ is inconsistent with the "top-down" neutrino dominated scheme in which superclusters form first: $z_{sc} > z_{galaxies}$. (2) Numerical simulations of the nonlinear "pancake" collapse taking into account dissipation of the baryonic matter show that at least 85% of the baryons are so heated by the associated shock that they remain ionized and unable to condense, attract neutrino halos, and eventually form galaxies [45]. (3) The neutrino picture predicts [46] that there should be a factor of ~5 increase in M_{tot}/M_{lum} between large galaxies ($M_{tot} \sim 10^{12} M_\odot$) and large clusters ($M_{tot} \gtrsim 10^{14} M_\odot$), since the larger clusters, with their higher escape velocities, are able to trap a considerably larger fraction of the neutrinos. As we discussed in Sec. 2.1, although there is evidence that M/L increases with M, the ratio of total to luminous mass $M/M_{lum} \approx 14$ for galaxies with large halos and for rich clusters [7,8]. (4) Both theoretical arguments [47] and data on Draco [48,49] imply that dark matter dominates the gravitational potential of dwarf spheroidal galaxies. The phase-space constraint [50] then sets a lower limit [49] $m_\nu > 500$ eV, which is completely incompatible with the cosmological constraint Eq. (10). (Note that for neutrinos as the DM in spiral galaxies, the phase space constraint implies $m_\nu > 30$ eV.)

These problems, while serious, may not be fatal for the hypothesis that neutrinos are the dark matter. It is possible that galaxy density does not closely correlate with the density of dark matter, for example because the first generation of luminous objects heats nearby matter, thereby increasing the baryon Jeans mass and suppressing galaxy formation. This could complicate the comparison of nonlinear simulations [42] with the data. Also, if dark matter halos of large clusters are much larger in extent than those of individual galaxies and small groups, then virial estimates would underestimate mass on large scales and the data could be consistent with M/M_{lum} increasing with M_{lum}. But it is hard to avoid the constraint on z_{sc} from streaming velocities in the linear regime [44] except by assuming that the local group velocity is abnormally low. And the only explanation for the high M/L of dwarf spheroidal galaxies in a neutrino-dominated universe is the rather ad hoc assumption that the dark matter in such objects is baryons rather than neutrinos. Of course, the evidence for massive halos around dwarf spheroidals is not yet solid.

6. Galaxy Formation with Warm DM

Suppose the dark matter consists of an elementary particle species X that interacts much more weakly than neutrinos. The X's decouple thermally at a temperature $T_{Xd} \gg T_{\nu d}$ and their number density is not thereafter increased by particle annihilation at temperatures below T_{Xd}. With the standard assumption of conservation of entropy per comoving volume, the X number density today n_X^o and mass m_X can be calculated in terms of the effective number of helicity states of interacting bosons (B) and fermions (F), $g = g_B + (7/8)g_F$, evaluated at T_{Xd} [51]. These are plotted in Fig. 3, assuming the "standard model" of particle physics. The simplest grand unified theories predict $g(T) \approx 100$ for T between 10^2 GeV and $T_{GUT} \sim 10^{14}$ GeV, with possibly a factor of two increase in g beginning near 10^2 GeV due to N = 1 supersymmetry partner particles. Then for T_{Xd} in the enormous range from ~ 1 GeV to $\sim T_{GUT}$, $n_X^o \sim 5 g_X \,\text{cm}^{-3}$ and correspondingly $m_X \simeq 2\Omega h^2 g_X^{-1}$ keV [52], where g_X is the number of X helicity states. Because of free streaming (see Fig. 2), such "warm" DM particles of mass $m_X \sim 1$ keV will cluster on a scale $\sim M_{P\ell}^3 m_X^{-2} \sim 10^{12} M_\odot$, the scale of large galaxies such as our own [27,53,54].

6.1 Candidates for Warm DM

What might be the identity of the warm DM particles X? It was initially [27] suggested that they might be the $\pm\frac{1}{2}$ helicity states of the gravitino $\tilde{G}$, the spin 3/2 supersymmetric partner of the graviton G. The gravitino mass is related to the scale of supersymmetry breaking by $m_{\tilde{G}} = (4\pi/3)^{\frac{1}{2}} m_{SUSY}{}^2 m_{P\ell}^{-1}$, so $m_{\tilde{G}} \sim 1$ keV corresponds to $m_{SUSY} \sim 10^6$ GeV. This now appears to be phenomenologically dubious, and supersymmetry models with $m_{SUSY} \sim 10^{11}$ GeV and $m_{\tilde{G}} \sim 10^2$ GeV are currently popular [55]. In such models, the photino $\tilde{\gamma}$, the spin $\frac{1}{2}$ supersymmetric partner of the photon, is probably the lightest R-odd particle, and hence stable. But in supersymmetric GUT models $m_{\tilde{\gamma}} \sim 10\, m_{\tilde{g}}$, and there is a phenomenological lower bound on the mass of the gluino $m_{\tilde{g}} > 2$ GeV [56]. The requirement that the photinos almost all annihilate, so that they do not contribute

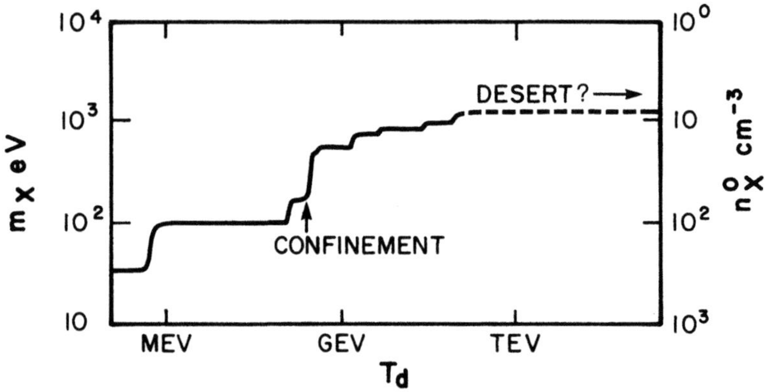

Figure 3. The mass m_X and present number density of warm dark matter particles X assuming the standard particle physics model with no entropy generation. The mass scales as $h^2\Omega$.

too much mass density, implies that $m_{\tilde{\gamma}} \gtrsim 2\,\mathrm{GeV}$ [34,57], and they become a candidate for cold rather than warm dark matter.

A hypothetical right-handed neutrino ν_R could be the warm DM particle [58], since if right-handed weak interactions exist they must be much weaker than the ordinary left-handed weak interactions, so $T_{\nu_R d} \gg T_{\nu d}$ as required. But particle physics provides no good reason why any ν_R should be light.

Thus there is at present no obvious warm DM candidate elementary particle, in contrast to the hot and cold DM cases. But our ignorance about the physics above the ordinary weak interaction scale hardly allows us to preclude the existence of very weakly interacting light particles, so we will consider the warm DM case, mindful of Hamlet's prophetic admonition

There are more things in heaven and earth, Horatio,
Than are dreamt of in your philosophy.

6.2 Fluctuation Spectrum

The spectrum of fluctuations δ_ν at late times in the hot DM model is controlled mainly by free streaming; $\delta_\nu(M)$ is peaked at $\sim M_{J,\nu}$, Eq. (12), for any reasonable primordial fluctuation spectrum. This is not the case for warm or cold DM.

The primordial fluctuation spectrum can be characterized by the magnitude of fluctuations just as they enter the horizon. It is expected that no mass scale is singled out, so the spectrum is just a power law

$$\delta_{DM,H} = \left(\frac{\delta\rho_{DM}}{\rho_{DM}} \right)_H = \kappa \left(\frac{M_{DM}}{M_o} \right)^{-\alpha} \tag{13}$$

Furthermore, to avoid too much power on large or small mass scales requires $\alpha \approx 0$ [59], and to form galaxies and large scale structure by the present epoch without violating the upper limits on both small [26] and large [60] scale (quadrupole) angular variations in the microwave background radiation requires $\kappa \sim 10^{-4}$. Eq. (13) corresponds to $|\delta_k|^2 \propto k^n$ with $n = 6\alpha + 1$. The case $\alpha = 0$ $(n = 1)$ is commonly referred to as the Zeldovich spectrum.

Inflationary models predict adiabatic fluctuations with the Zeldovich spectrum [38]. In the simplest models κ is several orders of magnitude too large, but it is hoped that this will be remedied in more realistic - possibly supersymmetric - models [61].

The important difference between the fluctuation spectra δ_{DM} at late times in the hot and warm DM cases is that $\delta_{DM,warm}$ has power over an increased range of masses, roughly from 10^{11} to $10^{15} M_\odot$. As for the hot case, the lower limit, $M_X \sim M_{P\ell}^{3} m_X^{-2}$, arises from the damping of smaller-scale fluctuations by free streaming. In the hot case, the DM particles become nonrelativistic at essentially the same time as they become gravitationally dominant, because their number density is nearly the same as that of the photons. But in the warm case, the X particles become nonrelativistic and thus essentially stop free streaming at $T \sim m_X$, well before they begin to dominate gravitationally at $T_{eq} \approx 6\Omega h^2$ eV. The subscript "eq" refers to the epoch when the energy density of massless particles equals that of massive ones:

$$z_{eq} = \frac{\Omega \rho_c c}{4\sigma T_0^4 (1+\gamma)} = 2.47 \times 10^4 \, \Omega h^2 \left(\frac{1.681}{1+\gamma}\right) \theta^{-4}. \tag{14}$$

We assume here that there are n_ν species of very light or massless neutrinos, and $\gamma \equiv \rho_\nu^0/\rho_\gamma^0 = (7/8)(4/11)^{4/3} n_\nu$ $(= 0.681$ for $n_\nu = 3)$, $\theta \equiv T_0/2.7\mathrm{K}$, and σ is the Stefan-Boltzmann constant. During the interval between $T \sim m_X$ and $T \sim T_{eq}$, growth of δ_{DM} is inhibited by the "stagspansion"[5] phenomenon (also known as the Meszaros [62] effect), which we will discuss in detail in the section on cold DM. Thus the spectrum δ_{DM} is relatively flat between M_X and

$$M_{eq} = \frac{4\pi}{3} \left(\frac{c t_{eq}}{1+z_{eq}}\right)^3 \rho_c \Omega = 2.2 \times 10^{15} \, (\Omega h^2)^{-2} M_\odot \,. \tag{15}$$

Fluctuations with masses larger than M_{eq} enter the horizon at $z < z_{eq}$, and thereafter δ_{DM} grows linearly with $a = (1+z)^{-1}$ until nonlinear gravitational effects become important when $\delta_{DM} \sim 1$. Since for $\alpha = 0$ all fluctuations enter the horizon with the same magnitude, and those with larger M enter the horizon later in the matter-dominated era and subsequently have less time to grow, the fluctuation spectrum falls with M for $M > M_{eq}$: $\delta_{DM} \propto M^{-2/3}$. For a power-law primordial spectrum of arbitrary index,

$$\delta_{DM} \propto M^{-\alpha - 2/3} = M^{-(n+3)/6}, \quad M > M_{eq}. \tag{16}$$

This is true for hot, warm, or cold DM. In each case, after recombination at $z_r \approx 1300$ the baryons "fall in" to the dominating DM fluctuations on all scales larger than the baryon Jeans mass, and by $z \approx 100$,

$\delta_b \approx \delta_{DM}$ [63].

In the simplest approximation, neglecting all growth during the "stagspansion" era, the fluctuation spectrum for $M_X < M < M_{eq}$ is just $\delta_{DM} \propto M^{-\alpha} = M^{-(n-1)/6} = M^{-(n_{eff}+3)/6}$, where $n_{eff} = n-4$; i.e., the spectrum is flattened by a factor of $M^{2/3}$ compared to the primordial spectrum. The small amount of growth that does occur during the "stagspansion" era slightly increases the fluctuation strength on smaller mass scales: $n_{eff} \approx n-3$. Detailed calculations of these spectra are now available [39,53].

6.3 Which Formed First, Galaxies or Superclusters?

For $\alpha \gtrsim 0$, $\delta_X(M)$ has a fairly broad peak at $M \sim M_X$. Consequently, objects of this mass - galaxies and small groups - are the first to form, and larger-scale structures - clusters and superclusters - form later as $\delta_X(M)$ grows toward unity on successively larger mass scales. For a particular primordial spectral index α, one can follow Peebles [64,65] and use the fact that the galaxy autocovariance function $\xi(R) \approx 1$ for $R = 5h^{-1}$, together with the (uncertain) assumption that the DM is distributed on such scales roughly like the galaxies, to estimate when the galaxies form in this scenario. For $\alpha = 0$, $z_{galaxies} \sim 4$, which is consistent with the observed existence of quasars at such redshifts. But superclusters do not begin to collapse until $z < 2$, so one would not expect to find similar Lyman α absorption line redshifts for quasars separated by $\sim 1h^{-1}$ Mpc perpendicular to the line of sight [66]. Indeed, Sargent et al. [67] found no such correlations. This is additional evidence against hot DM.

6.4 Potential Problems with Warm DM

The warm DM hypothesis is probably consistent with the observed features of typical large galaxies, whose formation would probably follow roughly the "core condensation in heavy halos" scenario (68,7,69]. The potentially serious problems with warm DM are on scales both larger and smaller than M_X. On large scales, the question is whether the model can account for the observed network of filamentary superclusters enclosing large voids [43,20]. A productive approach to this question may require sophisticated N-body simulations with $N \sim 10^6$ in order to model the large mass range that is relevant [70]. We will discuss this further in the next section in connection with cold DM, for which the same question arises.

On small scales, the preliminary indications that dwarf spheroidal galaxies have large DM halos [47-49] pose problems nearly as serious for warm as for hot DM. Unlike hot DM, warm DM is (barely) consistent with the phase space constraint [48-50]. But since free streaming of warm DM washes out fluctuations δ_X for $M \lesssim M_X \sim 10^{11} M_\odot$, dwarf galaxies with $M \sim 10^7 M_\odot$ can form in this picture only via fragmentation following the collapse of structures of mass $\sim M_X$, much as ordinary galaxies form from superclusters fragmentation in the hot DM picture. The problem here is that dwarf galaxies, with their small escape velocities $\sim 10\,km\,s^{-1}$, would not be expected to bind more than a small fraction of the X particles, whose typical velocity must be $\sim 10^2 km\,s^{-1}$ (~ rotation velocity of spirals). Thus we expect M/M_{lum} for dwarf galaxies to be much smaller than for large galaxies - but the indications are that they are comparable [47-49]. Understanding dwarf galaxies may well be crucial for unravelling the mystery of the identity of the DM. Fortunately, data on Carina, another dwarf spheroidal companion of the Milky Way, is presently being analyzed [71].

7. Galaxy Formation with Cold DM

Damping of fluctuations by free streaming occurs only on scales too small to be cosmologically relevant for DM which either is not characterized by a thermal spectrum, or is much more massive than 1 keV. We refer to this as cold DM.

7.1 Cold DM Candidates

Quantum chromodynamics (QCD) with quarks of nonzero mass violates CP and T due to instantons. This leads to a neutron electric dipole moment that is many orders of magnitude larger than the experimental upper limit, unless an otherwise undetermined complex phase θ_{QCD} is arbitrarily chosen to be extremely small. Peccei and Quinn [72] have proposed the simplest and probably the most appealing way to avoid this problem, by postulating an otherwise unsuspected symmetry that is spontaneously broken when an associated pseudoscalar field - the axion [73] - gets a nonzero vacuum expectation value $\langle\phi_a\rangle \sim f_a e^{i\theta}$. This occurs when $T \sim f_a$. Later, when the QCD interactions become strong at $T \sim \Lambda_{QCD} \sim 10^2\,MeV$, instanton effects generate a mass for the axion $m_a = m_\pi f_\pi / f_a \approx 10^{-5}\,eV (10^{12}\,GeV/f_a)$. Thereafter, the axion contribution to the energy density is [74] $\rho_a = 3 m_a T^3 f_a^2 (M_{P\ell} \Lambda_{QCD})^{-1}$. The requirement $\rho_a^o < \rho_c \Omega$ implies that $f_a \lesssim 10^{12}\,GeV$, and $m_a \gtrsim 10^{-5}\,eV$.[6] The longevity of helium-burning stars implies [75] that $m_a < 10^{-2}\,eV$, $f_a > 10^9\,GeV$. Thus if the hypothetical axion exists, it is probably important cosmologically, and for $m_a \sim 10^{-5}$ eV gravitationally dominant. (The mass range 10^{9-12} GeV, in which f_a must lie, is also currently popular with particle theorists as the scale of supersymmetry [55] or family symmetry breaking, the later possibility connected with the axion [76].)

Two quite different sorts of cold DM particles are also possible. One is a heavy stable "ino", such as a photino [57] of mass $m_{\tilde{\gamma}} > 2$ GeV as discussed above. By a delicate adjustment of the theoretical parameters controlling the $\tilde{\gamma}$ mass and interactions, the $\tilde{\gamma}$'s can be made to almost all annihilate at high temperatures, leaving behind a small remnant that, because $m_{\tilde{\gamma}}$ is large, can contribute a critical density today [34].

The second possibility may seem even more contrived: a particle, such as a ν_R, that decouples while still relativistic but whose number density relative to the photons is subsequently diluted by entropy generated in a first-order phase transition such as the Weinberg-Salam symmetry breaking [52]. (Recall that the m_X bound in Fig. 3 assumes no generation of entropy.) More than a factor $\sim 10^3$ entropy increase would over dilute $\eta = n_b/n_\gamma$, if we assume η was initially generated by GUT baryosynthesis; correspondingly, $m_X \lesssim 1$ MeV, and $M_X \gtrsim 10^6 M_\odot$.

Actually, it is not clear that we have a good basis to judge the plausibility of any of these DM candidates, since in no case is there a fundamental explanation - or, even better, a prediction - for the ratio $\omega \equiv \rho^o_{DM}/\rho^o_{lum}$, which is thought to lie in the range $10 \lesssim \omega \lesssim 10^2$. Two fundamental questions about the universe which the fruitful marriage of particle physics and cosmology has yet to address are the value of ω and of the cosmological constant Λ. (We have here assumed $\Lambda = 0$, as usual.)

7.2 "Stagspansion"

Peebles [65] has calculated the fluctuation spectrum for cold DM, with results that are well approximated by the expression

$$|\delta_k|^2 = k^n(1 + \alpha k + \beta k^2)^{-2},$$
$$\alpha = 6\,\theta^2\,h^{-2}\,\text{Mpc},\ \beta = 2.65\,\theta^4\,h^{-4}\,\text{Mpc}^2,\ \theta = T_o/2.7\text{K}. \qquad (17)$$

This calculation neglects the massless neutrinos; we find qualitatively similar results with their inclusion [77]. For an adiabatic Zeldovich ($n = 1$) primordial fluctuation spectrum, the spectrum of rms fluctuations in the mass found within a randomly placed sphere,[7] $\delta M/M$, is relatively flat for $M < 10^9 M_\odot$, and steepens to $\delta M/M \propto M^{-2/3}$ ($n = 1$, reflecting the primordial spectrum) for $M >> M_{eq}$. Our results [77] for $\delta M/M$ are shown in Fig. 4, normalized in such a way that $\delta M/M = 1$ at $R = 8 h^{-1}$ Mpc [65].

The flattening of the spectrum for $M < M_{eq}$ is a consequence of "stagspansion", the inhibition of the growth of δ_{DM} for fluctuations which enter the horizon when $z > z_{eq}$, before the era of matter domination. In the conventional formalism [32,64,78] - synchronous gauge, time-orthogonal coordinates - the fastest growing adiabatic fluctuations grow $\propto a^2$ when they are larger than the horizon. When they enter the horizon, however, the radiation and charged particles begin to oscillate as an acoustic wave with constant amplitude (later damped by photon diffusion for $M < M_{SILK}$), and the neutrinos free stream away. As a result, the main source term for the growth of δ_{DM} disappears, and once the fluctuation is well inside the horizon δ_{DM} grows only as [62; 64 pp 56-59]

$$\delta_{DM} \propto 1 + \frac{3a}{2a_{eq}} \tag{18}$$

until matter dominance ($a = a_{eq}$); thereafter, $\delta_{DM} \propto a$. Based on Eq.(18), it has sometimes been erroneously remarked [53,79] (also by the present authors [54], alas) that there is only a factor of 2.5 growth in δ_{DM} during the entire stagspansion era, from horizon crossing until matter dominance. There is actually a considerable amount of growth in δ_{DM} just after the fluctuation enters the horizon, since $d\delta_{DM}/da$ is initially large and since the photon and neutrino source terms for the growth of dark matter fluctuations do not disappear instantaneously. (See reference 77 for details.) This explains how $(\delta M/M)_{DM}$ can have grown by a factor ~ 30 larger at $10^9 M_\odot$ than at M_{eq}, and it also explains how galaxies can form at $z \approx 10$ even though $\delta M/M = 1$ for $M \sim 10^{15} M_\odot$ at the present time.[8]

7.3 Galaxy Formation

When δ reaches unity, nonlinear gravitational effects become important. The fluctuation separates from the Hubble expansion, reaches a maximum radius, and then contracts to about half that radius (for spherically symmetric fluctuations), at which point the rapidly changing gravitational field has converted enough energy from potential to kinetic for the virial relation $\langle PE \rangle = -2\langle KE \rangle$ to be satisfied. (For reviews see [80] and [64].)

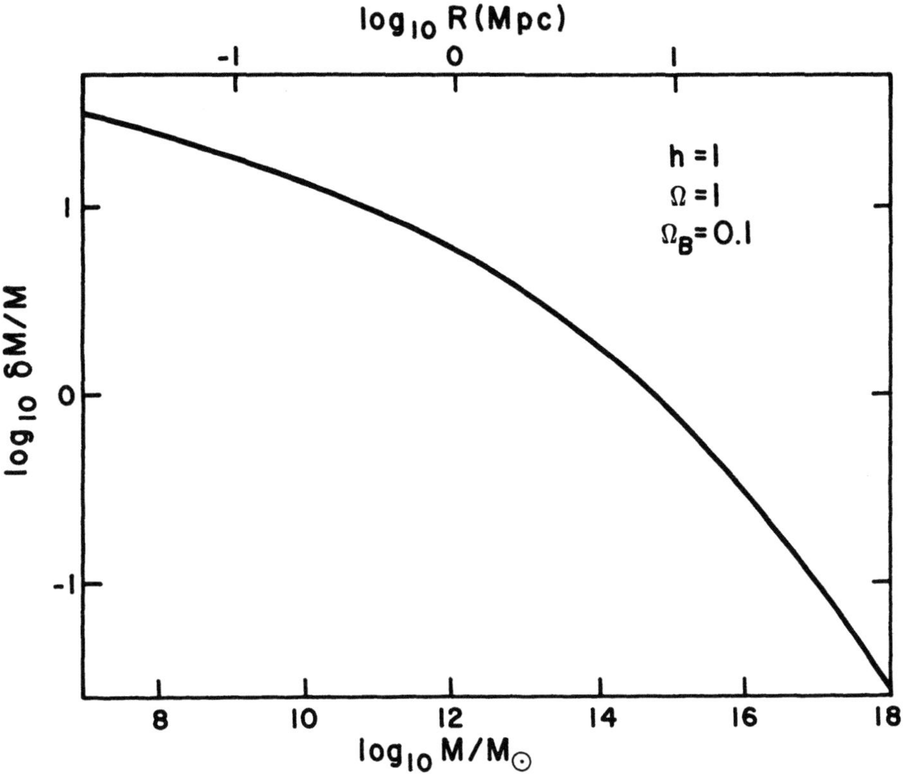

Figure 4. The logarithm of the r.m.s. mass fluctuations ($\log_{10}\delta M/M$) within a randomly placed sphere of radius R in a cold DM universe. The curve is normalized at R = 8 Mpc and assumes an initial Zeldovich (n = 1) fluctuation spectrum.

Although small-mass fluctuations are the first to go nonlinear in the cold DM picture, pressure effects inhibit baryons from falling into such fluctuations if $M < M_{J,b}$. More importantly, even for $M > M_{J,b}$, the baryons are not able to contract further unless they can cool by emitting radiation. Without such mass segregation between baryons and DM, the resulting structures will be disrupted by virialization as fluctuations that contain them go nonlinear [68]. Moreover, successively larger fluctuations will collapse relatively soon after one another if they have masses in the flattest part of the $\delta M/M$ spectrum, i.e., (total) mass $\lesssim 10^9\ M_\odot$.

Gas of primordial composition (about 75% atomic hydrogen and 25% helium, by mass) cannot cool significantly unless it is first heated to $\sim 10^4$K, when it begins to ionize [82]. Assuming a primordial Zeldovich spectrum normalized so that at the present time, $\delta M/M = 1$ at $R = 8\,h^{-1}$ Mpc [65], the smallest protogalaxies for which the gas is sufficiently heated by virialization to radiate rapidly and contract have total mass $M \sim 10^9\,M_\odot$ [82]. One can also deduce an upper bound on galaxy masses by requiring that the cooling time be shorter than the dynamical time [81]; this upper bound is $M \lesssim 10^{12}\,M_\odot$ [82]. These limits are illustrated in Fig. 5 where we have plotted the baryonic density versus temperature for virialized protogalaxies resulting from an initial Zeldovich fluctuation spectrum [82]. Only in protogalaxies for which the cooling time is short compared to the dynamical time can the baryons dissipate and contract. This dissipation leads to higher baryonic densities and somewhat higher temperatures. The collapse of fluctuations having mass $> 10^{13}\,M_\odot$ leads to clusters of galaxies in this picture. In clusters, only the outer parts of member galaxy halos are stripped off; the inner baryonic cores continue to contract, presumably until star formation halts dissipation [7].[9]

7.4 Potential Problems with Cold DM

Dwarf galaxies with heavy DM halos are less of a problem in the cold than in the hot or warm DM pictures. There is certainly plenty of power in the cold DM fluctuation spectrum at small masses; the problem is to get sufficient baryon cooling and avoid disruption. Perhaps dwarf spheroidals are relatively rare because most suffered disruption.

The potentially serious difficulties for the cold and warm DM pictures arise on very large scales, where galaxies are observed to form filamentary superclusters with large voids between them [20,43]. These features have seemed to some authors to favor the hot DM model,

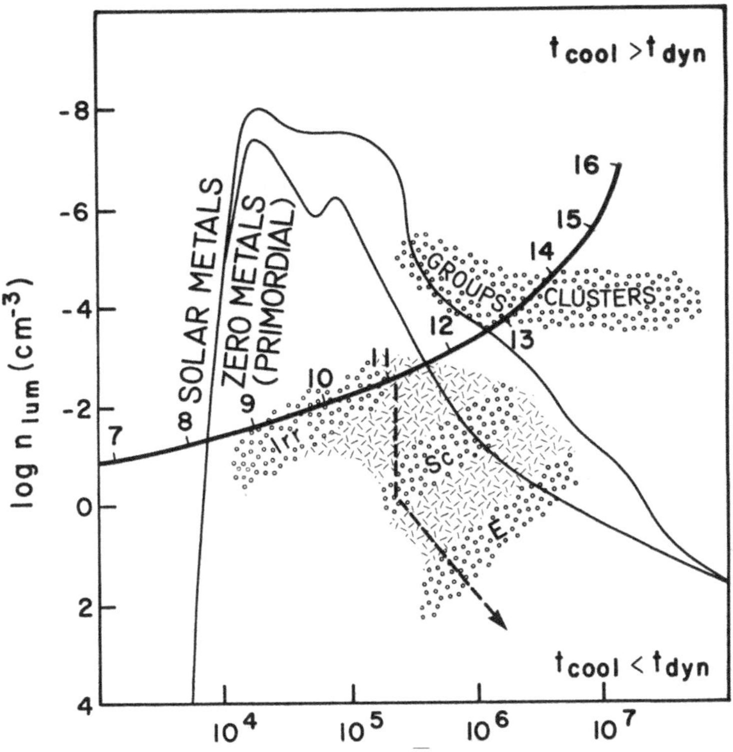

Figure 5. The (virialized) baryonic density versus temperature as perturbations having mass M_{tot} become nonlinear [82]. This curve assumes an initial Zeldovich spectrum, $\Omega h^2 = 1$, and $\Omega_b/\Omega = 0.07$. The region in which baryons can cool within a dynamical time is indicated by $t_{cool} < t_{dyn}$; also shown are positions of observed galaxies, groups, and clusters of galaxies [7]. The dashed line represents a possible evolutionary path for dissipating baryons.

apparently for two main reasons: (1) it is thought that formation of caustics of supercluster size by gravitational collapse requires a fluctuation power spectrum sharply peaked at the corresponding wavelength, and (2) the relatively low peculiar velocities of galaxies in superclusters are seen as evidence for the sort of dissipation expected in the baryonic shock in the "pancake" model. Recent work by Dekel [84] suggests, however, that nondissipative collapse fits the observed features of superclusters. Results from N-body simulations with $N \sim 10^6$ [70] will soon show whether broad fluctuation spectra lead to filaments.

8. Summary and Reflections

Although only very tentative conclusions can be drawn on the basis of present information, it is our impression that the hot DM model is in fairly serious trouble. Maybe that is mainly because it has been the most intensively studied of the three possibilities considered here.

Probably the greatest theoretical uncertainty in all three DM pictures concerns the relative roles of heredity vs. environment. For example, are elliptical galaxies found primarily in regions of high galaxy density, and disk galaxies in lower density regions, because such galaxies form after the regions have undergone a large-scale dissipative collapse which provides the appropriate initial conditions, as in the hot DM picture? Or is it because disks form relatively late from infall of baryons in an extended DM halo, which is disrupted or stripped in regions of high galaxy density? An exciting aspect of the study of large scale structure and DM is the remarkable recent increase in the quality and quantity of relevant observational data, and the promise of much more to come.

Perhaps even more remarkable is the fact that this data may shed important light on the interactions of elementary particles on very small scales. Fig. 6 is redrawn from a sketch by Shelley Glashow which recently was reproduced in *The New York Times Magazine* [85]. Glashow uses the snake eating its tail - the uroboros, an ancient symbol associated with creation myths [86] - to represent the idea that gravity may determine the structure of the universe on both the largest and smallest scales. But there is another fascinating aspect to this picture. There are left-right connections across it: medium-small-to-medium-large, very-small-to-very-large, etc. Not only does electromagnetism determine structure from atoms to mountains [87], and the

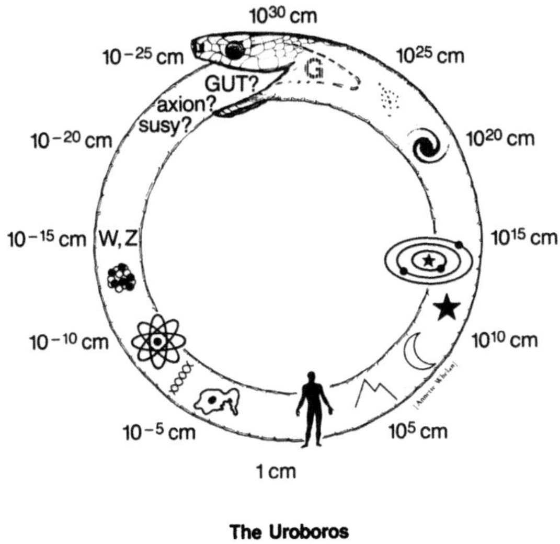

Figure 6. Physics Uroboros (after Glashow [85]).

strong and weak interactions control properties and compositions of stars and solar systems. The dark matter, which is gravitationally dominant on all scales larger than galaxy cores, may reflect fundamental physics on still smaller scales. And if cosmic inflation is to be believed, cosmological structure on scales even larger than the present horizon arose from interactions on the seemingly infinitesimal grand unification scale.

9. Acknowledgments

Our interest in the subject of dark matter grew out of collaboration with Heinz Pagels [27,54]. In preparing this paper we benefited from conversations with N. Abrams, J.R. Bond, A. Dekel, M. Davis, G. Efstatiou, C. Frenk, D. Lin, J. Silk, A. Szalay, M. Turner, S. White, and especially from extensive discussions with S. Faber at Santa Cruz and M. Rees at the Moriond Conference and subsequently. We received partial support from NSF grants and from the Santa Cruz Institute for Partical Physics.

10. Endnotes

1. The galaxies found in rich ($\gtrsim 10^3$ galaxies), well virialized clusters are mainly elliptical (E) or lenticular (SO), containing essentially no gas or young, bright stars. The roughly ten times as many galaxies not lying in rich clusters are mainly spiral (S) and SO, with a few Es and irregulars (I). S galaxies have higher L/M than Es mainly because their disks contain gas and short-lived bright stars [7]. For an excellent brief introduction to galaxies, see Fall [9].
2. A galaxy's peculiar velocity is its deviation from uniform Hubble expansion $\vec{v} = H_0\vec{r}$.
3. This discussion is approximate. Since neutrino decoupling and e^+e^- annihilation so nearly coincide, there is actually a little heating of the neutrinos too [33].
4. It is also possible that the DM is heavy stable neutrinos with mass $\gtrsim 2$ GeV, almost all of which would have annihilated [34]. This is a possible form of cold DM, discussed below.
5. In economic "stagflation", the economy stagnates but the economic yardstick inflates. The behavior of δ_{DM} during the "stagspansion" era is analogous: $\delta_{DM} \approx$ constant but a is expanding. We suggest here the term stagspansion rather than stagflation for this phenomenon since it occurs during the ordinary expansion era rather than during a possible very early "inflationary" (de Sitter) era.

6. One might worry that such a light particle could give rise to a force that at short distances $(10^{-5}\,\text{eV})^{-1} \sim 2$ cm would be much stronger than gravity. But because the axion is pseudoscalar, its nonrelativistic couplings to fermions are $\sim \vec{\sigma} \cdot \vec{p}$.

7. One calculates δ_k initially. In order to discuss mass fluctuations it is more convenient to use $\delta M/M$ than $\delta\rho/\rho$, the Fourier transform of δ_k [65]. Note that there is a simple relationship between $|\delta\rho/\rho|^2$ and $|\delta_k|^2$ only for a power law fluctuation spectrum $|\delta_k|^2 \propto k^n$.

8. Thus the Zeldovich spectrum is perfectly compatible with galaxy formation in a universe filled with cold DM, despite a recent claim to the contrary [79].

9. The model recently presented by Peebles [83] differs from that sketched here mainly in Peebles' assumption that there is sufficient cooling from molecular hydrogen for baryon condensation to occur rapidly even on globular cluster mass scales.

REFERENCES

1. S.M. Faber and J.S. Gallagher, Ann. Rev. Astron. and Astrophys. 17, 135 (1979).
2. V.C. Rubin, W.K. Ford, N. Thonnard, and D. Burstein, Astrophys. J. 261, 439 (1982); D. Burstein, V.C. Rubin, N. Thonnard, and W.K. Ford, Astrophys. J. 253, 70 (1982).
3. K.A. Innanen, W.E. Harris, and R.F. Webbink, Astron, J. 88, 338 (1983).
4. D.N.C. Lin and D. Lynden-Bell, Mon. Not. R. Astr. Soc. 198, 707 (1982).
5. W.H. Press and M. Davis, Astrophys. J. 259, 449 (1982); M. Davis and P.J.E. Peebles, Astrophys. J. 267, 465 (1983).
6. S.D.M. White MNRAS in press (1983).
7. S.M. Faber, in *Astrophysical Cosmology* (eds. H. Bruck, G. Coyne and M. Longair) pp. 191-234 (1982).
8. J.E. Gunn, ibid. pp. 233-260.
9. S.M. Fall, in S.M. Fall and D. Lynden-Bell, eds., *The Structure and Evolution of Normal Galaxies*, Cambridge Univ. Press (1981).
10. W. Forman and C. Jones Ann. Rev. Astron. & Astrophys. 20, 547 (1982).
11. R.P. Kirschner, A. Oemler, and P.L. Schechter, Astron. J. 84, 951 (1979).
12. P.J.E. Peebles, *Xth Texas Symposium on Relativistic Astrophysics*, Ann. N.Y. Acad. Sci. 375, 157 (1981).
13. M. Davis and P.J.E. Peebles, Ann. Rev. Astron. Astrophys. (1983, in press).
14. L. Hart and R.D. Davies, Nature 297, 191 (1982).
15. G.L. Hoffman and E.E. Salpeter, Astrophys. J. 263, 485 (1982).
16. R.G. Kron, Yerkes Observatory preprint (1983).

17. E.J. Wampler, C.M. Gaskell, W.L. Burke, and J.A. Baldwin, Astrophys. J. (in press, 1983); E.J. Wampler, private communication.

18. H.S. Murdoch, MNRAS 202, 987 (1983).

19. Dolgov and Ya. B. Zeldovich, Rev. Mod. Phys. 53, 1 (1981).

20. J.H. Oort, Ann. Rev. Astron. Astrophys. 21 (in press, 1983).

21. B. Binggeli, Astron. Astrophys. 107, 338 (1982).

22. D.J. Hegyi and K.A. Olive, U. Mich. preprint (1982); D.J. Hegyi, Rencontre de Moriond, Astrophysics (1983).

23. G. Steigman in *ν81* (1981) and in C. Heusch. ed., *Particles and Fields 1981: Testing the Standard Model* (A.I.P. Conf. Proc., No. 81) pp. 548-571; D.N. Schramm and G. Steigman, Astrophys. J. 241, 1 (1981).

24. J.R. Bond, G. Efstatiou and J. Silk, Phys. Rev. Lett. 45, 1980 (1980).

25. A.G. Doroshkevich, M.Yu. Khlopov, R.A. Sunyaev, A.S. Szalay and Ya.B. Zeldovich, *Proc. Xth Texas Symposium on Relativistic Astrophysics*, Ann. N.Y. Acad. Sci. 375 32 (1981); H. Sato, ibid 43; and ref. therein.

26. R.B. Partridge, Astrophys. J. 235, 681 (1980).

27. H.R. Pagels and J.R. Primack, Phys. Rev. Lett. 48, 223 (1982).

28. J. Carr, Comments on Astrophys. 7, 161 (1978).

29. R. Canizares, Astrophys. J. 263, 508 (1982).

30. C. Lacey, Rencontre de Moriond Astrophysics (1983).

31. G.B. Gelmini, S. Nussinov and M. Roncadelli, Nucl. Phys. B209, 157 (1982).

32. S. Weinberg, *Gravitation and Cosmology*, 534 (Wiley, 1972).

33. D.A. Discus, E.W. Kolb, A.M. Gleeson, E.C.G. Sudarshan, V.L. Teplitz and M.S. Turner, Phys. Rev. D26, 2694 (1982).

34. B.W. Lee and S. Weinberg, Phys. Rev. Lett. 39, 165 (1977).

35. V.A. Lyubimov, et al., Phys. Lett. 94B, 266 (1980).

36. F. Boehm, *Proc. Fourth Workshop on Grand Unification*, in press.

37. G.S. Bisnovatyi-Kogan and I.D. Novikov, Sov. Astron. 24, 516 (1980).

38. G. Gibbons, S. Hawking, and S. Siklos, eds., *The Very Early Universe* (Cambridge Univ. Press) and ref. therein (1983).

39. J.R. Bond and A.S. Szalay, Proc. *ν81*, 1 pp 59; Astrophys. J., in press (1983).

40. A. Melott, Mon. Not. R. Astr. Soc. 202, 595 (1983) and ref. therein.

41. A.A. Klypin and S.F. Shandarin, preprint (1982).

42. C. Frenk, S.D.M. White and M. Davis, Astrophys. J., in press (1983); S.D.M. White, C. Frenk and M. Davis, Rencontre de Moriond, Astrophysics (1983).

43. Ya.B. Zeldovich, J. Einasto and S.F. Shandarin, Nature 300, 407 (1982) and refs. therein.

44. N. Kaiser, Berkeley preprint (1983).

45. J.R. Bond, Rencontre de Moriond, Astrophysics (1983); P.R. Shapiro, C. Struck-Marcell and A.L. Melott, preprint (1983).

46. J.R. Bond, A.S. Szalay and S.D.M. White, Nature 301, 584 (1983).

47. S.M. Faber and D.N.C. Lin, Astrophys. J. (Lett.) 266, L17 (1983).

48. M. Aronson, Astrophys. J. (Lett.) 266, L11 (1983).

49. D.N.C. Lin and S.M. Faber, Astrophys. J. (Lett.) 266, L21 (1983).

50. S.D. Tremain and J.E. Gunn, Phys. Rev. Lett. 42, 407 (1971).

51. G. Steigman, Ann. Rev. Astron. Astrophys. 14, 339 (1976).

52. J.R. Primack, *Particles and Fields 2*, (Plenum) (1983).

53. J.R. Bond, A.S. Szalay and M.S. Turner, Phys. Rev. Lett. 48, 1636 (1982).

54. G.R. Blumenthal, H. Pagels and J.R. Primack, Nature 299, 37 (1982).

55. C.A. Savoy, Rencontre de Moriond, Elementary Particles, (1983).

56. J. Ellis, private communication (1983).

57. H. Goldberg, Phys. Rev. Lett. 50, 1419 (1983).

58. K.A. Olive and M.S. Turner, Phys. Rev. D25, 213 (1982).

59. E.R. Harrison, Phys. Rev. D1, 2726 (1970); P.J.E. Peebles and J.T. Yu, Astrophys. J. 162, 815 (1970); Ya.B., Zeldovich, Mon. Not. R. Astr. Soc. 160, 1P (1972).

60. P.M. Lubin, G.L. Epstein and G.F. Smoot, Phys. Rev. Lett. 50, 616 (1983); D.J. Fixin, E.S. Cheng and D.T. Wilkinson, Phys. Rev. Lett. 50, 620 (1983).

61. K.A. Olive, Rencontre de Moriond, Astrophysics (1983).

62. M. Guyot and Ya.B. Zeldovich, Astron. Astrophys. 9, 227 (1970); P. Meszaros, Astron. Astrophys. 37, 225 (1974).

63. A.G. Doroshkevich, Ya.B. Zeldovich, R.A. Sunyaev and M. Yu Khlopov, Sov. Astron. Lett. 6, 252 (1980); A.D. Chernin, Sov. Astron. 25, 14 (1981).

64. P.J.E. Peebles, *The Large Scale Structure of the Universe*, Princeton Univ. Press (1980).

65. P.J.E. Peebles, Astrophys. J. (Lett.) 263, L1 (1982); Astrophys. J. 258, 415 (1982).

66. A. Dekel, Astrophys. J. (Lett.) 261, L13 (1982).

67. W.L.W. Sargent, P. Young and D.P. Schneider, Astrophys. J., 256, 374 (1981).

68. S.D.M. White and M. Rees Mon. Not. R. Astr. Soc. 183, 341 (1978).

69. J. Silk, Nature 301, 574 (1983).

70. M. Davis, G. Efstatiou, C. Frenk and S.D.M. White, private communication (1983).

71. P. Schechter, private communication (1983).

72. R. Peccei and H. Quinn, Phys. Rev. Lett. 38, 140 (1977).

73. S. Weinberg, Phys. Rev. Lett. 40, 223 (1978); F. Wilczek, Phys. Rev. Lett. 40, 279 (1978).

74. L. Abbott and P. Sikivie, Phys. Lett. 120B, 133 (1983); M. Dine and W. Fischler, Phys. Lett. 120B, 137 (1983); J. Preskill, M. Wise and F. Wilczek, Phys. Lett. 120B, 127 (1983); J. Ipser and P. Sikivie, Phys. Rev. Lett. 50, 925 (1983).

75. D. Dicus, E. Kolb, V. Teplitz and R. Wagoner, Phys. Rev. D18, 1829 (1978); M. Fukugita, S. Watamura and M. Yoshimura, Phys. Rev. Lett. 48, 1522 (1982).

76. F. Wilczek, preprint NSF-ITP-82-100 (1982).

77. G.R. Blumenthal and J.R. Primack, in preparation.

78. W.H. Press and E.T. Vishniac, Astrophys. J. 239, 1 (1980).

79. M. Turner, F. Wilczek, and A. Zee, Phys. Lett. 125, 35 (1983).

80. J.R. Gott, Ann. Rev. Astron. Astrophys. 15, 235 (1977).

81. M.J. Rees and J.P. Ostriker, Mon. Not. R. astr. Soc. 179, 541 (1977).

82. G.R. Blumenthal, S.M. Faber, J.R. Primack and M. Rees, in preparation.

83. P.J.E. Peebles, Rencontre de Moriond, Astrophysics (1983).

84. A. Dekel, Astrophys. Jl, 264, 373 (1983).

85. T. Ferris, *The New York Times Magazine*, Sept. 26, 1982, pp 38.

86. E. Neumann, *Origins and History of Consciousness*, Princeton Univ. Press (1954).

87. V.F. Weisskopf, *Knowledge and Wonder*, Hinemann (1962).

LATE EVOLUTION OF ADIABATIC FLUCTUATIONS

A.S. Szalay[1] and J.R. Bond[2,3]

[1] Department of Atomic Physics, Eotvos University, Budapest
[2] Institute of Astronomy, Cambridge
[3] Institute for Theoretical Physics, Stanford University

We classify massive stable collisionless relics of the Big Bang into three categories of dark matter: hot, with damping mass about supercluster scale; warm, with damping mass of galactic or cluster scale; and cold, with negligible damping. The first objects that form in universes dominated by hot and warm relics are pancakes. Coupled one-dimensional N-body and Eulerian hydrodynamical simulations follow the nonlinear evolution of pancakes, the separation of baryons from dark matter via shock formation and the evolution of the shocked gas by conduction as well as by cooling. Only ~10-20% of the gas cools sufficiently to fragment on sub-galactic scales in neutrino-dominated hot theories. Cooling is efficient for warm relics. In all cases, the typical fragment size is $\sim 10^9$-10^{10} $M_\odot$. Electrons in the hot gas created by the pancake shocks can upscatter photons in the microwave background radiation, causing spectral distortions. Angular differences in these distortions lead to temperature fluctuations which are on the edge of observability, and can be used as a test of the pancake scenario.

I. CLASSIFICATION OF DARK MATTER CANDIDATES

Stable collisionless relics of the Big Bang are perhaps the most attractive candidates for the dark matter. Bond and Szalay (1983) have classified the possibilities into three basic types defined by their background velocity dispersion: relics may be hot, warm or cold. The canonical example of a hot relic is a massive neutrino, with velocity dispersion $6\ (m_\nu\ /\ 30\ \text{eV})^{-1}(1+z)\ \text{km s}^{-1}$. However, any particle which is massive, stable and decouples when relativistic at an epoch when the temperature of the universe was $\lesssim T_{qh} \sim 200$ MeV is a hot particle. Any particle which decouples above T_{qh} is warm, with a velocity dispersion $0.085\ (100/g(T_d))^{1/3}\ (1\ \mathrm{keV}/m_x)\ \mathrm{km\ s^{-1}}$. Here we call our hypothetical collisionless relic X, and m_x is its mass, $g(T_d)$ is the effective number of degrees of freedom at the X-decoupling temperature, T_d. The quark-hadron phase transition temperature, T_{qh}, is the approximate decoupling boundary between warm and hot since above T_{qh} the number of relativistic species present is large due to all the liberated quark-antiquark pairs. At the neutrino decoupling temperature, $T_d \sim 1$ MeV, g is only 10.75, whereas g ~ 100 in the minimal Weinberg-Salam theory. Near grand unification energies, $\sim 10^{15}$ GeV, g ~ 160 in the minimal Georgi-Glashow (1974) SU(5) theory. Supersymmetric theories increase g by only a factor of about two.

The relationship between the density parameter of relativistic decouplers and their mass is: $\Omega_x = 1.1\ h^{-2}\ (100/g(T_d))(m_x/1\ \mathrm{keV})$, where h is Hubble's constant in units of 100 km $\mathrm{s^{-1}}$ $\mathrm{Mpc^{-1}}$. For g=10.75, we get the usual 20-100 eV mass needed for one species of massive neutrino to close the universe. The variation corresponds to the range h = 0.5 to 1. For $T_d > T_{qh}$, and for minimal theories, $m_x \sim 200 - 2000$ eV is the mass required for $\Omega = 1$. This mass could be much larger if: (1) there is no large plateau in g(T) - i.e. no desert; (2) significant entropy generation occurs after T_d - Ω_x scales inversely with the entropy amplification factor. The first option could be restricted since the number of stable relativistic neutrino species at the time primordial helium generation cannot be much greater than three to avoid overproduction (Shvartsman 1969, Olive et al. 1981); this could translate into a constraint on the number of leptoquark families. The second option suffers from constraints on the allowable entropy generation after baryon synthesis.

Hot particles have a damping scale arising from the constructive effects of gravitational attraction on large scales and the destructive effects of their random velocity on small scales: $M_d = 3.4\ m_p^{\,3}/m_x^{\,2}$ (Bond, Efstathiou and Silk 1980), where $m_p = 1.22\times10^{22}$ MeV is the Planck mass; this is of supercluster scale. Warm particles damp below

$$M_d = 0.11\ (m_p^{\,3}/m_x^{\,2})\ \ (100/g(T_d))^{4/3} \qquad (1)$$

(Bond, Szalay and Turner 1982, Bond and Szalay 1983), which is either the scale of galaxies or of clusters depending upon h. If m_x is very large, due to either (1) or (2), then the damping scale can be very small, and the particles are effectively cold. Indeed, we define cold particles to be those with almost no velocity dispersion, and which thus have $M_d \sim 0$. Collisionless relics which decouple when they are nonrelativistic are examples. Preskill, Wise and Wilczek (1982) have recently pointed out, that oscillations of "classical" fields, i.e. of boson vacuum expectation values can lead to a large time-averaged energy density, as well as to a rapidly fluctuating part. They claim that spatial fluctuations in the field have energy density growth identical to that of nonrelativistic decouplers.

The canonical hot particle is the massive neutrino. Another candidate is the Majoran, a goldstone boson whose raison d'etre is to generate neutrino masses via spontaneous symmetry breaking (Georgi, Glashow and Nussinov 1981, Gelmini, Nussinov and Roncadelli 1982). All background $\nu\bar{\nu}$'s would annihilate into a sea of Majorans - the temperature of which would be higher than that of the background photons. If they are massive, then $m \leq 10$ eV is required in order to have $\Omega \leq 1$. This implies uncomfortably large damping masses; and as we shall see, very little gas cooling in a Majoran-dominated universe. Suggestions for warm relics have included the gravitinos and photinos of supersymmetric theories, and righthanded neutrinos. Any of these could also decouple when nonrelativistic, and thus be cold. Heavy neutral leptons of the sort discussed by Lee and Weinberg (1977), primordial black holes and monopoles are other cold relic candidates. If strings form in phase transitions in the very early universe, and if they primarily exist as loops of subgalactic dimensions (Kibble 1983), galaxy formation in string-dominated universes will effectively follow the cold scenario. The model for classical field oscillation is provided by the axion (Preskill, Wise and Wilczek 1982).

In all cases in which relics form the dark matter, a remarkable coincidence is required - namely that Ω_x and Ω_B are not too dissimilar. We know that Ω_B cannot be too small, or else cooling on any scales would have been too inefficient. This is an anthropic argument which rules out extreme variations of Ω_B. If $\Omega \sim 1$ is required as a consequence of inflation (or simplicity), and m_x is given by particle physics, then H_o would be adjusted so that the $\Omega_X \sim m_X h^{-2}$ relation is enforced for relativistic decouplers; the coincidence $\Omega_X \sim \Omega_B$ implies $m_X \sim m_N s^{-1}$, where $s \sim 10^9$ is the entropy per baryon and m_N is the nucleon mass. In a simple model of baryon generation this becomes $m_X \sim m_N m_P m_{VB}^{-1} \times \alpha_{GUT} \varepsilon_{CP}$, where m_{VB} is the mass of the intermediate vector boson responsible for baryon generation, α is the fine structure constant at GUT energies, ε is a CP-violating parameter. Why should such quantities be interrelated in this manner? The case of nonrelativistic decouplers requires perhaps an even more stringent restriction upon the particle physics, namely that the freeze-out temperatures for the reactions which create X's must be - within some narrow range - a prescribed fraction of m_X ($15 \lesssim m_X/T_{fx} \lesssim 50$ for $1\ \text{GeV} \lesssim m_X \lesssim 10^{15}$ GeV).

2. COOLING SCALE AND PANCAKES

We have seen that the $m_p^3 m_X^{-2}$ damping scale applies for hot and warm particles; compare this with the mass scale of stars which is set by the combination $m_p^3 m_N^{-2} = 1.2\ M_o$. Another scale at high mass enters into the determination of the fluctuation spectrum: the horizon mass at equipartition between relativistic and nonrelativistic constituents, which occurs at $z_{eq} = 25000\ \Omega h^2$ when the photon temperature is $5.8\ \Omega h^2$ eV

$$M_{Heq} = 0.2\ m_p^3\ T_{\gamma eq}^{-2} = 10^{16}\ (\Omega h^2)^{-2}\ M_o$$

An initially scale-free density spectrum evolves in the linear phase to one in which there is a sharp damping cutoff at masses smaller than M_d, a strong flattening between M_d and M_{Heq} (Peebles 1982, Bond, Szalay and Turner 1982, Bond and Szalay 1983). It has been conventional to associate the appearance of voids and strings in the galaxy distribution with a large damping cutoff. An important unresolved issue is whether the shoulder below M_{Heq} is sufficient to generate such structure. In any case, for warm and hot particles, the first structures to become nonlinear will be on the scale M_d, will collapse preferentially along one axis, becoming highly asymmetric, and result in shock formation in the central regions (Zeldovich 1970, Sunyaev and Zeldovich 1972). The first structures to collapse in the cold scenario may also be asymmetric and lead to shocks; however, instead of a smooth collective inflow, the shocks may be more localized, arising from cloud-cloud collisions.

Binney (1977), Rees and Ostriker (1977) and Silk (1977) have demonstrated how galaxy masses may be related to a cooling scale. It is instructive to go through this exercise to demonstrate what must be done to get cooling in larger structures - from which galaxies ultimately arise. The Rees and Ostriker (1977) development yields

$$M_{cool} \sim \frac{m_p^{\,3}}{m_N^{\,2}} \frac{\alpha^5\, m_p}{(m_e m_N)^{1/2}} \quad \frac{\Omega_B}{\Omega} \sim \; 2 \times 10^{10} \; \frac{\Omega_B}{\Omega} \quad M_\odot \tag{2}$$

The ingredients which go into obtaining this scale are as follows. A virialized homogeneous sphere cools via bremsstrahlung faster than free-fall if its temperature satisfies:

$$T/m_e < M\, m_p^{-3}\, m_N^{2}\; (m_N\, m_e)^{1/2}\, m_p^{-1}\, \alpha^{-3}\, \Omega\, \Omega_B^{-1} \quad .$$

However, the temperature $T \sim M^{2/3}(1+z_t)$ depends not only upon the mass, but also upon the epoch of turn-around, z_t. Indeed, if z_t is too large, Compton cooling will replace bremsstrahlung. In any case, stability can never be regained if T falls in the helium-hydrogen recombination cooling regime. Since the ionization energy of the helium is $2\alpha^2 m_e$, and the characteristic temperature for helium recombination is some fraction of this, we obtain the scale M_{cool}. There are three ways to increase M_{cool}:(1) raise z_t into the Compton cooling epoch,
(2) utilize central condensation so that T can be lower in the central regions since the matter has less far to fall before shocking;
(3) stretch the sphere into an oblate configuration so again the gravitational acceleration is less. Effects (2) and (3) operate in pancakes; z_t is constrained by limits on the temperature fluctuations in the microwave background, hence (1) cannot be pushed too far.

3. PANCAKE SHOCK CALCULATIONS

This work is described more fully in Bond, Centrella, Szalay and Wilson (1983) , hereafter BCSW. Here, we outline the methods and give the main results. A pancake collapse similar to our runs has recently been computed by Shapiro, Struck-Marcell and Melott (1983).

We ignored the effects of random velocity dispersions of the collisionless relics since these redshift away as the universe expands and are small relative to the gravitationally-induced velocities at the time of pancaking (Bond, Szalay and White 1983). For such cold initial conditions the Zeldovich (1970) solution describes the deviations of the particle positions from their initial values. The formula is exact in the linear regime, and in one dimension until caustic formation. It also describes the early nonlinear phases of 3D evolution rather well.

The distribution of the principal eigenvalues, λ_i, of the deformation tensor describes the 3D patterns which first appear in the nonlinear evolution of a density fluctuation spectrum with a damping cutoff. The overdensity is related to these eigenvalues by

$$1 + \delta = (1-b\lambda_1)^{-1} (1-b\lambda_2)^{-1} (1-b\lambda_3)^{-1} \quad . \tag{3}$$

where $b = (1+z)^{-1}$, in the $\Omega=1$ models we are most concerned with. The principal axes are always ordered so that $\lambda_1 \geq \lambda_2 \geq \lambda_3$. Therefore, by definition, collapse is most rapid along the 1-axis; and caustics, where $\delta \to \infty$, occur at λ_1 -maxima. The surfaces of λ_1-maxima are generally curved, with curvature at most of the order of the damping scale. Different surfaces intersect at points of degeneracy of the 1 and 2-axes. These topological structures are catalogued by Arnold, Shandarin and Zeldovich (1982). In the neighbourhood of every λ_1- maxima surface, the flows are essentially one-dimensional: we expect our 1D simulations to describe the post-caustic evolution of these regions rather well. Directions 2 and 3 may continue undergoing transverse expansion which differs little from the Hubble expansion initially. However, within a few Hubble times of the collapse redshift, transverse flows toward the 1-2 degeneracy lines (strings) may become important.

Neutrinos are followed by direct N-body simulation of their equations of motion. The gravitational field equations reduce - for these nonrelativistic particles - to Poissons equation, except that the source is the overdensity relative to the background. This couples the N-body to the gas-dynamical code, which utilizes Eulerian hydrodynamics in comoving space. We solve transport equations for the following quantities: (1) Baryon number (2) Momentum - artificial viscosity is used to treat shocks (3) Matter energy, including ionic and electronic contributions. Artificial viscosity provides the shock heating. Energy losses arise from bremsstrahlung, Compton cooling, and He and H recombination cooling. Flux-limited conduction is included: transfer primarily occurs via electron-electron collisions, although ion-ion collision are also incorporated. (4) Ion energy. This transport equation includes ion-ion conduction and a term describing the relaxation of the ionic temperature towards the electron temperature via ion-electron collisions. It is the ions which are shock heated and the electrons which cool. The ion and electron temperatures therefore differ - especially near the shock front. The equation of state is that of an ideal gas with ionization fraction determined by the balance of collisional ionization with recombination. We do not explicitly calculate the He ionization fraction; it is however included in the cooling rate. Complications arise since the pancakes are generally optically thick in their central regions to Ly α and to radiation **above the** Ly edge. Still, lower energy radiation escapes, allowing continual rapid cooling.

We began our calculations in the linear regime. The initial phase of the collapse just demonstrates the validity of the Zeldovich-solution - even for the gas, which has a small pressure. However, near the redshift of caustic formation, z_c, the central density rises rapidly and the gas pressure builds up due to the adiabatic compression. The shock forms approximately at the point where the incoming ram pressure equals this gas pressure (Sunyaev and Zeldovich 1972), i.e. near the sonic point. The collisionless relics begin to separate from the gas after this. The neutrinos form density spikes at the edges of the well known phase space spirals which appear in the ID simulations. The gas is confined by the ram pressure, much of it to a region smaller than 10 kpc. The spread in neutrinos is more than an order of magnitude greater at these times.

Consider the temperature profile evolution for a 25 Mpc pancake. Conduction clearly transports energy from the hot exterior inward, thereby flattening the temperature gradient. The profiles drop precipitously toward 1 eV and hydrogen recombination below T ~ 100 eV. By z = 4.8 , 9% of the gas has cooled below this temperature; by z = 3, 11% has cooled with conduction, 14% without. It takes conduction time to operate: a 54 Mpc pancake begins with only 4% of the gas cold; an inward-eating conduction front breaks through the center by z = 3.8, resulting in an essentially flat profile by z = 3. For small wavelength runs, the temperatures achieved are lower, conduction is relatively unimportant, and a large fraction of the gas cools.

The major feature of our numerical runs is that the pressure is constant behind the front, and is equal in magnitude to the ram pressure. The ram pressure can be determined from the Zeldovich solution exterior to the shocked region.

4. COOLING FRACTIONS AND FRAGMENT SIZES

The fraction of gas which has **cooled below** 100 eV by z=3 for $z_c=5$ collapses is q_c. Below 100 eV, cooling down to 1 eV occurs rapidly; 100 eV is also the virial temperature for a typical galactic halo. With only bremsstrahlung cooling included, and transverse Hubble expansion assumed:

$$q_c = A\,(hL_{10})^{-1}\,(1 + B\,\Omega_B\,h\,(hL_{10})^2)^{1/3} \qquad (4)$$

$$A = 0.27\left(\frac{1+z_c}{6}\right)^{3/10}\left(\frac{4}{1+z}\right)^{4/5}$$

$$B = 1.0\left(\frac{1+z_c}{6}\right)^{1/10}\left(\frac{1+z}{4}\right)^{12/5}\left(\frac{1-(\frac{1+z}{1+z_c})^{1/10}}{0.04}\right)$$

A and B are only weakly dependent upon z_c, assuming wa are not in the Compton cooling regime, but are relatively sensitive to z. In BCSW, we also give an approximate solution with conduction included. Both expressions agree well with our full numerical runs. The comoving size of the cooled region can be computed from the cooled fraction:

$$d \simeq 4.7\,h^{-1}\,q_c\,(hL_{10})^{-1}\ \text{kpc},$$

which turns out to be much smaller than the usual 3D Jeans length. It is more appropriate to consider the longitudinal dimension of the fragmenting region fixed at d; the transverse dimension of the most rapidly growing perturbation is then given by the Sunyaev-Zeldovich length:

$$L_{SZ} = 77\,(1+z)^{-1}\,(q_c L_{10})^{-1}\ \text{kpc}.$$

The fragments are thus initially extremely elongated, but can rapidly undergo transverse collapse at the freefall rate due to the efficiency of cooling. The mass of the fragments is

$$M_{SZ} = 5\times10^8\,(\Omega_B h)^{-1}\,(hL_{10}\,q_c)^{-1}\,M_\odot \qquad (5)$$

According to Eq.(5), M_{SZ} is only weakly dependent upon the pancake mass; for $\Omega_B \sim 0.1$, we have $\sim 10^{10} M_o$ representing the typical fragment scale.

5. CBR DISTORTONS FROM THE HOT PANCAKE GAS

The upscattering of the microwave background photons by hot electrons is described by the Kompaneets equation for the photon distribution. Its solution is frequency independent in the Rayleigh-Jeans part of the spectrum, and the spectral distortions are given by the CBR temperature fluctuation (Zeldovich and Sunyaev 1969):

$$dT/T = -2y = -2 \int n_e \sigma_T (kT_e/mc^2)\, dr \qquad (6)$$

The integral is taken along the line of sight, T_e, n_e and m are the electron temperature, density and mass, and σ_T is the Thomson cross section.

The distortions themselves are difficult to measure. Our goal is to predict angular fluctuations arising from the statistical dispersion along the line of sight in this integral, since this is observable. Radio experiments of small beam size probe only small separations; they measure integral quantities along the line of sight. Typical experiments involve either beam switching through some angle, with the beam pair swept around a part of the sky, or imaging of a given region.

Pancakes are isobaric in their shocked regions, hence $n_e T_e \sim P$ is constant, so our line integral becomes a simple sum over all pancakes. Along the line of sight there will be many pancakes, and each will have a different collapse redshift, length scale, orientation, etc. To calculate the contribution in detail requires the knowledge of the appropriate probability distributions. Lacking this, we use a simple model to estimate the expected magnitude of the effect. We assume all pancakes form at the same epoch, z_p, with the same length, and are evenly spaced in comoving length along the line of sight. The number of pancakes along this line (N) is a random variable with the mean

$$\langle N\rangle = (600/hL_{10}) \; [1 - (1+z_p)^{1/2}]$$

up to factors of order one, which depend upon the precise distribution of spacing. Here L_{10} = L/10 Mpc is the comoving wavelength of the collapsing region, z_p is the redshift of pancaking. We assume the statistics of N are Poisson, with $\sqrt{N}$ fluctuations. We further assume, that the inclination angle is statistically independent of the other variables, and is randomly oriented. Then the dispersion of y will depend upon the relative dispersion in the number of pancakes along the line of sight ($\sim 1/\sqrt{N}$) and on the relative dispersion of the value $\langle 1/\cos i \rangle$, the projected relative thickness, averaged over the N pancakes. Angular fluctuations in the temperature will arise as a consequence of the variance about the mean as we go from one line of sight to another, as long as the two directions are statistically independent. This will be valid, provided their angular separation is much greater than the angular size associated with the transverse dimension of a pancake ($\Theta > 1^{\circ} > \Theta_L$). We have assumed that the transverse and initial longitudinal scales are about the same. In Table I. we give more accurate evaluations of dT/T as a function of L and z_p for $\Omega=1$, $\Omega_B=0.1$, and h=1 for these large-angle fluctuations.

L \ z_p	1.0	2.0	3.0	5.0	7.0
20	7.1E-07	3.0E-06	7.1E-06	2.1E-05	4.5E-05
40	4.0E-06	1.7E-05	4.0E-05	1.2E-04	2.5E-04
60	1.1E-05	4.7E-05	1.1E-04	3.3E-04	7.0E-04
80	2.3E-05	9.6E-05	2.3E-04	6.9E-04	1.4E-03
100	4.0E-05	1.7E-04	4.0E-04	1.2E-03	2.5E-03

Table I. The dependence of dT/T on large $(>1^{\circ})$ angular scales on the pancake length scale in Mpc and on the redshift of pancaking z_p.

Since pancakes form a cell-like structure in the universe, they are highly correlated at small angular separations. In this case consider a 'beam-switch' differential measurement. We can only get a contribution from the pancake if it intersects with one beam but not with the other. The only obvious way to achieve this is that we have either a pancake edge-on, nearly tangential to the line of sight, or we have a branching point of three pancakes. The total averaged contribution will be proportional to $d(\Theta,z)/L$, therefore on small angles we expect dT/T to be rising as Θ up to the large angle limit.

A rewiev of the present observational situation was given by Partridge (1981), further references therein. The overall limits on dT/T for scales from a few arc minutes to a few degrees are about 10^{-4}.

However, by imaging techniques, there is a nonvanishing chance, that one may see signals at 3σ or higher above the rms background, with a rather sharp boundary: a clear signature of a pancake.

Even though our numbers are order of magnitude estimates, it is clear that the dT/T coming from hot gas is very close to present upper limits. With more sophisticated treatment one may obtain stronger limits on the characteristic scale and on the epoch of galaxy formation in the pancake picture. The inclusion of transverse contraction leading to filaments and clusters will be important. The estimates presented here can already be used to rule out extremely large pancakes, which form at early epochs.

6. DISCUSSION

If we assume the density fluctuations are initially adiabatic, the nature of the dark matter determines how the large scale structure first arises. If cold relics dominate, the theory is a variant of the hierarchical clustering picture developed by White and Rees (1978) for the isothermal picture with dark matter. Peebles (1983) and Primack (1983) discuss the many positive aspects of this picture. However, it is difficult to see (1) how superclusters and large voids arise, (2) how Ω can be one - if indeed it is one. If warm relics dominate, cooling is no problem, but again (1) and (2) do not come naturally. Further, dwarf irregulars and ellipticals would have to arise via fragmentation in the warm scenario, and if M_d is of cluster scale, so would all the galaxies.

In neutrino-dominated models, (1) and (2) could naturally follow. On the other hand, building a theory with 90% of the gas too hot to immediately condense on galaxies could be a problem. The X-ray emitting gas in rich clusters cannot have a mass much larger than about 2 times that in galaxies of the clusters (Ku et al. 1982). However, since rich clusters may arise at the points of branching points of pancakes, and their gas has T ~ 7 keV, significantly hotter, than than pancake gas, we should not identify the two sorts of hot gas. Intergalactic gas with T > 10 eV and one-tenth of the critical density cannot be ruled out (Sherman 1982). Indeed, it may be desirable to have most of the baryons unclustered on galactic scales. Thus the neutrino theory may well survive this cooling problem.

REFERENCES

Arnold, V. I., Shandarin, S.F. and Zeldovich, Ya.B. 1982, Geophys. Astrop. Fluid Dynamics 20, 111.

Binney, J. 1977, Ap.J. 215, 483.

Bond, J.R.,Efstathiou, G. and Silk, J. 1980, Phys. Rev. Lett. 45, 1980.

Bond, J.R. and Szalay A.S. 1983, Ap.J. , in press.

Bond, J.R., Szalay, A.S. and Turner, M.S. 1982, Phys. Rev. Lett. 48, 1636.

Bond, J.R., Centrella, J., Szalay, A.S. and Wilson, J.R. 1983, M.N.R.A.S. to be published.

Bond, J.R., Szalay, A.S. and White, S.D.M. 1983, Nature 301, 584.

Gelmini, G.B., Nussinov, S. and Roncadelli, M. 1982, preprint MPI-PAE 37182

Georgi, H. and Glashow, S.L. 1974, Phys. Rev. Lett. 32, 438.

Georgi, H., Glashow, S.L. and Nussinov, S. 1981, Nucl. Phys. B 193, 297.

Kibble, T.W. 1983, private communication.

Ku, W.H.M. et al. 1982, M.N.R.A.S. 202.

Lee, B.W. and Weinberg, S. 1977, Phys. Rev. Lett. 39, 165.

Olive, K.A., Schramm, D.N., Steigman, G., Turner, M.S. and Yang, J. 1981, Ap.J. 246, 557.

Partridge, B. 1981, Proc. Int. School on Cosmology, Erice, Italy, p. 121. edited by B.J.T. Jones, Reidel, Dordrecht.

Peebles, P.J.E., 1982, Ap.J. 258, 415.

Preskill, J., Wise, M.B. and Wilczek, F. 1982. Harvard preprint HUTP-82/A048.

Primack, J. 1983, this volume.

Rees, M.J. and Ostriker, J. 1977, M.N.R.A.S. 179, 541.

Shvartsman, V.F. 1969, Sov.Physics. JETP Lett. 9, 184.

Shapiro, P.R., Struck-Marcell, C. and Melott, A.L. 1983, preprint.

Sherman, R.D. 1982, Ap.J. 256, 370.

Silk, J. 1977, Ap.J. 211, 638.

Sunyaev, R.A. and Zeldovich, Ya.B. 1972, Astron. Astrophys. 20, 189.

Zeldovich, Ya.B. 1970, Astron. Astrophys. 5, 84.

EXPERIMENTAL SEARCHES FOR SUPERSYMMETRIC PARTICLES*

G.L. Kane
Randall Laboratory of Physics
University of Michigan
Ann Arbor, MI 48109

ABSTRACT

In this review, recent limits on masses of supersymmetric partners of quarks, leptons, and gauge bosons are analyzed, and most experiments so far proposed to find such states in the next few years are examined, with emphasis on experiments that can be done now (especially gluino pair production in hadron reactions, $e^+e^- \to \tilde{\gamma}_2\tilde{\gamma}_1$, and $W^\pm$ decay). The implications of heavy photinos are considered. An attempt is made to systematize the terminology so that names and symbols denote the properties of the states and are consistent.

1. Introduction

Among the ideas which are currently promising to explain, at least in part, the Standard Model and why it works, and to lead beyond the Standard Model, Supersymmetry is certainly the most popular.[1] And it is popular in spite of (because of?) a total absence of any experimental evidence which can be interpreted, even indirectly, as an indication that nature is Supersymmetric. Although finding any experimental evidence for Supersymmetry would have an extraordinary impact on particle physics, and although (as we will see below) it is not obvious how to look for most signatures of supersymmetry, very little effort has gone into this area. Fortunately, a number of ways exist to find evidence for Supersymmetry if it is present.

If nature were Supersymmetric, it could be on a level where no direct experimental evidence would appear, but it could also happen that the Supersymmetry is directly observable on our energy scale. My approach will be to examine the (essentially model independent) minimal alternative that nature is explicitly supersymmetric and that each of the usual quarks, leptons, gauge bosons, and Higgs bosons has its superpartner particle, displaced one half unit in spin but otherwise identical in quantum numbers to the familiar particle. The states are listed in Table I.

Table I

Standard Particles	Supersymmetric Partners
u_L, u_R, d_L, ...	$\tilde{u}_L$, $\tilde{u}_R$, $\tilde{d}_L$, ...
e^-_L, e^-_R, μ^-_L,...	$\tilde{e}^-_L$, $\tilde{e}^-_R$, $\tilde{\mu}^-_L$, ...
ν_e, ν_μ, ...	$\tilde{\nu}_e$, $\tilde{\nu}_\mu$, ...
-·-·-·-·-·-	-·-·-·-·-·-
g	$\tilde{g}$
$W^\pm$	$\tilde{W}^\pm$
γ, Z^o	$\tilde{\gamma}$, $\tilde{Z}^o$
$\begin{pmatrix} H_1^+ \\ H_1^o \end{pmatrix}$, $\begin{pmatrix} H_2^o \\ H_2^- \end{pmatrix}$	$\tilde{H}_1^+$, $\tilde{H}_1^o$, $\tilde{H}_2^o$, $\tilde{H}_2^-$

The set of standard particles (quarks, leptons, gauge bosons, Higgs) is given on the left. On the right are the superpartners, written in terms of the weak isospin eigenstates. Those superpartners above the dot-dashed line have spin zero, those below have spin 1/2.

Since $\tilde{\gamma}, \tilde{Z}°$, $\tilde{H}_1^o$, $\tilde{H}_2^o$ all have spin 1/2 they can mix, so the mass eigenstates will be linear combinations of the weak eigenstates. The mass eigenstates will be denoted as $\tilde{\gamma}_1, \tilde{\gamma}_2, \tilde{h}_1, \tilde{h}_2$, and called photinos, ziggsinos, or neutral higgsinos, depending on whether their couplings are photon-like, intermediate, or Higgs-like. Similarly, $\tilde{W}^{\pm}$ and $\tilde{H}_{1,2}^{\pm}$ will mix, and the mass eigenstates will be denoted $\tilde{w}_1^{\pm}$, $\tilde{w}_2^{\pm}$. The charged mass eigenstates will be called winos, wiggsinos, or charged higgsinos depending on whether their couplings are W-like, intermediate, or Higgs-like. The collective set of neutral, uncolored, spin 1/2 superpartners will be called "neutralinos", and the collective set of charged, uncolored, spin 1/2 superpartners called "charginos". When it is convenient to refer to a generic neutralino (any one of $\tilde{\gamma}_1, \tilde{\gamma}_2, \tilde{h}_1^o, \tilde{h}_2^o$) the designation $\tilde{y}^o$ will be used, and for a generic chargino, $\tilde{w}^{\pm}$.

If an unbroken Supersymmetry held, the partners would have the same masses as the familiar particles, and they would not have escaped detection. It is customary to assume that the necessary breaking of Supersymmetry makes the superpartners heavier but does not affect their couplings or other properties, and that is presumably a good assumption for our purposes. One practical advantage of a Supersymmetric theory is that all the coupling strengths of the superpartners are determined in terms of those of the standard particles (essentially by replacing pairs of standard particles by their superpartners in any vertex). Thus, given a superpartner with some (unknown) mass but specific electric charge, weak isospin, and color one can see how it couples to quarks, gluons, and leptons and compute its production rates at any accelerator as a function of its mass.

In the simplest form of the theory the number of superpartners at a vertex is always even, so all superpartners will decay eventually into the lightest one. I will assume that is the photino, although it could be a higgsino or a scalar neutrino[4] -- then some modifications in detailed conclusions would occur, but not major changes.

What is needed for progress is experimental information, so I will concentrate on how to search for superpartners. As will be seen, that is sometimes subtle; most superpartners have detectable signatures, but often not ones that would be noticed without an explicit search. Background problems can be serious. Interestingly, some of the superpartners (gluinos, photinos and whiggsinos) could have been produced in experiments but not yet detected because the usual experimental cuts and procedures are not appropriate for the new states.

For some states [mainly scalar partners of leptons (sleptons) and partners of gluons (gluinos)] good lower limits exist on their masses. That is, if they were not heavier than a certain mass they would probably already have been detected. I will summarize these. Otherwise the emphasis will be on how a positive signal for supersymmetry might be detected at present machines and those of the next few years. Table II is a summary of the situation, meant to give a broad overview. At the time of writing it is not known whether there will exist a hadron collider beyond the FNAL 2 TeV $\bar{p}p$ Tevatron I. A "desertron" with $\sqrt{s} > 10$ TeV and high luminosity can easily produce and detect any colored hadrons (here, gluinos and squarks) up to at least 1 TeV in mass and perhaps more; it is unlikely we will have such a machine running in less than a decade so I have not explicitly included it in Table II. Two intermediate energy machines have been proposed, CBA (pp, $\sqrt{s}$=0.8 TeV, $\mathcal{L} \approx 10^{33}/\text{cm}^2$ sec) at Brookhaven, and DC ($\bar{p}p$, $\sqrt{s}$=4 TeV, $\mathcal{L} \approx 10^{31}/\text{cm}^2$ sec) at Fermilab. The latter proposal is quite new, and there has not been time to study in detail the sensitivity it can achieve, so I have quoted numbers from CBA where the studies have already been done. Studies are needed because it is not just a matter of production rates (which are easily calculated) -- one must compare with backgrounds, make cuts, etc. At its full luminosity one expects DC to do somewhat better at reaching high gluino masses than CBA, while for the important squark-photino final state CBA may do a little better. Some comparison can be made from the information in the Snowmass Summer Study[3], especially Figure 4 of the report of R. Palmer et al., which allows comparison of various bellwether experiments for machines of various $\sqrt{s}$, $\mathcal{L}$.

If supersymmetry were a spontaneously broken global symmetry, a Goldstone particle (a fermion called a Goldstino) would exist, coupled to every particle and its superpartner. Then various additional decays would be allowed, while production properties are unaffected. Goldstinos could also[5] be pair produced or produced associated with superpartners. I will not include the consideration of these in the present summary, because of space limitations; their role in gluino searches is described in Ref. 5, and they will be fully discussed in a forthcoming more complete review.[6]

It has been customary to assume photinos were rather light, with masses less than a few GeV. Recently Goldberg[7] has emphasized that cosmological arguments combined with the Majorana nature of the photino imply that photino masses may be rather large, up to of order 25 GeV; they might still be the lightest supersymmetric partners. Various superpartner masses are correlated -- heavier sleptons implies heavier photinos. I will proceed here by giving the discussion for relatively light photinos, up to a few GeV (implying sleptons less than about 50 GeV in mass) and adding qualifying remarks for the situation with heavier photinos.

Various indirect methods exist to learn about restrictions on supersymmetric partners. Since these can never give a positive indication of supersymmetry, I will not cover them here. The relevant literature can be traced from the review of Fayet.[1] The status of models can be studied from the talk of Polchinski in these proceedings.

Sometimes I will call various possibilities "excluded" by experiment. It should always be kept in mind that various experimental assumptions and cuts go into each reported result, and that it is very hard to find something for which one is not looking. Thus repeated examination of data for signals of new physics is very worthwhile.

Now let us consider the entries in Table II, explaining them one row at a time.

Table II

SUSY Partners	Present Limits	Soon		about 1989
$\tilde{\ell}^{\pm}$	$\tilde{m} \gtrsim 16$ GeV	~20 GeV	$[e^+e^- \to \tilde{\ell}^+\tilde{\ell}^-]$ $[W^{\pm} \to \tilde{\ell}^{\pm}\bar{\tilde{\nu}}?]$	~45 GeV
$\tilde{\nu}$	$m(\tilde{\nu}_\tau)+m(\tilde{\nu}_\ell) \gtrsim m_\tau$		$[e^+e^- \to \tilde{\nu}\tilde{\nu}]$	
$\tilde{q}$	$\tilde{m} \gtrsim 15$ GeV??		$[e^+e^- \to \tilde{q}\bar{\tilde{q}}]$ $\bar{p}p, pp \to \tilde{q}\tilde{\gamma}X$	~40 GeV ~125 GeV
$\tilde{g}$	$\tilde{m} > 2$-5 GeV	~25 GeV	$[\bar{p}p, pp \to \tilde{g}\tilde{g}X]$	~140 GeV
$\tilde{\gamma}$	cosmology		$[e^+e^- \to \gamma\tilde{\gamma}\tilde{\gamma}?]$	
$\tilde{\gamma}_2, \tilde{h}^0_1, \tilde{h}^0_2$	----	$m(\tilde{\gamma}_2)+m(\tilde{\gamma}_1) \lesssim 40$GeV	$[e^+e^- \to \tilde{\gamma}_2\tilde{\gamma}_1]$	$m(\tilde{\gamma})_2+m(\tilde{\gamma}_1) \lesssim 95$GeV
$\tilde{w}^{\pm}_1, \tilde{w}^{\pm}_2$	$\tilde{m} \gtrsim 16$ GeV???	20 GeV	$[e^+e^- \to \tilde{w}^+_i\tilde{w}^-_i]$	~40 GeV
		$m(\tilde{w})+m(\tilde{y}) < 50$ GeV	$[W^{\pm} \to \tilde{w}^{\pm}_i\ \tilde{y}]$ $(\tilde{y}=\tilde{\gamma}_1, \tilde{\gamma}_2, \tilde{h}^0)$	$m(\tilde{w})+m(\tilde{y}) \lesssim 70$GeV

Charged Scalar Leptons (sleptons) $\tilde{\ell}^{\pm}$

The charged sleptons $\tilde{e}, \tilde{\mu}, \tilde{\tau}$ are straightforward to produce at e^+e^- colliders, up to the beam energy in mass. All are produced by direct channel $\gamma, Z°$ as for any charged scalar, with $\beta^3/4$ units of R for each of $\tilde{\ell}^{\pm}_L$, $\tilde{\ell}^{\pm}_R$, and a $\sin^2\theta$ production distribution.

If photinos are light, $\tilde{\ell}^{\pm}$ decay rapidly, with essentially 100% branching ratio,

$$\tilde{\ell}^{\pm} \to \ell^{\pm}\tilde{\gamma}.$$

The $\tilde{\gamma}$ interacts too weakly to be detected in a normal collider detector, so it escapes. Thus on average 1/2 of the energy is missing. The ℓ^+ and ℓ^- are not collinear. Consequently the signature is quite good. There is some background from $\tau^+\tau^-$ events but the branching ratios reduce the background, and $\gamma\gamma$ events can be cut away.

All present detectors have results.[8] Combining them one can exclude $\tilde{e}$ and $\tilde{\mu}$ up to about 16 GeV and $\tilde{\tau}$ to 15.3 GeV, including[9] a retroactive analysis of SPEAR data at the lower end. One can do a little better on $\tilde{e}$, by singly producing[10] $\tilde{e}+\tilde{\gamma}$, giving $\tilde{m}_e>19.5$ GeV.[11]

If photinos are not light one can still say something. Long-lived or stable charged hadrons are excluded[12] for m<14 GeV, so if $\tilde{\gamma}$ is too heavy for $\tilde{\ell}^{\pm}$ to decay quickly via $\ell\tilde{\gamma}$, that is the limit. If $\tilde{m}_\gamma$ is less than about 10 GeV, the leptons from $\tilde{\ell}^{\pm}\to\ell^{\pm}\tilde{\gamma}$ would probably still have been seen, though experimenters need to examine that question with their cuts in mind. For $\tilde{m}_\gamma\gtrsim 10$ GeV, but $\tilde{m}_{\ell^\pm}>\tilde{m}_\gamma$, the leptons from $\tilde{\ell}^{\pm}$ decay probably would not have been detected, so the limits on $\tilde{m}_{\ell^\pm}$ do not hold -- i.e., it appears that a 12 GeV $\tilde{\ell}^{\pm}$ and a 10 GeV photino would contradict no accelerator data. [For the $\tilde{\mu}$ there is a restriction[13] of order 15 GeV from g-2, but this is close enough so that it could probably be made acceptable.] The decay lifetime is still quite short, with a width at least in the MeV range unless $\tilde{\ell}^{\pm}$ and $\tilde{\gamma}$ are nearly degenerate.

Even if $\tilde{\gamma}$ were heavier than $\tilde{\ell}$, it is very unlikely that there could be a charged stable supersymmetric partner, since cosmological arguments suggest there would then be too much energy density in charged matter. There are several possibilities under consideration: (a) It could be that sneutrinos were the lightest stable partner[4]. Then $\tilde{\gamma}\to\tilde{\nu}\nu$ via one loop, and $\tilde{\ell}\to\ell\tilde{\nu}\nu$ with a lifetime on the weak scale. (b) If sneutrinos can get a vacuum expectation[14] value, then photinos will decay to normal particles via one loop, $\tilde{\gamma}\to\gamma\nu$. Then $\tilde{\ell}\to\ell\nu\gamma$. In such cases the final state signatures, including perhaps the decay of a long-lived charged object, are accessible at e^+e^- colliders and are generally somewhat different from standard signatures. Some relevant data is mentioned in the photino section.

Scalar Neutrinos (sneutrinos) $\tilde{\nu}$

The present limits on sneutrinos are not very strong. The best one can do so far[15] is to observe that there would be modifications to τ decay properties if the decay $\tau\to\ell\tilde{\nu}_\ell\tilde{\nu}_\tau$ could occur, as in Fig. 1. Depending on the wino mass, this gives limits which qualitatively imply $m(\tilde{\nu}_\tau)+m(\tilde{\nu}_\ell)\gtrsim m_\tau$ for $\ell=e,\mu$.

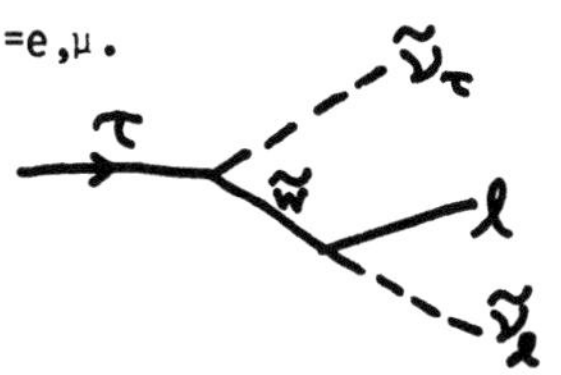

Fig. 1

There are two ways to look for more massive $\tilde{\nu}$.

(a) one can have $e^+e^- \to \tilde{\nu}\bar{\tilde{\nu}}$

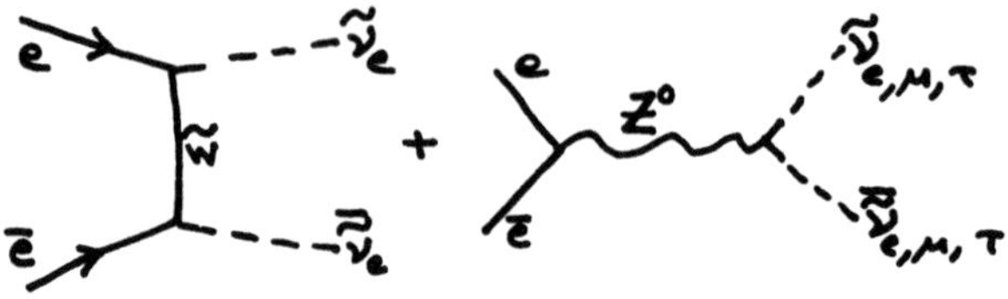

Fig. 2

as in Fig. 2. This will occur if $m(\tilde{\nu})$ is not too large, with the contribution of Fig. 2 giving $\sigma<1$ pb. On the Z°, the branching ratio for $Z^\circ \to \tilde{\nu}_x \bar{\tilde{\nu}}_x$ is 3% times phase space. The crucial question for detection is the signature, which requires a knowledge of the $\tilde{\nu}$ decay branching ratios. This has been studied by Barnett, Lackner, and Haber recently.[16]

One decay is via a loop, $\tilde{\nu}\to\nu\tilde{\gamma}$. An example is shown in Fig. 3a.

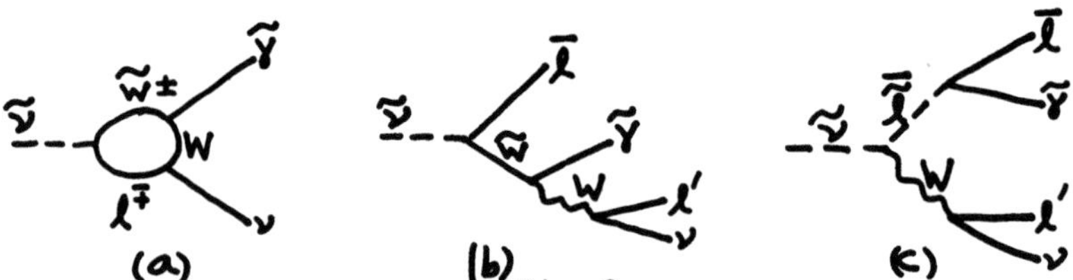

Fig. 3

Alternatively, one could have $\tilde{\nu}\to\nu\tilde{\gamma}\ell'\bar{\tilde{\ell}}$ with a number of contributions, two of which are shown in Fig. 3b,c. In general one can consider a number of alternatives -- e.g. the $\tilde{w}$ or the $\tilde{\ell}$ in Fig. 3b,c could be light enough to be physical particles, in which case those contributions clearly dominate. If $\tilde{w},\tilde{\ell}$ are heavy, the conclusion of Ref. 16 is that the neutral mode $\tilde{\nu}\to\nu\tilde{\gamma}$ will dominate, but for certain ranges of parameters ($m(\tilde{w})$ or $m(\tilde{e})$ not too large) the branching ratio for $\nu\tilde{\gamma}\ell'\bar{\ell}$ can be as large as 50%. If $BR(\nu\tilde{\gamma}\ell'\bar{\ell})$ is large the signature is quite good, with only a very soft pair of leptons produced by one $\tilde{\nu}$ decay and only missing neutrals from the other. Under these conditions, unstable light $\tilde{\nu}$ could be found at present e^+e^- machines.

If such a signature were detected, it could be distinguished from other kinds of new physics by details of angular distributions, and by comparing e^+e^-, $\mu^+\mu^-$, μe modes -- here all are present, with extra electrons because of the w exchange diagram in Fig. 2, whereas for $\tilde{\gamma}_2\tilde{\gamma}_1$ final states (see below) one can have e^+e^- and $\mu^+\mu^-$ but not μe.

If the $\nu\tilde{\gamma}$ mode totally dominstes, one can say very little here. Eventually, neutrino pair counting experiments could put a limit on the $\tilde{\nu}$ mass, and the absence of the branching ratio for $\nu\tilde{\gamma}\bar{\ell}\ell'$ constrains parameters.

(b) The second possibility is

$$W^{\pm} \rightarrow \tilde{\ell}^{\pm}\,\tilde{\nu}_{\ell}$$

which can occur if it is kinematically allowed. The rate can be large, with

$$\frac{\Gamma(W\rightarrow\tilde{\ell}\tilde{\nu}_{\ell})}{\Gamma(W\rightarrow\ell\nu)} = \frac{1}{2}\left\{[1-(m(\tilde{\ell})+m(\tilde{\nu}))^2/m_W^2]\;[1-(m(\tilde{\ell})-m(\tilde{\nu}))^2/m_W^2]\right\}^{3/2}$$

The signature is not too bad[16]. Assuming light photinos, presumably $\tilde{\ell}\rightarrow\ell\tilde{\gamma}$ and the photino escapes. If $\tilde{\nu}\rightarrow\nu\tilde{\gamma}$, only the ℓ can be detected, and it has an energy spectrum which depends on $m(\tilde{\ell})$, $m(\tilde{\nu})$. If $\tilde{\nu}\rightarrow\nu\tilde{\gamma}\bar{\ell}\ell'$ one has three soft charged leptons. The escaping particles give missing E_T, $\not{p}_T$, and the single or three leptons are isolated, not in hadron jets and not accompanied by hadron jets.

The background from W decays and from QCD effects involving heavy quark semileptonic decays will be serious and needs to be studied.

<u>Scalar quarks (squarks) $\tilde{q}$</u>

Squarks can be looked for at both e^+e^- and hadron colliders. At e^+e^- machines the situation is somewhat similar to sleptons, but less good because (a) there is 1/9 or 4/9 the rate because of the smaller electric charge (in addition to the 1/4 for scalars), and (b) presumarly $\tilde{q}\rightarrow q\tilde{\gamma}$ (with the same caveats for heavy $\tilde{\gamma}$) so one has non-collinear jets with missing energy (since the photinos escape). This is certainly detectable but much harder to see without careful study. At the present time it is not clear to me that there exist any firm limits on squark masses, although limits in the 10-15 GeV range could probably be imposed with present data (or a squark signal detected).

At hadron colliders one can have

$$\bar{p}p, pp \rightarrow \tilde{q}\tilde{\bar{q}}X$$

with reasonable cross sections, but combining rate and signature considerations perhaps the best way to search[17] is for

$$\bar{p}p, pp \to \tilde{q}\tilde{\gamma}X$$

via the elementary processes shown in Fig. 4. Then the signature is

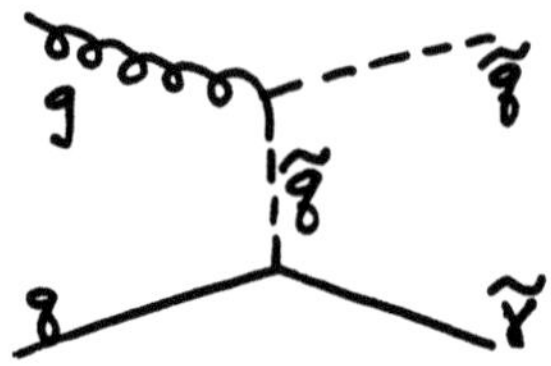

Fig. 4

large missing $\not{p}_T$ from the escaping $\tilde{\gamma}$, with a single quark jet at large p_T from $\tilde{q} \to q\tilde{\gamma}$. A full analysis in terms of useful variables has been done by Littenberg, Paige and collaborators, including background effects. They conclude that at the proposed CBA machine one can detect $\tilde{q}$ of mass 100 GeV and perhaps higher, with 125 GeV conceivably feasible. Signal/noise studies have not been done for other facilities.

Photinos ($\tilde{\gamma}$)

There are no direct experimental constraints on photinos, nor is there any compelling argument about what mass they should have. Cosmological arguments analagous to the usual ones for neutrinos suggest[18] the mass should either be less than about 100 eV (actually, that the sum of all light particles masses, neutrinos + photinos + ..., should be less than about 100 eV) or above a GeV; the latter range has been studied quantitatively by Goldberg[7], who calculated a lower limit of about 2 GeV for scalar fermion masses $m(\tilde{f})$ below about 50 GeV and then increasing rapidly with $m(\tilde{f})$ to about 15 GeV at $m(\tilde{f}) \simeq m_W$, as shown in (his) Figure 5. Theoretically, the photino starts out massless before the SUSY

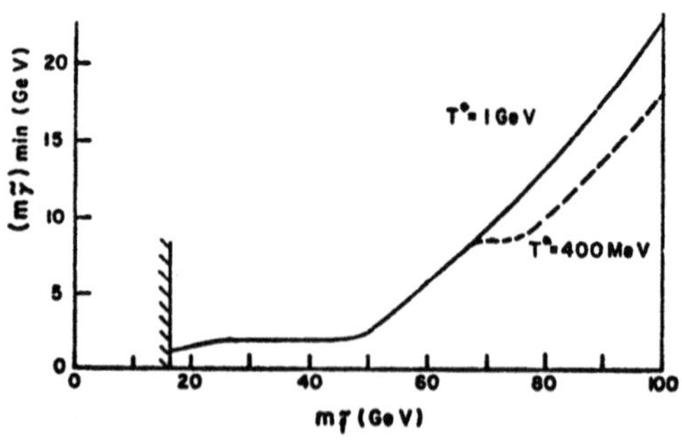

Fig. 5

is broken, and can get mass directly or radiatively. In models radiative contributions to the mass are often related to gluino masses by $m(\tilde{\gamma}) \simeq k\alpha m(\tilde{g})/\alpha_s$ where k is a group factor of order unity. But in some models[19] the expectation is $m(\tilde{\gamma})=m(\tilde{g})$ as all gauginos get a common mass. If photinos are heavy, many of the experimental tests of SUSY and ways to find SUSY particles get more difficult and subtle.

One direct test may be possible. Photinos can be produced in e^+e^- (or $q\bar{q}$) annihilation as in Fig. 6. It has been suggested[20] that the

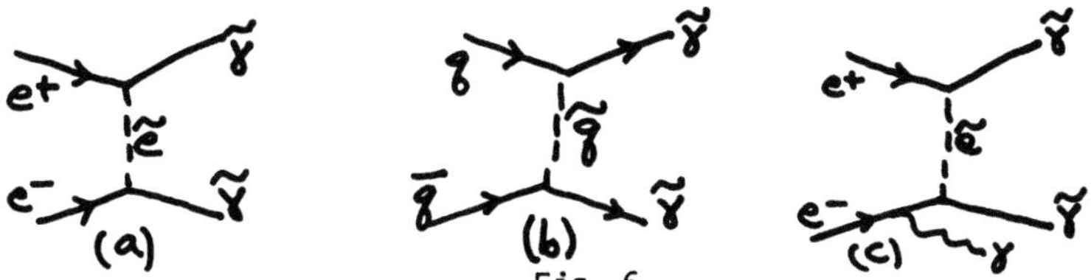

Fig. 6

search analagous to neutrino counting is viable, and it may be attempted[21] in the near future. The rate for Fig. 6c is large large enough to detect if $m(\tilde{e})$ is less than about 40 GeV. At future hadron colliders Fig. 6b will give an interesting signature, and a gluon jet can be radiated from one of the quark lines to provide a recoil trigger.

If photinos decay to photon + anything, it is possible to search by using $e^+e^- \to \tilde{\gamma}\tilde{\gamma} \to \gamma\gamma X$. The mode $\tilde{\gamma} \to \gamma\tilde{G}$ is discussed in Ref. 20, 18, and $\tilde{\gamma} \to \gamma\nu$ in Ref. 14 and the slepton section above. A search with no candidate events based on 7.1 pb^{-1} at 34 GeV/c has been reported in Ref. 22.

Gluinos ($\tilde{g}$)

Gluinos are the partners of gluons and consequently they are a color octet of electrically neutral particles. When the SUSY is unbroken they are degenerate with the gluons, at zero mass. In models they acquire masses ranging from rather small ones below a GeV, to hundreds of GeV. Typical models give 15-100 GeV for $m_{\tilde{g}}$.

If photinos are lighter than gluinos, the gluinos will decay via Fig. 7

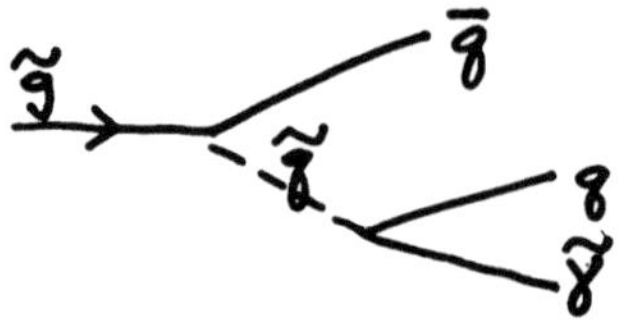

Fig. 7

with a lifetime (for light $\tilde{\gamma}$)

$$\tau_{\tilde{g}} \approx 10^{-6} \left(\frac{m_\mu}{\tilde{m}_g}\right)^5 \left(\frac{\tilde{m}_q}{m_W}\right)^4 \text{ sec.}$$

Gluinos, being colored, are produced by coupling to gluons in hadron collisions. The cross section for pair-producing gluinos is about an order of magnitude larger than that for quarks of the same mass.

If gluinos were very light they would be produced with mb cross sections, and then travel an observable distance, having been shielded by gluons (or conceivably[23] $q\bar{q}$ in a color octet) to give a new, long-lived neutral (or charged) hadron. It is unlikely[5,24] such states would have escaped detection.

If $\tilde{m}_g \gtrsim 1$ GeV, one can do a perturbative QCD calculation[5] for the production cross section of $\tilde{g}$, and obtain σ as a function of $\tilde{m}_g$. If gluinos are not detected, one assumes it is because the mass is large enough so that the production cross section is too small. [The mass of the hadron -- e.g. gluinoball -- contains some constituent mass, presumably of order a GeV, as well as the current algebra gluino mass.] Such arguments exclude $\tilde{m}_g \lesssim 1$ GeV.

For 1 GeV $\lesssim \tilde{m}_g \lesssim$ 6-8 GeV, beam dump experiments with $E_{beam} \lesssim 1$ TeV are relevant. Gluino pairs are produced in the dump. The $\tilde{g}$ decay as in

Fig. 7. The photinos travel downstream until they interact in the beam dump detectors. The photinos interact[25] as in Fig. 8.

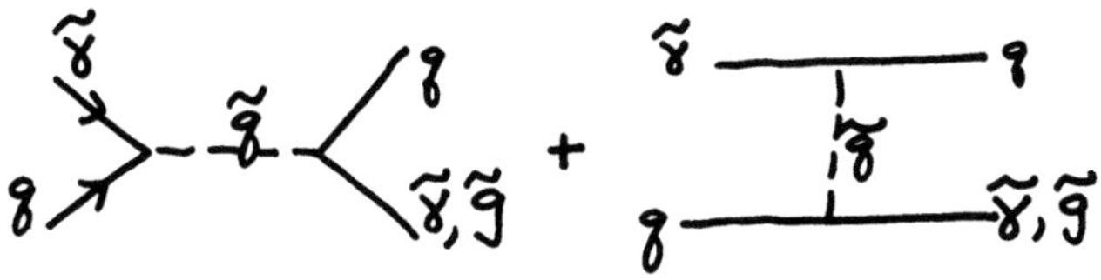

Fig. 8

Which contributions dominate depends on the photino energy and on the masses. In all cases, there are extra candidates for neutral current events in the detector because no charged lepton appears. [If such candidates appear, the y, E_{vis} distributions are different from those for neutrinos.] The absence of such events allows an experiment to set limits on the gluino mass; the limits depend on $\tilde{m}_q$ since the photino interaction probability varies as $\tilde{m}_q^{-4}$. The FNAL beam dump experiment[26] has the limits shown in Fig. 9, and the CHARM experiment[27] has similar limits, slightly less stringent since their detector was further from the dump.

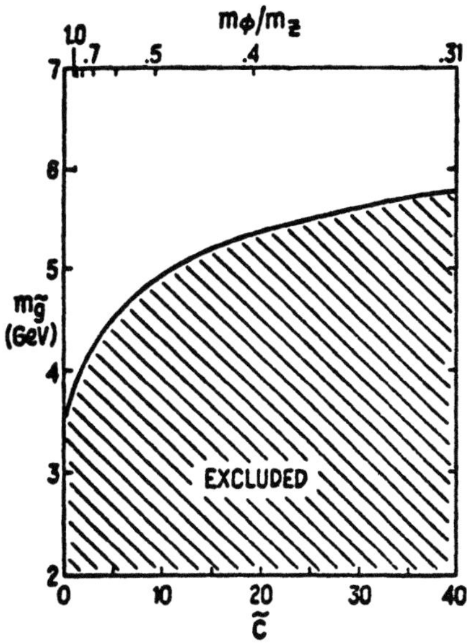

Fig. 9

If photinos are heavy, the gluinos do not decay, and there is no limit beyond the absence of stable new hadrons, under the assumptions discussed in the introduction.

At hadron colliders more massive gluino pairs can be produced. The signature is considerable missing $\vec{p}_T$ because of the escaped $\tilde{\gamma}$'s. An analysis has been done by Paige, Littenberg and collaborators[17] to make this quantitative, and to see the effect of detector efficiency. Basically they define variables such as $x_E = -\vec{p}_T \cdot \vec{p}_T{}' / |\vec{p}_T|^2$ where $\vec{p}_T$, $\vec{p}_T{}'$ are simply the transverse momenta of the particles in in opposite hemispheres, whether clear jets are visible or not. If normal QCD jets were formed, one tends to find $\vec{p}_T \approx -\vec{p}_T{}'$, so $x_E \approx 1$. For gluinos there is missing $\vec{p}_T$ so $|\vec{p}_T| \neq |\vec{p}_T{}'|$ and the clumps are at an angle, so x_E is less than unity, and almost flat down to $x_E \sim 0$. Then a cut at $x_E = 1/2$ eliminates almost all QCD background, but only about 2/3 of the signal. QCD background with $x_E < 1$ often arises from events with a heavier quark and a semileptonic decay, so a cut to eliminate events with a hard charged lepton further enhances the $\tilde{g}$ sample. Monte Carlo studies based on these procedures suggest that the SPS collider can produce and detect gluinos for $\tilde{m}_g < 25$ GeV, FNAL can reach $\tilde{m}_g \lesssim 90$ GeV, and CBA can reach $\tilde{m}_g \lesssim 140$ GeV. Very crudely, future high energy hadron colliders with luminosities of $10^{30}/\mathrm{cm}^2\mathrm{sec}$ can reach $x \approx 2\tilde{m}_g/\sqrt{s} \lesssim 0.1$ and those with luminosities of $10^{33}/\mathrm{cm}^2\mathrm{sec}$ can reach $x \lesssim 1/3$.

Although gluinos, being colored, but having no electric charge or weak isospin, are most easily produced by hadrons, they occur[28] with a reasonable signature in Z° decay at one loop, as in Fig. 10.

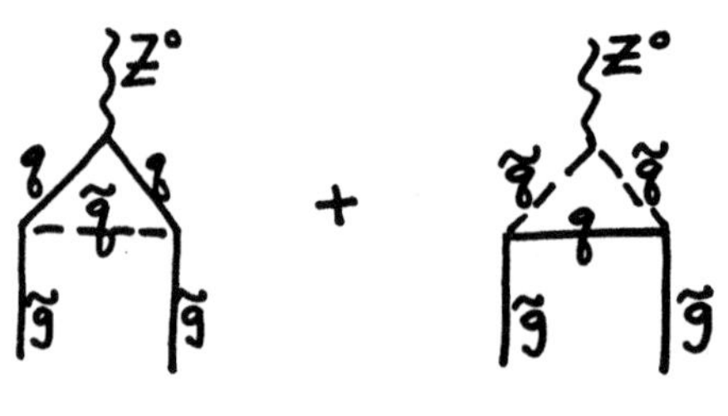

Fig. 10

The branching ratio depends on m_t, $\tilde{m}_q$, and $\tilde{m}_g$. For typical values one gets $BR \lesssim 10^{-5}$, which is detectable though not easily. Given the expected chronological order of new machines, it is not likely that gluinos will be discovered this way, but if they are found with $\tilde{m}_g \lesssim 40$ GeV this is a nice check on the theory, and the branching ratio is sensitive to the scalar quark spectrum. In particular, every theoretical model has a definite prediction for BR($Z^\circ \to \tilde{g}\tilde{g}$).

Neutral, Spin 1/2, Color Singlet, SUSY Partners (zino, ziggsino, neutral higgsino, photino; $\tilde{z}$, $\tilde{\gamma}_2$, $\tilde{h}^\circ$, $\tilde{\gamma}_1$)

As shown above in Table I, there will be several neutral spin 1/2 SUSY partners, associated with the Z°, the photon, and the two neutral Higgs bosons H_1, H_2. [In addition, in many models a singlet boson often denoted Y is added for technical reasons, and its SUSY partner adds a fifth state to this set; I will not include it in this discussion.] The photino was discussed separately above since it is more familiar, but it should really be considered in the context of the full set of neutral spin 1/2 SUSY partners --- I will call them "neutralinos". The crucial consideration is that they all have the same charge and spin, and they are coupled by the (SU(2) breaking) mass generating mechanism, so the mass eigenstates are mixtures of the weak eigenstates. They have been discussed in Ref. 29, and studied extensively in Ref. 30, 31.

The weak eigenstates are denoted $\tilde{Z}^\circ, \tilde{\gamma}, \tilde{H}_1^0, \tilde{H}_2^0$. The mass eigenstates will have couplings that are mixed gauge-like couplings and Higgs-like couplings and are denoted $\tilde{z}^\circ$ for the heaviest, $\tilde{\gamma}_1$ for the lightest state with mainly photon-like couplings, $\tilde{h}_1^0$ for the lightest state with mainly Higgs-like couplings, and $\tilde{\gamma}_2^0$ for the state of intermediate mass -- the couplings of $\tilde{\gamma}_2$ will be part Higgs-like and part gauge-like (it is usually dominantly Higgs-like since $\tilde{z}^\circ$ and $\tilde{\gamma}_1$ are more gauge-like, so it might be more appropriate to use $\tilde{h}_2^0$ for its name, but it is its gauge-like part which leads to its phenomenological importance so we will stick to $\tilde{\gamma}_2$ as a useful mnemonic name). The neutralinos will be denoted collectively by $\tilde{y}^0$ when it is useful to refer to them as a group or without specifying precisely which state is meant.

A neutralino with essentially gauge couplings is called a zino, one with essentially Higgs-like couplings is called a neutral higgsino, one with strongly mixed couplings a ziggsino, and one with photon-like couplings a photino.

The masses of these neutralinos are (as all SUSY partner masses) very model dependent. Typically the $\tilde{z}^\circ$ mass is greater than m_Z, the $\tilde{\gamma}_1$ mass is 0.5-5 GeV, the $\tilde{h}^0$ and $\tilde{\gamma}_2$ masses in the 20-70 GeV range.

There are two especially good ways to look for neutralinos, both possibly applicable in the near future.

(a) In essentially all SUSY models the decay

$$W^{\pm} \to \tilde{w}_1^{\pm}\, \tilde{\gamma}^{\circ}$$

will occur, with a potentially detectable branching ratio. I will consider it in detail in the next section, since a discussion of the charged spin 1/2 states is also required.

(b) Specific to the neutralinos is one very hopeful place to look for SUSY partners[30],

$$e^+e^- \to \tilde{\gamma}_2\, \tilde{\gamma}_1$$

as shown in Fig. 11.

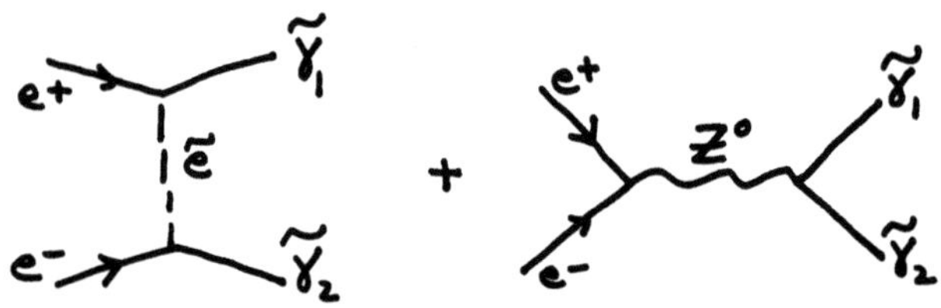

Fig. 11

Some models have $\tilde{m}_{\gamma_1} + \tilde{m}_{\gamma_2} < 35$ GeV, so this could be detected in the near future at PETRA/PEP. Almost all models have $\tilde{m}_{\gamma_1} + \tilde{m}_{\gamma_2} < 100$ GeV, so this will be effectively (though not rigorously) a crucial test of whether supersymmetry is relevant to physics on the weak scale.

The cross section for $\tilde{\gamma}_2\tilde{\gamma}_1$ production is of the form

$$\sigma \simeq \pi(\lambda\lambda')^2\, \alpha^2\, s/\tilde{m}_e^4$$

for $s \lesssim \tilde{m}_e^2$. The factors λ,λ' arise at the vertices because of the mixing of the weak eigenstates into the mass eigenstates -- typically $0.1 \lesssim \lambda\lambda' \lesssim 0.7$. In models, $\tilde{m}_e$ may be of order $m_{W/2}$ or less, up to or above m_W.

Thus typically

$$\sigma \gtrsim 0.1 \text{ pb}.$$

Since e^+e^- colliders are currently collecting more than pb^{-1}/day, they can be producing more than an event a week of $\tilde{\gamma}_2\tilde{\gamma}_1$.

The signature is quite good.[30] $\tilde{\gamma}_2$ can be seen to decay to any $f\bar{f}$ pair plus $\tilde{\gamma}_1$, as in Fig. 12.

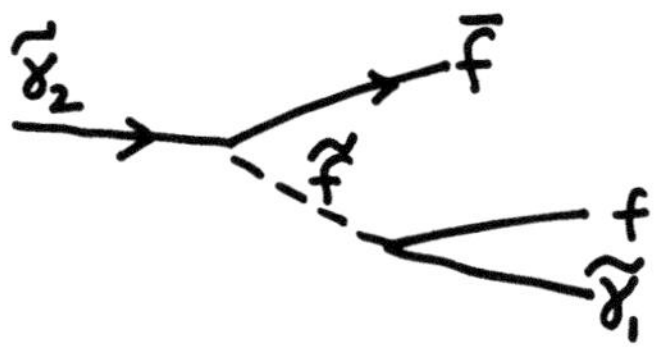

Fig. 12

where $f\bar{f}$ is e^+e^-, $\mu^+\mu^-$, $\tau^+\tau^-$, $q\bar{q}$. Since the primary and secondary $\tilde{\gamma}_1$ both escape, at least 2/3 of the energy of the event is missing, and there is a lepton pair or two jets with large $\vec{p}_T$ imbalance. [There is no standard model background for such events. Similar events could arise from $\tilde{\nu}\bar{\tilde{\nu}}$ production or production of a fourth generation massive neutrino, either of which would be as exciting to detect; one can distinguish what kind of new physics is occurring by various properties of the branching ratios, energy distributions, etc.]
Thus,

$$e^+e^- \rightarrow \tilde{\gamma}_2\tilde{\gamma}_1$$

provides one of the most promising ways to produce, detect, and study SUSY partners, either soon or at the future colliders. At high luminosity hadron colliders the same analysis holds and even larger $\hat{s}$ can be reached for $q\bar{q} \rightarrow \tilde{\gamma}_2\tilde{\gamma}_1$, so one will eventually have essentially a definitive test of any supersymmetric approach which produces particles on the weak scale. At CBA one could produce an event a weak for constituent cross sections down to about 0.01 pb and for masses of $\tilde{\gamma}_1+\tilde{\gamma}_2$ up to well over 100 GeV.

Charged, Spin 1/2, Color Singlet SUSY Partners (wino, wiggsino,charged higgsino; $\tilde{W}_1^{\pm}, \tilde{W}_2^{\pm}$)

The charged, spin 1/2 SUSY partners of W^+ and of H_1^+ are a pair of weak eigenstates that are mixed by mass terms in general, forming two mass eigenstates -- called "charginos". The mass eigenstates will have couplings that are a mixture of gauge-like couplings and Higgs-like couplings. I will denote the charginos as $\tilde{W}_1^{\pm}, \tilde{W}_2^{\pm}$. When a chargino has essentially gauge couplings it is called a wino, when it has essentially Higgs couplings it is called a charged higgsino, and when it has strongly mixed couplings it is called a wiggsino. The mass eigenstates are collectively denoted $\tilde{w}^{\pm}$. They have been discussed in References 29-35.

The chargino mass matrix will have the form

$$\begin{matrix} \tilde{w}^{+} \tilde{H}_1^{+} \end{matrix} \begin{pmatrix} M & g_2 v_2 \\ g_2 v_1 & \mu \end{pmatrix} \begin{matrix} \tilde{w}^{-} \\ \tilde{H}_2^{-} \end{matrix}$$

where $v_i = \langle 0|H_i^0|0\rangle$, $m_W = g_2\sqrt{(v_1^2+v_2^2)/2}$. The matrix is labeled in terms of two-component weak eigenstates. It may be that $v_1 \neq v_2$. M is a gaugino mass term and μ is from a Higgs sector coupling -- both are unknown and must originate in SUSY breaking physics. The eigenstates of the mass matrix are $\tilde{w}_1, \tilde{w}_2$. One of the mass eigenvalues can be smaller than m_W, and it will be smaller[32] if M=0 or if μ=0 [then the determinant of the mass matrix is $g_2^2 v_1 v_2$ in magnitude, and by comparison with m_W one can see one eigenvale is smaller], but it could easily happen that both eigenvalues are greater than M_W. In a number of present models, one eigenvalue is typically $M_W/3$ or $M_W/2$, as has been emphasized in Ref. 32, 33.

If the above mass matrix is mainly off-diagonal (M, μ small), then the eigenstates are Dirac particles

$$\tilde{w}^{\pm}_{1,2} = \begin{pmatrix} \tilde{W}^{+} \\ \tilde{H}^{-*} \end{pmatrix}, \quad \begin{pmatrix} \tilde{H}^{+} \\ \tilde{W}^{-*} \end{pmatrix}$$

with mixed couplings, i.e. wiggsinos. If a diagonal term is large compared to m_W, then the eigenstates are mainly wino or higgsino,

$$\tilde{w}^{\pm}_{1,2} = \begin{pmatrix} \tilde{W}^{+} \\ \tilde{W}^{-*} \end{pmatrix}, \quad \begin{pmatrix} \tilde{H}^{+} \\ \tilde{H}^{-*} \end{pmatrix} \quad ,$$

with gauge-like or Higgs-like couplings.

With these explanations, one can examine the production and decay of the charginos. One should think of the above Dirac eigenstates in the same way as one views an electron as

$$e^{-} = \begin{pmatrix} e_L^{-} \\ e_R^{+*} \end{pmatrix},$$

with e_L, e_R as degenerate in mass but different in weak interaction properties (SU(2) doublet for e_L, singlet for e_R). We speak of separate production of e_L, e_R by the weak interactions, while the electromagnetic interactions do not distinguish them. Thus above one can have separate production of upper or lower pieces of $\tilde{w}_i^{\pm}$.

The two most interesting ways to produce charginos soon are

(a) $e^+e^- \to \tilde{w}^+\tilde{w}^-$ (Ref. 29, 30, 31, 33)

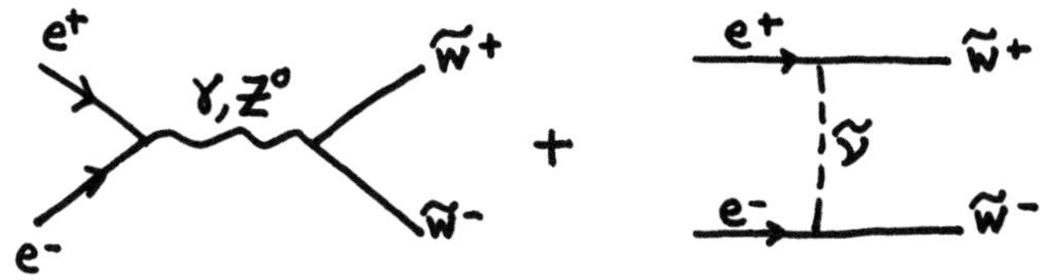

Fig. 13

(b) $\bar{p}p$, $pp \to W^{\pm}X$ $\;\;\hookrightarrow \tilde{w}^{\pm}\tilde{\gamma}^{0}$ (Ref. 30-33)

For (a), the cross section has the unit of R for production of a charged fermion via a photon, plus the Z° and $\tilde{e}$ contributions. The coupling to a Z° is large whether $\tilde{w}$ is wino or higgsino, and the $\tilde{e}$ contribution could be significant for lighter $\tilde{e}$.

The signatures [30,31,33] can be seen from the possible $\tilde{w}$ decays.

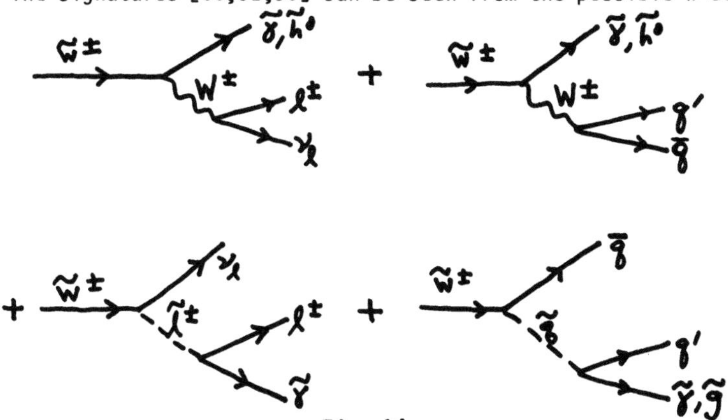

Fig. 14

The relative rates of these contributions depend on the ratio of vacuum expectation values v_1/v_2, on $\tilde{m}_\ell$, $\tilde{m}_q$, $\tilde{m}_g$, $\tilde{m}_\gamma$ etc. In crudest approximation all modes could be comparable in size.

Since $\tilde{w}^{\pm} \to \nu_\ell \ell^{\pm}\tilde{\gamma}$ looks like the decay of a heavy lepton if $\tilde{\gamma}$ is light, people have tended to say that $\tilde{w}$ is excluded for $\tilde{m}_W < 18$ GeV by PETRA and PEP data. However, because of the changes if $\tilde{\gamma}$ is massive, and various details of the SUSY couplings, published data does not provide firm limits on $\tilde{w}^{\pm}$ masses, and they still could be found[30] below 18 GeV. The essential point is that often enough energy is carried off undetected in these decays so that the events in question would not have passed typical experimental cuts. The charged lepton spectrum becomes much softer than for a typical heavy lepton decay.

At higher $\sqrt{s}$ one can use the forward-backward asymmetry and the rate on the Z° to distinguish[29,30,31] among winos, wiggsinos, and higgsinos. For a heavy lepton $g_V^2+g_A^2=0.25$, for a wino $g_V^2+g_A^2=2.43$, for a higgsino $g_V^2+g_A^2=0.31$, and for a wiggsino $g_V^2+g_A^2=1.37$. Intermediate mixing is possible, of course. The asymmetry[36] is shown below [(a) is a normal lepton, (c) either a pure wino or pure higgsino, (d) a wiggsino with $\tilde{w}_L=\tilde{w}^-$. A wiggsino with $\tilde{w}_L=\tilde{H}^-$ gives the opposite sign for (d), and curves numbered 1,2,3 correspond to m=0,20 GeV, 40 GeV for production of a massive pair. Forward and backward are defined by integrating over 0<|cosθ|<0.8. The reduction is caused by threshold effects and by helicity flip of massive states giving interference effects.] One can see that SLC, LEP will be very well suited to untangle many of the properties of SUSY partners, both charginos and neutralinos.

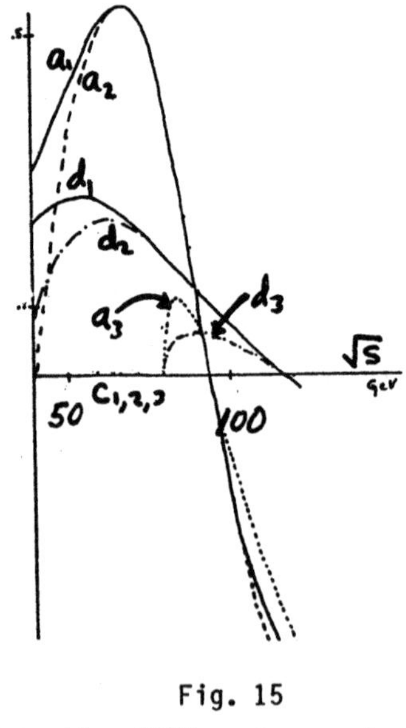

Fig. 15

Next consider detecting SUSY partners via W decay,

$W^{\pm} \to \tilde{w}^{\pm}\tilde{y}^{o}$

Calculating the branching ratio is somewhat subtle. The interaction Lagrangian contains terms ($\tilde{m}_{w_2} < \tilde{m}_{w_1}$, $v_2 < v_1$)

$$\mathcal{L} \sim W_{\mu}^{+} \{e\, \tilde{\gamma}_1\, \gamma^{\mu} R\, \tilde{w}_2^{-} - \frac{g_2}{\sqrt{2}} \frac{v_2}{v} \tilde{\gamma}_2 \gamma^{\mu} L\, \tilde{w}_2^{-}$$

$$+ \frac{g_2}{\sqrt{2}} \frac{v_1}{v} \tilde{h}^{o}\, \gamma^{\mu} L\, \tilde{w}_2^{-} + \frac{g_2}{\sqrt{2}} \bar{e}\, \gamma^{\mu} L\, \nu + \ldots\} \; .$$

Thus one expects three different decays if all are allowed by phase space,

$$W^{\pm} \to \tilde{w}_2 + \tilde{\gamma}_1, \tilde{\gamma}_2, \tilde{h}^{o} \; .$$

If all final masses were small, the sum of partial branching rates if

$$\Gamma(W \to \tilde{w}\tilde{y})/\Gamma(W \to e\nu) < \frac{e^2 + \frac{g_2^2}{2}\left(\frac{v_1^2+v_2^2}{v^2}\right)}{g_2^2/2} = 1 + 2\sin^2\theta_W \approx 1.45.$$

However, one expects a significant reduction by phase space. For W decay into final state masses m,μ the phase space factor is

$$\left[1-(m^2+\mu^2)/2M_W^2 - (m^2-\mu^2)^2/2M_W^4\right] \left[(1-(m+\mu)^2/M_W^2)\,(1-(m-\mu)^2/M_W^2)\right]^{1/2} \; .$$

The signatures[30,31,33] are quite different, and depend on which of the $\tilde{w}$ decay modes dominate. For $\tilde{w}\,\tilde{\gamma}_1$ the $\tilde{\gamma}_1$ will escape, giving the large missing $\vec{p}_T$ siganture. $\tilde{w}$ may decay to $\ell^{\pm}\nu\tilde{\gamma}_1$, $\bar{q}q'\tilde{\gamma}'$, or $\bar{q}q'\tilde{g}$. The $\bar{q}q'\tilde{\gamma}$ events have been discussed in detail as a good signature (the "Zen" events) in Ref. 31. For the other modes both states decay, so one has $\tilde{w}$ decays on one side and $\tilde{\gamma}_2$ or $\tilde{h}^{o}$ decays on the other. There are various cases, all with some missing $\vec{p}_T$, and with some charged leptons or quark jets.

Because of the dilution of the signal by branching ratio effects (several decays with various signatures), and because of background effects [these have been studied in some detail by Littenberg and Paige[17]-- they find that with strong cuts, which reduce the signal, one can get ordinary standard model background that mimics the signal to the level where signal-to-noise is somewhat better than unity] it will not be easy to find a signal for $W \to \tilde{w}\tilde{y}$ unless we are very lucky about which modes occur and which branching ratios dominate. Nevertheless, it is such a good opportunity that experimenters should search mode-by-mode for a possible signal.

Acknowledgements

I am grateful to S. Raby, H. Haber, J. Hagelin, J. Leveille, and especially J.-M. Frere, for stimulating and instructive conversations and collaborations.

References

[1] For recent reviews of the theory, history, motivation, and models, see P. Fayet, Proceedings of the 21st International High Energy Physics Conference, Paris 1982, ed. P. Petiau and M. Porneuf, and Ref. 2. For an earlier review of experimental aspects, see I. Hinchliffe ed L. Littenberg, Ref. 3.

[2] J. Polchinski, these proceedings.

[3] Proceedings of the 1982 DPF study on Future Facilities, Snowmass, ed. R. Donaldson, R. Gustafson, and F. Paige.

[4] J. Hagelin, G.L. Kane, and S. Raby, in preparation.

[5] G.L. Kane and J.P. Leveille, Phys. Lett. 112B, 227 (1982).

[6] H.E. Haber and G.L. Kane, in preparation.

[7] H. Goldberg, Northeastern preprint NUB-2592 (1983).

[8] A useful review is K.H. Lau, SLAC-PUB-3001, Oct. 1982. See also H. Behrend et al., Phys. Lett. 114B, 287 (1982) CELLO Collaboration; W. Bartel et al., Phys. Lett. 114B, 211 (1982) JADE Collaboration; D. Barber et al., Phys. Rev. Lett. 45, 1904 (1981) MARK J Collaboration; R. Brandelik et al., Phys. Lett. 117B, 365 (1982) TASSO Collaboration; C.A. Blocker et al., Phys. Rev. Lett. 49, 517 (1982) MARK II; D. Ritson, XXI International Conference on High Energy Physics, Paris, July 1982, MAC.

[9] F.B. Heile et al., Nucl. Phys. B138 (1978) 189.

[10] M.K. Gaillard, L. Hall, and I. Hinchliffe, Phys. Lett. 116B 279 (1982).

[11] See Hinchliffe and Littenberg, Ref. 3, p. 248, for the Mark II result.

[12] B. Adeva et al., MARK J Collaboration, Phys. Rev. Lett. 48 967 (1982).

[13] P. Fayet, in "Unification of the Fundamental Particle Interactions (1980) ed, by. S. Ferrara, J. Ellis, and P. Van Nieuwenhuizen, p.587. See also, J.A. Grifols and A. Mendez, Barcelona preprint UABFT-84 (revised), and R. Barbieri and L. Maiani, Phys. Lett. 117B (1982) 203.

[14] J.-M. Frere and G.L. Kane, to be published. This question has also been discussed by Aulakh and Mohapatra, and is under consideration by Gaillard, Hall, and Suzuki and by Polchinski and Wise.

[15] G.L. Kane and W. Rolnick, in preparation.

[16] R.M. Barnett, K. Lackner, and H.E. Haber, SLAC-PUB-3066 (1983) and SLAC-PUB-3105 (1983).

[17] L. Littenberg and F. Paige and collaborators, private communication and Ref. 3.

[18] N. Cabibbo, G. Farrar, and L. Maiani, Phys. Lett. 105B (1981) 155.

[19] E. Cremmer, P. Fayet, and L. Girardello, Phys. Lett. 122B (1983) 41.

[20] P. Fayet, Phys. Lett. 117B (1982) 460; J. Ellis and J. Hagelin, Phys. Lett. 122B (1982) 303.

[21] D. Burke, private communictaion.

[22] H.J. Behrend et al., CELLO collaboration, Phys. Lett. 123B, 127 (1983).

[23] T. Goldman, Phys. Lett. 78B 110 (1978).

[24] G. Farrar and P. Fayet, Phys. Lett. 76B (1978) 575; 79B (1978) 442; P. Fayet, Phys. Lett. 78B (1978) 417; See also P.R. Harrison and C.H. Llewellyn Smith, Oxford preprint, 1982.

[25] P. Fayet, Phys. Lett. 86B (1979) 272.

[26] R.C. Ball et al., Univ. of Michigan preprint UM HE 82-21, Submitted to the XXI International Conference on High Energy Physics, Paris, July 1982.

[27] F. Bergsma et al., Phys. Lett. 121B, 429 (1983).

[28] G.L. Kane and W. Rolnick, Nucl. Phys. B217 (1983) 117.

[29] J. Ellis and Graham G. Ross, Phys. Lett. 117B 397 (1982).

[30] J.-M. Frere and G.L. Kane, Univ. of Michigan preprint UM TH 83-2, Feb. 1983.

[31] J. Ellis, John S. Hagelin, D.V. Nanopoulos, and M. Srednicki, SLAC-PUB-3094, April 1983.

[32] S. Weinberg, Phys. Rev. Lett 50 (1983) 387.

[33] R. Arnowitt, A.H. Chamseddine, and Pran Nath, Phys. Rev. Lett. 50 232 (1983).

[34] P. Fayet, Ecole Normale Superieure preprint LPTENS 83/16.

[35] Luis Alvarez-Gaumi, J. Polchinski, and Mark B. Wise, Harvard preprint HUTP-83/A063.

[36] The curves for the massive case are due to calculations by J.-M. Frere, private communication.

LOW ENERGY SUPERGRAVITY: THE MINIMAL MODEL

Joseph Polchinski

Lyman Laboratory of Physics
Harvard University
Cambridge, Massachusetts 02138

ABSTRACT

This is a review of recent models in which $N = 1$ supergravity plays a role in the physics at the weak interaction scale. I focus on models with the minimal low energy particle content - the non-supersymmetric standard model, plus an additional Higgs, plus the superpartners of these. My contribution to this was done in collaboration with M. Wise and L. Alvarez-Gaumé.

In the past few years there has been a proliferation of models in which the known particles have superpartners at or below the weak interaction scale. The idea of these models is to cancel the large radiative corrections to the Weinberg-Salam Higgs scalar mass and cure the naturalness problem [1]. Since the world is not supersymmetric, any such model must have a source of supersymmetry breaking. The existing models fall neatly into classes according to the way that the known particles couple to the supersymmetry (SS) breaking. In the case of spontaneous SS breaking the quantity of interest is g_{fbG}, the Yukawa coupling between a fermion, its bosonic partner, and the Goldstone fermion (Goldstino).

In the Fayet models [2], one has

$$g_{fbG} = e' = O(1) \qquad \sqrt{f_G} \sim 10^2\,\mathrm{GeV} . \tag{1}$$

The couplings of the quarks and leptons to the Goldstino are tree level Yukawa couplings, related by SS to a new U(1) gauge coupling e'. The quantity f_G is the Goldstino decay constant, which measures the overall scale of SS breaking; f_G^2 is the scale of typical contributions to the vacuum energy density. The splitting between superpartners is given by

$$m_f^2 - m_b^2 \sim f_G \, g_{fbG} \; . \tag{2}$$

This is the Goldberger-Trieman relation for broken SS. It says simply that the splitting in any supermultiplet is the overall scale of SS breaking times the strength with which that supermultiplet couples to the Goldstino. Since the left-hand side of (2) is $\sim 10^2$ GeV in any model which tries to cure the naturalness problem, f_G must be taken to be larger as we consider models with smaller g_{fbG}.

The Fayet models were very promising, but suffer from the defect that it is exceedingly difficult to make a model of this type with the correct vacuum, with SS broken and color unbroken [3,4]. The difficulty of making an acceptable model in which g_{fbG} is a tree level coupling has led to models in which the coupling to the Goldstino is less direct. Several groups [3,5,6] studied models in which g_{fbG} was absent at tree level but induced at some number of loops by radiative corrections. In this case one has

$$g_{fbG} \sim \left(\frac{g^2}{4\pi}\right)^n \leqslant 10^{-2} \qquad \sqrt{f_G} \geqslant 10^3\,\text{GeV} \tag{3}$$

where g is a generic coupling. A limit of this idea is the case where the coupling to the Goldstino comes about only from loops of superheavy fields [7,8]

$$g_{fbG} \sim \frac{M_W}{M_{GUT}} \left(\frac{g^2}{4\pi}\right)^n \leqslant 10^{-16} \qquad \sqrt{f_G} \geqslant 10^{10}\,\text{GeV} \; . \tag{4}$$

This possibility was suggested by the inverted hierarchy models [9], in which it was found to happen accidentally and automatically [10].

This last coupling is so small that it is comparable to the supergravitational coupling between the Goldstino and ordinary matter

$$g_{fbG} \sim \frac{M_W}{M_P} \sim 10^{-17} \tag{5}$$

(M_P = Planck scale). Thus, in models with supersymmetry broken at 10^{10} GeV or above, the effects of gravity cannot be neglected [11,12,13]. Indeed, one could imagine that the renormalizable couplings between the Goldstino and ordinary fields are entirely absent, and only those induced by supergravity remain. This idea has attracted a great deal of attention, partly because it is exciting that gravity might play a key role in low energy physics, and partly because the resulting models are the simplest of all the realistic models of broken SS. It may still be that the correct model has both couplings (4) and (5) at competitive levels, but here we will focus on the purely gravitational coupling. In this case $\sqrt{f_G} \sim \sqrt{M_W M_P} \sim 10^{10-11}$GeV. The models then have three sectors:

(1) The "observable" sector contains the quarks, leptons, gauge and Higgs bosons, plus all the superpartners of these, as well as all the fields associated with grand unification.

(2) The "hidden" sector contains the Goldstino and its bosonic partner, and perhaps additional fields.

(3) Supergravity. The first and second sectors both couple to supergravity, but not to each other. Thus, ignoring supergravity, the physics we see, sector (1), would be entirely supersymmetric. Since supergravity is still a very weak force at the scale of particle physics, sectors (2) and (3) are almost invisible to us. The only effects which survive are those which involve the large scale f_G to compensate the small gravitational coupling. The net result is that the physics at ordinary energies look like sector (1) plus soft SS breaking terms induced by supergravity coupling to the hidden sector. We outline the analysis, which is given more fully in Refs. 12-14. The interesting tree level effects come, not from exchange of gravitons or gravitinos, but from exchange of auxiliary fields present when gravity is supersymmetrized. These have been summed into an effective scalar potential [12,15].

$$V(Z) = \exp(\kappa^2 \sum_i Z_i^* Z_i) \left\{ \sum_i \left| \frac{\partial f(Z)}{\partial Z_i} + \kappa^2 Z_i^* f(Z) \right|^2 - 3\kappa^2 |f(Z)|^2 \right\} + \tfrac{1}{2} \sum_a D_a^2(Z, Z^*) \ . \tag{6}$$

This messy expression involves the scalar fields Z_i, the superpotential f, and the gravitational coupling $\kappa^2 = 8\pi/M_P^2$. The D-terms have the usual globally SS form. Turning off gravity ($\kappa \to 0$) gives the usual global result

$$V(Z) = \sum_i \left| \frac{\partial f(Z)}{\partial Z_i} \right|^2 + \tfrac{1}{2} \sum_a D_a^2(Z, Z^*) \ . \tag{7}$$

In the present case of decoupled sectors [12],

$$f(Z) = g(X) + h(Y) \tag{8}$$

where the X_i are the observable fields and the Y_i the hidden fields. For the simplified case that there are only light fields $(O(M_W))$ and no superheavy fields, one obtains the effective scalar potential for the X by putting in the vacuum expectation values for the hidden fields Y and expanding in powers of κ:

$$V(X) = V_{SS}(X) + m_g^2 \sum X_i^* X_i + m_g \sum X_i \frac{\partial g(X)}{\partial X_i} + h.c. + m_g(A-3)g(X) + h.c. \quad (9)$$

Here, $V_{SS}(X)$ is the ordinary SS scalar potential, $m_g = (4\pi/3)^{\frac{1}{2}} f_G/M_P$ is the gravitino mass, and A is a parameter, O(1), depending on the VEV's in the hidden sector. The form of the hidden sector enters only through the two parameters m_g and A. One sees in addition to the supersymmetric potential certain SS breaking terms. (In the realistic case that there are superheavy fields, the analysis is more complicated [13], but the same breakings are found with slightly different coefficients.) The first is a positive contribution to the mass-squared of all the scalars: squarks, sleptons, and Higgs. The other terms are additional scalar breakings whose form is determined in terms of the superpotential. All the breaking terms are governed by the scale m_g. m_g must then be of order 10^2 GeV. It cannot be much smaller since it determines the masses of the scalar quarks and leptons, and it cannot be much larger since (9) gives an $O(m_g^2)$ contribution to the Higgs mass, and our purpose in introducing SS was to prevent any contributions to the Higgs mass larger than the weak scale.

Now we have to ask whether it can be correct to use the tree level potential, neglecting incalculable loop graphs such as Figure 1. The theory being studied, N = 1 supergravity with a more or less arbitrary

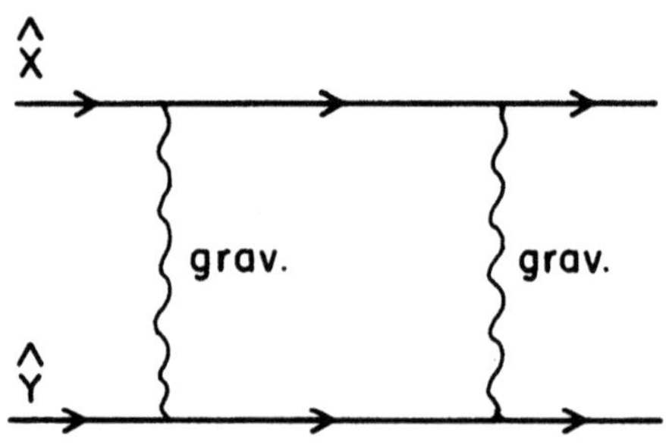

Figure 1

Gravitational loop graph. $\hat{X}$ is observable superfield and $\hat{Y}$ is hidden superfield.

matter system, has no special finiteness properties. To make sense of it we have to think of supergravity as an effective field theory. The graph of Figure 1 is roughly $\kappa^4\int d^4k/k^2$. When the loop momentum is small compared to the Planck scale, the graph makes sense and is small. At some scale Λ, presumably of order the Planck scale, our understanding of gravity breaks down and Figure 1 no longer makes sense. This corresponds to the loop in Figure 1 being very small in spacetime and gives rise to effective local operators such as $\int d^4\theta\hat{X}^\dagger\hat{X}\hat{Y}^\dagger\hat{Y}$ (the hat denotes a superfield). The coefficients of these operators should be of the order of the Planck scale to the appropriate dimensional power. What we do not know about the physics at the Planck scale corresponds to the inclusion in the Lagrangian of the most general such local operators consistent with the symmetries. This is entirely analogous to the use of the non-linear sigma model to represent low energy pion dynamics [16]. Note that the argument does not require that we know what the underlying dynamics at the Planck scale is, but we must speculate about its symmetries. To the extent that matter interactions and masses can be neglected, supergravity respects a very strong symmetry, a unitary transformation of all the matter fields into one another:

$$\hat{\phi}_i \rightarrow U_{ij}\,\hat{\phi}_j \tag{10}$$

If this is a good symmetry at the Planck scale, it greatly restricts the additional terms in the Lagrangian. Neglecting terms with extra derivatives, which are uninteresting at low energy, we must simply replace the renormalizable kinetic energy, $\int d^4\,\theta\sum_i \hat{\phi}_i^\dagger\,\hat{\phi}_i$, with the more general form

$$\int d^4\,\theta\kappa^{-2}\,d(\kappa^2\sum_i \hat{\phi}_i^\dagger\,\phi_i) \tag{11}$$

where d is an unknown function. The argument from (6) to (9) must be repeated with this more general kinetic term [13,14] and the result is that one finds the same three breakings as in (9) but the coefficients are now three unknown masses of the order of the gravitino mass:

$$V(X) = V_{SS}(X) + m_1^2\sum_i X_i^* X_i + m_2\sum_i X_i\frac{\partial g(X)}{\partial X_i} + \text{h.c.} + m_3\,g(X) + \text{h.c.} \tag{12}$$

The foregoing argument is due to S. Weinberg (private communication and [13]). It may, of course, be the case that the dynamics at the Planck scale respects a weaker symmetry than (10), and the SS breakings then involve more free parameters, but we will assume the strong symmetry (10) and study the consequences. One immediate consequence is that the second term in (12) gives an equal contribution, call it δm^2, to all scalar masses:

$$(\delta m^2)_{squark} = (\delta m^2)_{slepton} = (\delta m^2)_{Higgs} \tag{13}$$

Of course matter interactions distinguish different superfields and violate the unitary symmetry (10). To estimate this effect, dress Figure 1 with various matter interactions, such as a gauge boson attaching at two points to the X line. Such graphs give incalculable order α corrections to the unitary symmetry, from the region where all loop momenta are large. It is then consistent to assume that (10) is a good approximate symmetry provided all matter interactions remain weak at the Planck scale. There is a larger contribution, of order $\alpha\,\ell n(M_P/M_W)$, from the region where all gravitational lines are at large momentum but some non-gravitational loops are at smaller momenta. Since $\alpha\,\ell n(M_P/M_W)$ can be of order 1, these effects cannot be neglected. They can be calculated, fortunately, since they are essentially lower energy radiative corrections in the matter theory to local operators induced by gravity, and all such effects can be summed with the renormalization group. So the relation (13), like the SU(5) relation $m_b = m_\tau$, may hold at high energy but must be correctly renormalized at the weak scale.

At this point we are studying effectively globally SS theories, with explicit breakings having the form (12) at the Planck scale and then RNG improved at low energy. Now we have to find a way to break $SU(2)\times U(1)$. This is a puzzle since (12) includes a contribution to the Higgs mass-squared which is, at least before RNG correction, positive. We could try to get negative contributions to the Higgs mass from the other terms in (12). This is easier if the low energy Higgs sector contains additional fields beyond doublets, but there are fairly general arguments against this [8,17,18]: the breakings in (12) involving g(Y) tend to give such fields VEV's, spoiling the relation $\rho = M_W^2/M_Z^2\cos^2\theta_W = 1$, unless the fields are gauge singlets, and gauge singlets in the low energy theory are unnatural when $f_G \gg M_W^2$. With only doublets (we will assume two doublets, H with Y = -1 and H' with Y = +1, since there is no advantage to adding more) the most general Higgs potential from (12) is

$$\begin{aligned} V(H,H') = {} & (\mu^2+\delta m_H^2)H^*H + (\mu^2+\delta m_{H'}^2)H'^*H' - 2\mathrm{Re}(\mu BHH') \\ & + \tfrac{1}{8}e_1^2(H^*H - H'^*H')^2 + \tfrac{1}{8}e_2^2(H^*\vec{\sigma}H + H'^*\vec{\sigma}H')^2 \; . \end{aligned} \tag{14}$$

Here μ and B are parameters to be defined later, δm_H^2 and $\delta m_{H'}^2$ are the contribution of the second term in (12), the quartics are from the U(1) and SU(2) D terms and we have suppressed SU(2) indices. If

$$|\mu^2 B^2| > (\mu^2 + \delta m_H^2)(\mu^2 + \delta m_{H'}^2) \tag{15}$$

then the origin is a saddle point and SU(2) breaks, but if the unitary symmetry relation

$$\delta m_H^2 = \delta m_{H'}^2 \tag{16}$$

is not modified, then the potential is unbounded below along the line

$$H'_i = \varepsilon_{ij} H_j^* \tag{17}$$

where the quartics vanish, and the Higgs VEV slides out at least to M_{GUT}. So, we have a fairly strong indication that the low energy Higgs potential should contain only doublets, and that the relation (16) must receive radiative correction.

Since there are many more decades below the unification scale than above, most of the RNG correction comes about between M_{GUT} and M_W and is independent of the details of unification. We will focus on the case that the theory below the unification scale is minimal, with 3 generations, the SU(3) × SU(2) × U(1) gauge bosons, 2 Higgs doublets, and the superpartners of these. The most general Lagrangian with soft SS breaking [19] and this particle content (and we assume a discrete symmetry to prevent renormalizable B and L violation) is

$$\begin{aligned}\mathscr{L} = {} & \text{gauge invariant kinetic terms} + \\ & + \int d^2\theta\, g_E \hat{E}^c \hat{L} \hat{H} + g_D \hat{D}^c \hat{Q} \hat{H} + g_U \hat{U}^c \hat{Q} \hat{H}' \\ & + m_1 A_E g_E E^c L H + m_1 A_D g_D D^c Q H + m_1 A_U g_U U^c Q H' \\ & - \sum_{\text{scalars}} \delta m_i^2 X_i^* X_i + \mu \int d^2\theta \hat{H} \hat{H}' + m_1 \mu B H H' \\ & + \xi \int d^4\theta \hat{V}_1 - \tfrac{1}{2} \sum \tilde{m}_a \lambda_a \lambda_a + \text{h.c. where appropriate.}\end{aligned} \tag{18}$$

After the kinetic terms in (18), there is a supersymmetric term containing the Yukawa couplings which give mass to the quarks and leptons (a hat distinguishes a superfield from the corresponding scalar, and generation, SU(3), and SU(2) indices have been suppressed). The next term is a soft breaking from the m_2 and m_3 terms in (12). A_E, A_D, and A_U (for all generations) have a common value $A = (3m_2 + m_3)/m_1$ at high energy (defined to correspond to A in (9)). Next is a scalar mass-squared term, with the δm_i^2 having a common value $\delta m_i^2 = m_1^2 \sim m_g^2$ at high energy. The parameter μ is a supersymmetric Higgs mass, while μB is the related soft breaking (B, like A, is a parameter of order 1). The term ξ is the Fayet-Iliopoulos D-term, and is expected to be negligible. The final term is a gauge fermion Majorana mass for the SU(3), SU(2), and U(1) gauginos, which is a soft breaking of SS that we have not yet mentioned. It is produced at tree level if the nonrenormalizable term

$$\kappa \int d^2\theta \sum_i c_i \hat{Z}_i \hat{W}^\alpha \hat{W}_\alpha \tag{19}$$

is present in the Lagrangian, with $\hat{W}^\alpha$ the field strength superfield. (19) gives $\tilde{m}_1 = \tilde{m}_2 = \tilde{m}_3 = \tilde{m}$ at high energy. If the c_i are O(1), the gaugino mass $\tilde{m}$ is $\sim m_g$. However, (19) does not respect the unitary symmetry (10); thus it will not be produced by purely gravitational radiative corrections and the assumption (10) which allowed us to use the form (12) for the breakings requires the term (19) to be absent [20]. The gauge fermion mass $\tilde{m}$ then comes about through unitary-symmetry violating radiative corrections and is of order αm_g or a few GeV [17, 21]. On the other hand, the strong symmetry (10) is just a speculation, and we might also study weaker symmetries such as a unitary symmetry acting only on the observable fields. This symmetry would allow (19) while keeping the form (12) of the scalar breakings intact. So, we will also study the case that $\tilde{m}_a$ is of order m_g.

We wish to study the renormalization group for (18). The most interesting equation is that for the mass-squared of the Higgs which gives mass to the top quark:

$$\mu\frac{\partial}{\partial\mu}\begin{pmatrix}\delta m^2_{H'}\\ \delta m^2_{T_R}\\ \delta m^2_{T_L}\end{pmatrix} = \frac{g_T^2}{8\pi^2}\begin{pmatrix}3&3&3\\2&2&2\\1&1&1\end{pmatrix}\begin{pmatrix}\delta m^2_{H'}\\ \delta m^2_{T_R}\\ \delta m^2_{T_L}\end{pmatrix} + \frac{g_T^2 A_T^2 m_1^2}{8\pi^2}\begin{pmatrix}3\\2\\1\end{pmatrix} - \frac{2e_3^2\tilde{m}_3^2}{3\pi^2}\begin{pmatrix}0\\1\\1\end{pmatrix} \tag{20}$$

where we have neglected gauge couplings other than e_3 and Yukawa couplings other than that, g_T, for the top. A_T is the A parameter for the three scalar term $H'T_RT_L$, $\delta m^2_{T_{L,R}}$ are the mass-squared breakings for the scalar partners of the left- and right-handed top quark, and $\tilde{m}_3$ is the gluino mass. The sign of the first term is such as to make δm^2 smaller at low energy, and the 3:2:1 weighting has the effect that, even when $A_T = \tilde{m}_3 = 0$, the $\delta m^2_{H'}$ is driven negative for large enough g_T (that is, a heavy enough top quark). This is just what we need to break SU(2) × U(1). The last two terms, involving A_T and $\tilde{m}_3$, enhance this effect (the latter indirectly, by increasing $\delta m^2_{T_{L,R}}$, which then drive $\delta m^2_{H'}$ via the first term) reducing the needed value of g_T. Thus, the radiative corrections associated with a heavy top quark (or a heavy fourth-generation quark) is a plausible answer to the puzzle of why, in supersymmetry, the squarks and sleptons have positive mass-squared and the Higgs scalars negative. It was discovered first in global SS by several groups [3,5]. It was then noted that the same idea would work in local SS [22-24]. Several groups have studied the minimal model for various values of parameters [23-26]; see also Inoue *et al.* [5].

In (18) we have five unknown parameters, g_T, A, m_1^2, μ, and B, and we will include the possibility of large $\tilde{m}$ as a sixth. By choosing values for three of them and using the scale of SU(2) × U(1) breaking to fix the fourth (an overall scale), we can plot the physics as a function of two of them. Figure 2 shows the case $\tilde{m} \sim \mu \sim \mu B \sim 0$ as a function of $g_T(M_X)$ (and the resulting value of m_{TOP}) and $A(M_X)$. Figure 3 shows $\mu \sim \mu B \sim 0$, $\tilde{m} = m_1$, with the same axes. For large enough m_{TOP} (to the right of the line marked "∞"), SU(2) × U(1) is broken, and the necessary value of m_{TOP} is seen to be smaller when $|A|$ or $\tilde{m}$ is large. The contours mark the value of the lightest squark or slepton, which is also close to the value of m_1. For Figure 2, $\tilde{m} \sim 0$, all the squarks and sleptons are within 10-20 GeV of this value, while for Figure 3, $\tilde{m} = m_1$, the sleptons are all within 10-20 GeV of the given value and the squarks are approximately twice as heavy. The precise ordering of masses depends on small parameters. The lightest new scalar is typically in the range 70-200 GeV, though there is no actual upper limit if the parameters are adjusted to lie close enough to the line marked "∞". When Eq. (20) drives the SU(2) × U(1) breaking too well, the quadratics become negative along the line (17) where the quartics vanish. The potential is then unbounded below and the Higgs VEV becomes enormous; this is the region marked "potential unbounded". Actually it is possible to go a bit past the limit of classical stability and radiative corrections

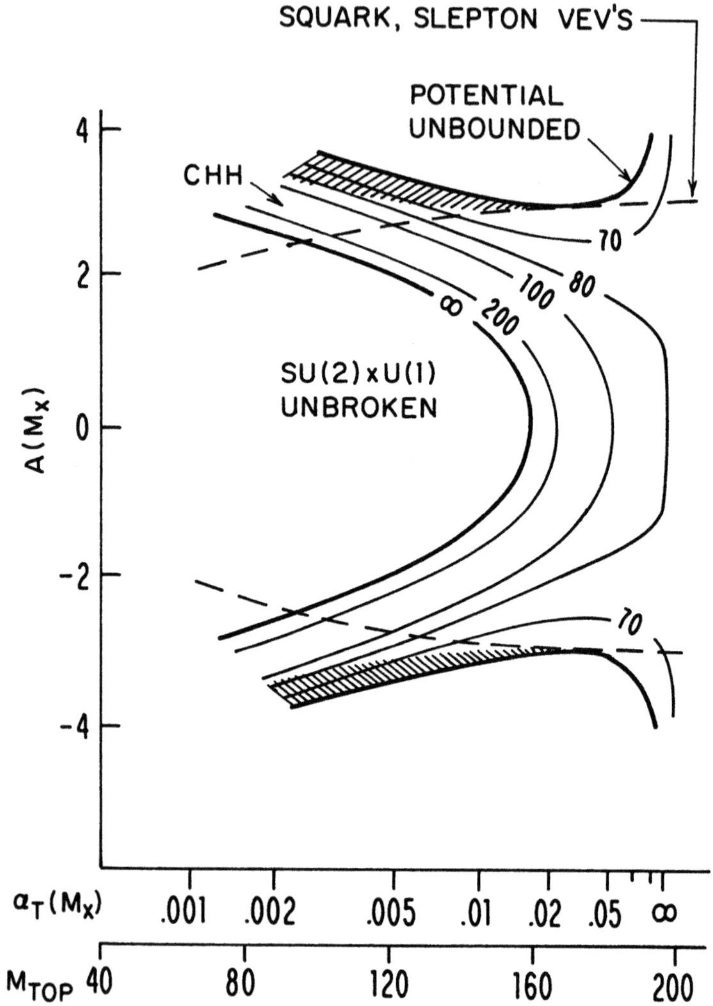

Figure 2

Physics as a function of $g_T(M_X)$ and $A(M_X)$ for $\tilde{m} \sim \mu \sim \mu B \sim 0$. The horizontal axis also gives the corresponding top quark mass. The contours give the value of the lightest squark or slepton. The horizontal dashed lines and the regions "potential unbounded", "squark and slepton VEV's", and "CHH" are discussed in the text. The shaded region is excluded from CHH by vacuum instability.

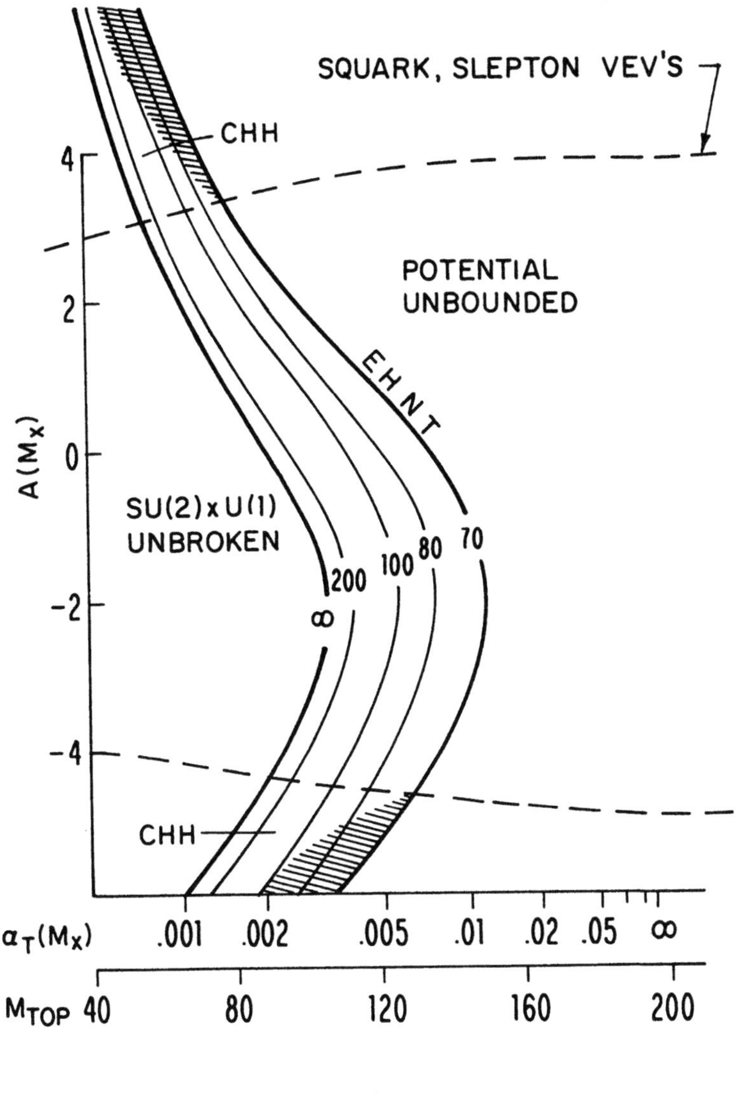

Figure 3

The same as Figure 2 with $\tilde{m} = m_1$, $\mu \sim \mu B \sim 0$. The region "EHNT" is also discussed in the text.

will still fix the Higgs VEV at an acceptable value [25]. This is the region of Figure 3 marked EHNT; its most notable feature is that the sleptons can be much lighter, limited only by the experimental bound of 18 GeV. Another dangerous region is $|A| > 3\text{-}4$ (marked with a dashed line) where new, lower minima develop with unacceptable squark and slepton VEV's [27]. In fact, this region is not completely forbidden, as tunneling to the bad vacua may be sufficiently slow on cosmological time scales [25,26]. There is then a narrow band of allowed parameters marked CHH in Figures 2 and 3 and extending onward. The notable feature here is the possibility of a very light top quark and stop squark [26]. Figures 2 and 3 are both for μ small. When μ is large, one sees in (14) three competing effects - the μ^2 terms tend to prevent $SU(2) \times U(1)$ breaking, the μB term tends to cause breaking in the dangerous direction (17), and the renormalization of $\delta m^2_{H'}$ tends to cause breaking in the safe direction $|H'| >> |H|$. Once μ is comparable to m_1, some tuning of μ and B is needed to cause $SU(2) \times U(1)$ to break while keeping the potential bounded along (17).

For $\tilde{m} \sim 0$ (Fig. 2), $m_{TOP} > 100$ GeV is needed in the region between the dashed lines, while for $\tilde{m} = m_1$ (Fig. 3), $m_{TOP} > 60$ GeV is needed and for $\tilde{m} >> m_1$, $m_{TOP} > 50$ GeV is needed [23,24]. The top quark can be lighter in the CHH region, or for μ large, though in the latter case μ and B must be quite finely adjusted. Also, Ibañez [28] has noted that if one adds to the minimal model fields with $SU(3) \times SU(2) \times U(1)$ content $(8,1)_0 + (1,3)_0 + (1,1)_2 + (1,1)_{-2}$, as might arise from some unified models, the effect on the RNG is such as to lower the necessary value of m_{TOP} to 30 GeV. Thus, there is no firm lower bound on the mass of the top quark needed to drive $SU(2) \times U(1)$ breakdown and the recently rumored value of 30-40 GeV is acceptable, in particular if $|A|$ is large. On the other hand, a rather larger value would have been expected from Figures 2 and 3 and is needed if A is near the value of $3-\sqrt{3}$ found in the simplest case [11-12], so it may be that a fourth-generation up quark is the source of $SU(2) \times U(1)$ breakdown. This works in much the same way, except that the fourth-generation down quark should not be too heavy or it will drive the other Higgs mass, δm^2_H, negative and make the potential unstable along (17).

There are too many unknown parameters in (18) to generalize about the expected spectrum. As noted earlier, the lightest new scalar will be rather heavy, 70-200 GeV, in the classically stable region of Figures 2 and 3, but can be lighter in the EHNT and CHH regions. The "ino" spectrum depends strongly on the unknown parameters μ and $\tilde{m}$. If either of these is small, the decay $W \rightarrow$ neutral ino + charged ino is allowed, and may be seen in the near future [20,21,29].

The last thing I would like to discuss is some indirect signatures of this model - the ρ-parameter, flavor changing neutral currents, and the neutron electric dipole moment. Various particles in this model appear in loop contributions to $\delta\rho = M_W^2/M_Z^2 \cos^2\theta_W - 1$. In the most extreme case (maximally split doublets, $m_{TOP} \sim 200$ GeV) one finds $\delta\rho = .02$ (.012 from top quark, .004 from top squark, .002 from all other scalars, and .002 from the W-ino)[17], which is still consistent with experiment but large enough to be interesting. More typically, however, one finds $\delta\rho \leqslant .01$. As far as flavor changing neutral currents, the unitary symmetry (10) guarantees that the breakings (12) satisfy the super-GIM mechanism, so that the quark and squark mass matrices are diagonal in the same basis and the gluino couplings are diagonal. The RNG gives calculable corrections to this. These have been studied in Ref. 30, where it was found that the contribution of the first two generations is unacceptably large if the squark and gluino masses are both $\leqslant 40$ GeV. The potentially larger contribution of the third generation depends on unknown mixing angles.

Note that Figure 2 is symmetric about $A = 0$, while Figure 3 is not. This is because the relative sign of m_1A and $\tilde{m}$ is a physical parameter. So, in fact, is the relative phase δ; the phase of $\tilde{m}$ can only be redefined by an R-transformation, and this changes the phase of m_1A as well. Thus, the low energy supergravity model has new sources of CP violation beyond the standard model [31]. Figure 4 shows a one-loop graph in which the new phase δ appears; a quark line splits into a squark-gluino loop. A factor of $\tilde{m}^*$ appears in the gluino line and a factor of $m_1A\langle H\rangle$ appears in the squark line. A photon couples in as well, so the whole graph gives rise to a neutron electric dipole moment which is of order

$$\frac{\alpha_3 e}{2\pi} \frac{m_{\tilde{g}}}{m_1^2} \sin\delta \qquad (21)$$

or $10^{-(22-23)} \sin\delta$ e-cm. Thus, $\sin\delta$ is already constrained to be $\leqslant 10^{-2}$. The origin of δ lies in the physics of the unification scale and the hidden sector. *A priori*, it could have been order 1, order α, or much smaller. The evidence of baryon asymmetry is that there are nontrivial phases beyond the KM phase, so we might hope to see a neutron electric

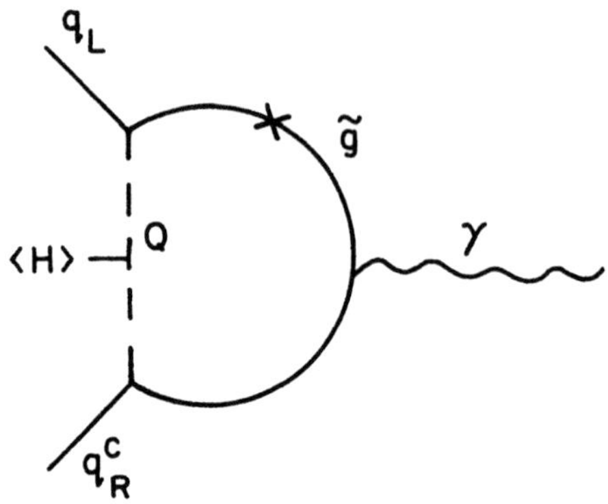

Figure 4

One-loop contribution to the neutron electric dipole moment. The cross on the gluino ($\tilde{g}$) line is a factor of $\tilde{m}^*$, and the insertion on the squark line is a factor of $m_1 A\langle H\rangle$.

dipole moment as the experimental limit is improved. Since the standard model suggests a moment of order 10^{-30} e-cm, this would be an indication of new physics, and perhaps supergravity.

In summary, we have looked at a simple question: does the minimal supersymmetric model, combined with the SS breakings characteristic of supergravity, lead to an acceptable theory? The answer is that it works quite well, provided that there is one rather heavy quark. In doing this I have been able to put off many hard questions, but I will remind you of what they are. We still need a good unified model. In particular, the Higgs doublets do not receive large radiative masses, but it still must be arranged that they are light at tree level while their color triplet partners, which mediate B violation, are heavy. There are several schemes for arranging this, but none are very attractive. A further problem is to get the right, SU(3) × SU(2) × U(1), vacuum. This is tied to the problem of an acceptable cosmology. Next, remember that the small number M_W/M_P, or f_G/M_P^2, is still an input, to be explained in the future. Finally, we have worked so hard to cure one naturalness problem, that of large radiative corrections to the Higgs mass, and our answer, spontaneously broken SS, has given no clue to the solution of that other naturalness problem, the cosmological constant. It may still be that a new approach is needed which answers both at once.

This research was supported in part by the National Science Foundation under Grant Number PHY82-15249.

References

[1] E. Witten, Nucl. Phys. B185 (1981) 513; M. Dine, W. Fischler, and M. Srednicki, Nucl. Phys. B189 (1981) 575; S. Dimopoulos and S. Raby, Nucl. Phys. B192 (1981) 353; N. Sakai, Z. Physik C11 (1981) 153; S. Dimopoulos and H. Georgi, Nucl. Phys. B193 (1981) 150.

[2] P. Fayet, Phys. Lett. 69B (1977) 489, and 84B (1979) 416.

[3] L. Alvarez-Gaumé, M. Claudson, and M. Wise, Nucl. Phys. B207 (1982) 16.

[4] L. Hall and I. Hinchliffe, Phys. Lett. 112B (1982) 351; S. Weinberg, Phys. Rev. D26 (1982) 287.

[5] J. Ellis, L. Ibañez, and G. Ross, Phys. Lett. 113B (1982) 283; K. Inoue, A. Kakuto, H. Komatsu, and S. Takeshita, Prog. Th. Phys. 67 (1982) 1889.

[6] M. Dine and W. Fischler, Phys. Lett. 110B (1982) 227; C. Nappi and B. Ovrut, Phys. Lett. 113B (1982) 175.

[7] T. Banks and V. Kaplunovsky, Nucl. Phys. B206 (1982) 45; S. Dimopoulos and S. Raby, Los Alamos preprint LA-UR-82-1982 (1982); M. Dine and W. Fischler, Nucl. Phys. B204 (1982) 346; R. Barbieri, S. Ferrara and D. V. Nanopoulos, Z. Physik C13 (1982) 267.

[8] J. Polchinski and L. Susskind, Phys. Rev. D26 (1982) 3661.

[9] E. Witten, Phys. Lett. 105B (1982) 267.

[10] T. Banks and V. Kaplunovsky, Ref. [7]. See also J. Polchinski, Phys. Rev. D27 (1983) 1320.

[11] B. Ovrut and J. Wess, Phys. Lett. 112B (1982) 347; R. Barbieri, S. Ferrara, D. V. Nanopoulos, and K. S. Stelle, Phys. Lett. 113B (1982) 219; J. Ellis and D. V. Nanopoulos, Phys. Lett. 116B (1982) 133.

[12] A. Chamseddine, R. Arnowitt, and P. Nath, Phys. Rev. Lett. 49 (1982) 970; R. Barbieri, S. Ferrara, and C. Savoy, Phys. Lett. 119B (1982) 343.

[13] L. Hall, J. Lykken, and S. Weinberg, Phys. Rev. D27 (1983) 2359; P. Nath, R. Arnowitt, and A. Chamseddine, Northeastern preprint NUB2579 (1982).

[14] S. Soni and H. A. Weldon, U. Penn. preprint (1983).

[15] E. Cremmer, S. Ferrara, L. Girardello, and T. Van Proeyen, CERN preprint TH-3343 (1982); J. Bagger and E. Witten, Phys. Lett. 118B (1982) 103; S. Weinberg, Phys. Rev. Lett. 48 (1982) 1776.

[16] S. Weinberg, Physica 96A (1979) 327.

[17] L. Alvarez-Gaumé, J. Polchinski, and M. Wise, Harvard preprint HUTP-82/A063, to be published in Nucl. Phys. B221.

[18] H. P. Nilles, M. Srednicki, and D. Wyler, Phys. Lett. 124B (1983) 337; A. B. Lahanas, Phys. Lett. 124B (1983) 341.

[19] L. Girardello and M. Grisaru, Nucl. Phys. B194 (1981) 65.

[20] S. Weinberg, Phys. Rev. Lett. 50 (1983) 387.

[21] R. Arnowitt, A. H. Chamseddine, and P. Nath, Phys. Rev. Lett. 50 (1983) 232.

[22] H. P. Nilles, Phys. Lett. 115B (1982) 193; L. Ibañez, Phys. Lett. 118B (1982) 73; J. Ellis, D. V. Nanopoulos, and K. Tamvakis, Phys. Lett. 125B (1983) 275.

[23] L. Ibañez, Madrid preprint FTUAM/82-8 (1982); L. Ibañez and C. Lopez, Madrid preprint FTUAM/83-1 (1983).

[24] L. Alvarez-Gaumé, J. Polchinski, and M. Wise, Ref. [17].

[25] J. Ellis, J. Hagelin, D. V. Nanopoulos, and K. Tamvakis, SLAC preprint 3042 (1983).

[26] M. Claudson, L. J. Hall, and I. Hinchliffe, LBL preprint 15948 (1983).

[27] J.-M. Frére, D. Jones, and S. Raby, Michigan preprint UMHE-82-58.

[28] L. Ibañez, Madrid preprint FTUAM/83/4 (1983).

[29] J.-M. Frére and G. Kane, Michigan preprint UM-TH-83-2; J. Ellis, J. Hagelin, D. V. Nanopoulos, K. Tamvakis, SLAC-PUB 3042; B. Grinstein, J. Polchinski, and M. Wise, CalTech preprint; A. Chamseddine, P. Nath, and R. Arnowitt, Northeastern preprint NUB2588.

[30] J. Donoghue, H. P. Nilles, and D. Wyler, CERN preprint 3583 (1983).

[31] W. Buchmüller and D. Wyler, Phys. Lett. 121B (1983) 321; J. Polchinski and M. Wise, Harvard preprint HUTP-83/A016; F. del Aguila, M. Gavela, J. Grifols, and A. Mendez, Barcelona preprint (1983).

SUPERUNIFICATION FROM ELEVEN DIMENSIONS*

M.J. Duff, B.E.W. Nilsson and C.N. Pope

The Blackett Laboratory, Imperial College,
London SW7 2BZ

ABSTRACT

We show how N=8 supersymmetry can break spontaneously to N=1 at the Planck scale via a Kaluza-Klein compactification of d=11 supergravity on the squashed seven-sphere. Features unique to Kaluza-Klein supergravity are (i) the massless gravitino of the N=1 phase comes from a massive N=8 supermultiplet, (ii) the scalars developing nonzero VEVs also belong to massive N=8 supermultiplets, (iii) parity remains unbroken when N=8 breaks to N=1.

Next we ask whether the resulting N=1 theory can provide a realistic SU(3)xSU(2)xU(1) unification and speculate that it might if some of the gauge bosons and fermions are composite as in the EGMZ model. In contrast to their model, however, we avoid unwanted helicities and problems with their non-compact E_7. Moreover, we suggest a scheme in which the electroweak SU(2)xU(1) is a subgroup of the d=11 general coordinate group but the strong SU(3) is a subgroup of the d=11 local Lorentz group and are not, therefore, to be combined into a GUT. The special properties of the seven-sphere also suggest a possible solution of the cosmological constant problem involving fermion condensates.

*Lecture delivered by M.J. Duff.

I. INTRODUCTION

Some time ago [1,2] it was pointed out that the field equations of N=1 supergravity in d=11 dimensions admit of vacuum solutions corresponding to the product of d=4 anti-de Sitter space times the seven-sphere; and that since S^7 admits 8 Killing spinors and since its isometry group is SO(8), this gives rise via a Kaluza-Klein mechanism to an effective d=4 theory with N=8 supersymmetry and local SO(8) invariance.

Since then there has been considerable progress in the understanding of S^7 compactification of d=11 supergravity [3→28] and in Sections 2 and 3 of this paper we bring the reader up to date, paying particular attention in Section 3 to the Higgs-Kibble, spontaneous symmetry - breaking interpretation [5] of the five different S^7 solutions which are presently known. This involves new results on the tower of massive N=8 supermultiplets. Whatever one's view of the physical relevance of extra dimensions and supergravity, the Kaluza-Klein S^7 approach to d=11 supergravity discussed in Sections 2 and 3 is mathematically sound and, few would dispute, beautiful.

Section 4, however, will be on much shakier ground because here we turn to the other theme of this paper and ask whether Kaluza-Klein supergravity stands any chance of describing low-energy particle physics phenomenology. Although major problems remain, the results are not all together discouraging.

In Section 5 we examine some rival approaches to S^7 supergravity and discuss the possible interpretation of some very recent results.

2. KALUZA KLEIN SUPERGRAVITY

Back in the 1920's Kaluza and Klein [29] tried to unify gravity with electromagnetism by giving an extra fifth dimension to spacetime. Nowadays, of course, we are more ambitious and would like to include the strong and weak interactions as well. The current renaissance in

Kaluza-Klein theories is based on the hope that this may be possible with a few more extra dimensions.

We have argued elsewhere [2,5,10] that the only viable Kaluza-Klein theory is supergravity and that the only way to do supergravity is via Kaluza-Klein. Some of the reasons are:

(i) The maximum dimension that one can consistently formulate a supersymmetric field theory is eleven [30]. This d=11 theory of Cremmer, Julia and Scherk [31] is unique and describes the interaction of gravity g_{MN} with the gravitino Ψ_M and a 3-index gauge field A_{MNP}. These fields have 44, 128 and 84 degrees of freedom respectively. The continuous symmetries are d=11 general covariance, local N=1 supersymmetry, local SO(1,10) Lorentz invariance, and abelian gauge invariance of A_{MNP}. There is also a discrete symmetry where one performs an odd number of space or time reflections and sends A_{MNP} to $-A_{MNP}$. (The signature is $-++++++++++$). This theory has only one parameter, the gravitational coupling, and supersymmetry forbids any tampering with the Lagrangian [65]. Either this is the theory of everything or its wrong: there are no half-measures!

(ii) The basic idea of Kaluza-Klein is that what we perceive to be internal symmetries in four dimensions are really spacetime symmetries in the extra dimensions. Carrying this logic to its ultimate conclusion would demand that <u>all</u> symmetries in Nature are spacetime symmetries. This extreme Kaluza-Klein philosophy is realized by d=11 supergravity both in the continous and discrete symmetries. For example charge conjugation in d=4 is parity in the extra seven dimensions [32]. (This extreme case would not be true of N>1 supersymmetry which would require d<11 dimensions).

(iii) Unless one introduces Yang-Mills fields already in the higher dimensional theory, which would contradict the fundamental spirit of Kaluza-Klein discussed in (ii), one cannot obtain light fermions in the d=4 theory by starting with the Dirac Lagrangian. Lichnerowicz's theorem [33] requires that one must start with the Rarita-Schwinger Lagrangian and this automatically requires supergravity. (For other, less attractive, ways out, see [2])

(iv) Modern approaches to higher dimensional theories demand that they exhibit 'spontaneous compactification'[34] i.e that the field equations admit of stable ground-state solutions of the form (d=4 spacetime) × (compact extra dimensions). Compactness ensures that the d=4 mass spectrum is discrete and also guarantees charge quantization. Higher dimensional theories where the extra dimensions are compact are indistinguishable from four dimensional theories with a very special mass spectrum and this is why Kaluza-Klein could be realistic despite the science-fiction overtones of extra dimensions. However, not all higher dimensional theories exhibit spontaneous compactification. In d=11 supergravity spontaneous compactification not only works but the dimension of spacetime (d=4) is now <u>output</u> rather than input.

To see this, consider the d=11 field equations [35] with $\langle\phi_M\rangle = 0$

$$R_{MN} - \frac{1}{2} g_{MN} R = \frac{1}{3}\left[F_{MPQR} F_N{}^{PQR} - \frac{1}{8} g_{MN} F^2\right] \tag{1}$$

$$\nabla_M F^{MPQR} = \frac{-1}{576} \epsilon^{PQRM_1..M_4M_5..M_8} F_{M_1..M_4} F_{M_5..M_8} \tag{2}$$

where M, N = 1....11 and $F_{MNPQ} = 4\partial_{[M} A_{NPQ]}$ is the field strength of the A_{MNP} potential. Denote by x^μ the coordinates of spacetime, whose dimension is not yet specified, and by y^m the coordinates of the extra dimensions. Maximal symmetry for the spacetime ground-state requires that we look for solutions of the form

$$\langle g_{\mu\nu}\rangle = \overset{o}{g}_{\mu\nu}(x) \qquad \langle F_{\mu\nu\rho\sigma}\rangle = \overset{o}{F}_{\mu\nu\rho\sigma}(x)$$

$$\langle g_{mn}\rangle = \overset{o}{g}_{mn}(y) \qquad \langle F_{mnpq}\rangle = \overset{o}{F}_{mnpq}(y) \tag{3}$$

with all components with mixed indices equal to zero. Moreover $\overset{o}{F}_{\mu\nu\rho\sigma}(x)$ must be an invariant tensor. This means that

$$\overset{o}{F}_{\mu\nu\rho\sigma} = 3\, m\, \varepsilon_{\mu\nu\rho\sigma} \tag{4}$$

where m is a constant. Hence, as pointed out by Freund and Rubin [36], we have singled out d=4 for spacetime since $F_{\mu\nu\rho\sigma}$ has rank four. For the moment let us also set $F_{mnpq} = 0$ for simplicity, then substituting (3) and (4) into (1) and (2), we find that (2) is trivially satisfied while (1) yields the product of a four dimensional Einstein spacetime with the usual Minkowski signature

$$R_{\mu\nu} = -12\, m^2\, g_{\mu\nu}, \tag{5}$$

which we take to be anti-de Sitter space (AdS) = SO(3,2)/SO(3,1) and a seven dimensional Einstein space with Euclidean signature

$$R_{mn} = 6\, m^2\, g_{mn} \tag{6}$$

and hence compact since $m^2 > 0$. There are still infinitely many solutions of (6) and, a priori, each could claim to be the ground state. We need some criterion for distinguishing the false vacua from the true vacuum. A necessary condition for the true vacuum is that it be stable, and one reason for stability would be an unbroken supersymmetry. For a supersymmetric vacuum, we require that $\langle\psi_M\rangle$ stay zero under a supersymmetry transformation, i.e. that $\langle\delta\psi_M\rangle = 0$ where

$$\delta\psi_M = D_M\varepsilon - \frac{i}{144}\,(\hat{\Gamma}^{NPQR}{}_M + 8\,\hat{\Gamma}^{PQR}\delta^N_M)\, F_{NPQR}\,\varepsilon \;\equiv \tilde{D}_M\varepsilon \tag{7}$$

For solutions of the type discussed above, we have

$$\bar{D}_\mu = D_\mu + m\,\gamma_\mu\,\gamma_5 \tag{8}$$

$$\bar{D}_m = D_m - \frac{m}{2}\Gamma_m \tag{9}$$

where we have used the decomposition of the d=11 Γ matrices

$$\hat{\Gamma}_A = (\gamma_\mu \otimes 1,\ \gamma_5 \otimes \Gamma_a) \tag{10}$$

The number of unbroken supersymmetries is thus given by the number of 'Killing spinors' on the d=7 manifold, i.e. the number of solutions to

$$\bar{D}_m\,\eta = 0 = (\partial_m - \frac{1}{4}\,\omega_m{}^{ab}\Gamma_{ab} - \frac{m}{2}\,e_m{}^a\Gamma_a)\,\eta \tag{11}$$

where $\eta(y)$ is an 8-component (commuting) spinor in d=7, and where $\omega_m{}^{ab}$ and $e_m{}^a$ are the spin connection and seibenbein of the ground-state solution to (6). Such Killing spinors satisfy the integrability condition.

$$\left[\bar{D}_m, \bar{D}_n\right]\eta = \frac{-1}{4}\,C_{mn}{}^{ab}\Gamma_{ab}\eta = 0 \tag{12}$$

where $C_{mn}{}^{ab}$ is the Weyl tensor. The subgroup of spin (7) generated by these linear combinations of the spin (7) generators Γ_{ab} corresponds to the holonomy group $\mathcal{H}$ of the generalized connection of (11). Thus the maximum number of unbroken supersymmetries, N, is equal to the number of spinors left-invariant by $\mathcal{H}$.

The maximally symmetric solution of (6) is given by the 'round' S^7 = SO(8)/SO(7) for which $C_{mn}{}^{ab} = 0$, $\mathcal{H} = 1$ and $N = N_{max} = 8$. However, there also exist other solutions which are topologically seven-spheres which deviate from the maximally symmetric geometry by being 'squashed', and solutions for which the $A_{mnp}(y)$ 'torsion' field is also non-zero. These vacua have symmetry group $G \subset SO(8)$ and $N<8$. To date, five S^7 solutions are known and are tabulated below.

TABLE 1 - KNOWN SOLUTIONS WITH S^7 TOPOLOGY

SOLUTION	HOLONOMY	SUPERSYMMETRY	GAUGE GROUP	REF.
I round	1	N=8	SO(8)	[1,2,10]
II left-squashed	G_2	N=1	SO(5)×SU(2)	[8]
III round + 'torsion'	SO(7)	N=0	SO(7)	[3]
IV right-squashed	G_2	N=0	SO(5)×SU(2)	[15]
V right-squashed + 'torsion'	SO(7)	N=0	SO(5)×SU(2)	[15]

This table suggests that these different solutions correspond to different phases of the same effective four-dimensional theory; solution I being the symmetric phase and solutions II,III,IV and V being spontaneously broken phases. This brings us to another attractive feature of the Kaluza-Klein theory based on N=1, d=11 supergravity:

V) The unique properties of S^7provide a Kaluza-Klein origin for the spontaneous breakdown of gauge symmetries, supersymmetries, and the discrete symmetries of C, P and CP.

The ability to find solutions with squashing [8,15], is due to the remarkable result (valid for spheres only of dimension 4k + 3, k≥1) that S^7 admits not one, but two Einstein metrics [37]. The ability to find solutions with 'torsion [3], is related to the equally remarkable result that, apart from group manifolds, S^7 is the only compact space to admit an 'absolute parallelism'[38]. More remarkable than both, in the authors' opinion, is that in d=11 supergravity these somewhat abstract results from differential geometry are translated into something familiar in the effective four-dimensional theory: the Higgs-Kibble mechanism [5].

3. HIGGS-KIBBLE INTERPRETATION

i) Vacuum I: The round S^7

Since the round S^7 admits 8 Killing spinors and since its isometry group is SO(8), the Kaluza-Klein mechanism will give rise to an effective d=4 theory with N=8 supersymmetry and local SO(8) invariance. The group SO(8) posesses the unique property of 'triality' i.e. there are three inequivalent eight dimensional representations with Dynkin labels (1,0,0,0), (0,0,0,1) and (0,0,1,0) to which we conventionally assign the names vector (8_v), spinor (8_s), and conjugate spinor (8_c) respectively. (We follow the notation of Slansky [39] except that s and v are interchanged). This triality will be crucial in what follows. In addition to the familiar massless N=8 supermultiplet of spins $\{2, 3/2, 1, 1/2, 0^+, 0^-\}$ with SO(8) content $\{1, 8_s, 28, 56_s, 35_v, 35_c\}$ this Kaluza-Klein theory also describes an infinite tower of massive, spin $\leqslant$ 2 supermultiplets with masses quantized in units of m, the inverse radius of S^7. They correspond to irreducible representations of OSp(4/8) but will be reducible under SO(8) [4]. They are obtained by taking the tensor product of the massless multiplet with Dynkin labels {(0,0,0,0), (0,0,0,1), (0,1,0, 0), (1,0,1,0), (2,0,0,0), (0,0,2,0)} and the representation (n,0,0,0) which correspond to eigenmodes of the scalar Laplacian on S^7, and which yield the massive gravitons. This done, we allow each massless spin 2 to eat a spin 1 and spin 0 to become a massive spin 2; each massless spin 3/2 to eat a spin 1/2 to become a massive spin 3/2; and each massless spin 1 to eat a spin 0 to become a massive spin 1. Of course, since spacetime is anti-de Sitter space, the quantity 'mass' must be defined accordingly. See [57, 41, 59]. For example, massless spin-3/2 require an apparent mass term and massless scalars must obey the conformal wave equation. Moreover, members of the same N=8 massive multiplet can have different de Sitter mass. The results are shown in Table 2.

TABLE 2. MASSIVE N=8 SUPERMULTIPLETS

SPIN	SO(8) REP.
2	(n,0,0,0)
3/2	(n,0,0,1) + (n-1,0,1,0)
1	(n,1,0,0) + (n-1,0,1,1) + (n-2,1,0,0)
1/2	(n+1,0,1,0) + (n-1,1,1,0) + (n-2,1,0,1) + (n-2,0,0,1)
0^+_-	(n+2,0,0,0) + (n-2,2,0,0) + (n-2,0,0,0)
0	(n,0,2,0) + (n-2,0,0,2)

For example, the lightest such multiplet has the SO(8) content $\{8_v, 8_c + 56_c, 56_v + 160_v, 160_c + 224_{vc}, 112_v, 224_{cv}\}$ and the next-to-lightest has the SO(8) content $\{35_v, 56_s + 224_{vs}, 28 + 350 + 567_v, 8_s + 160_s + 672_{vc} + 840_s, 1 + 294_v + 300, 35_s + 840'_s\}$. Note that the s and c type representations occur only for spins 3/2, 1/2, 0^- while spins 2, 1 and 0^+ have either no subscript or a v subscript. Note also that, although N=8, the maximum spin of these massive supermultiplets is 2 and not 4. This phenomenon of 'multiplet shortening' is discussed by Freedman and Nicolai [59].

The infinity of massive gravitinos is due to an infinity of spontaneously broken supersymmetries as may be seen by Fourier expansion of the d=11 supersymmetry parameter in harmonics on S^7

$$\varepsilon\,(x,y) = \sum_k \varepsilon_k\,(x)\;Y^k\,(y) \tag{13}$$

The lowest mode, which is 8-fold degenerate, corresponds to the unbroken N=8 supersymmetry and to the 8 massless gravitinos, whereas the remaining ε_k transform states of different mass into each other. Similar remarks apply to the other d=11 local symmetries of general covariance, local SO(1,10) and gauge invariance of A_{MNP}. In this way, one obtains an infinite dimensional non-compact super algebra which generalizes the finite-dimensional non-super algebras described by Salam and Strathdee. Note, moreover, that since the massive modes have masses quantized in units of m, the inverse radius of S^7, and since the SO(8) Yang-Mills coupling constant g obeys the Kaluza-Klein relation [27]

$$g^2 = 16\pi G m^2$$

where G is Newton's constant, the massless sector is in no sense a limiting case of the complete theory. Throughout this work, therefore, we shall always advocate that the massive states be retained.

As a matter of technical interest, however, one may ask whether the theory can be consistently truncated to include only the massless supermultiplet and if so whether this truncation coincides with N=8 gauged SO(8) theory of de Wit and Nicolai [42]. It has been conjectured that this is indeed the case [1,2,5,10]. Verifying this conjecture would require a complete non-linear analysis of the d=4 equations of motion obtained from d=11 and, to date, only the linearized theory has been obtained in full detail [9,10]. However, a crucial necessary condition is that the spacetime cosmological constant Λ must be related to the SO(8) gauge coupling constant g by the same formula [42]

$$4\pi G\Lambda = -3\,g^2$$

as obtains in the N=8 phase of the de Wit-Nicolai theory. Weinberg [43] has shown how the coupling constant may be calculated starting from pure Einstein gravity in higher dimensions. Applied to S^7 of radius m^{-1}, the formula gives $g^2 = 64\pi G m^2$ which, combined with $\Lambda = -12m^2$ from (5), yields

$$16\pi G\Lambda = -3g^2$$

which disagrees with the de Wit-Nicolai value and would disprove the conjecture. However, it turns out that the presence of the A_{MNP} field in d=11 supergravity leads to a modification of Weinberg's calculation [27]. When this is taken into account it changes the relation between g and m^2 to be $g^2 = 16\pi G m^2$ and hence to give precisely the de Wit-

Nicolai relation. This reinforces the conjecture and suggests that the 'hidden' symmetry of SO(7) discussed in Section 4 might be enlarged to the SU(8) of the de Wit-Nicolai theory. With $F_{\mu\nu\rho\sigma}$ given by (4), the criterion for unbroken supersymmetry is given by (11), and on the round S^7 there are 8 solutions denoted η^I (I = 1..8) where the index I corresponds to the 8_s of SO(8). Note, however, that by inserting a minus sign into (4), one obtains a new vacuum solution for which the criterion of unbroken supersymmetry is obtained by sending m to −m in (11). On the round S^7 there are still 8 solutions denoted $\eta^{I'}$(I' = 1..8) where the index I' now corresponds to the 8_c of SO(8). All of the above discussion again goes through provided the s and c labels are interchanged.

(ii) Vacuum II: The left-squashed S^7

The squashed S^7 has metric $ds^2 = c^2 e^a \otimes e_a$ (a = 1..7) where $c = (2 \times 7^{1/2} m \lambda)^{-1} (1 + 8 \lambda^2 - 2 \lambda^4)^{1/2}$ is chosen to give $R = 42 m^2$ and

$$e^0 = d\mu \ , \ e^i = \frac{1}{2} \sin \mu \ \omega_i, \ e^{\hat{i}} = \frac{1}{2} \lambda \left(\nu_i + \cos \mu \ \omega_i\right) \tag{14}$$

where i = 1,2,3 and $\hat{i}$ = 4,5,6 = $\hat{1},\hat{2},\hat{3}$, where

$$\nu_i = \sigma_i + \Sigma_i, \ \omega_i = \sigma_i - \Sigma_i \tag{15}$$

and where σ_i and Σ_i are left-invariant one-forms satisfying the SU(2) algebra

$$d\sigma_1 = - \sigma_2 \wedge \sigma_3 \quad , \ d\Sigma_1 = -\Sigma_2 \wedge \Sigma_3 \tag{16}$$

and cyclic permutations. The constant parameter λ describes the degree of distortion. The round S^7 solution of (6) corresponds to $\lambda^2 = 1$, and describes the coset space SO(8)/SO(7) for which $C_{mn}{}^{ab} = 0$, $\mathcal{H} = 1$ and $N = N_{max} = 8$. The left-squashed S^7 solution of (6) corresponds to $\lambda^2 = 1/5$ and describes the Einstein metric on the coset space Sp(2) × Sp(1)/Sp(1) × Sp(1) [15,20] where Sp(2) ≃ SO(5) and Sp(1) ≃ SU(2) , for which $C_{mn}^{ab} \neq 0$, $\mathcal{H} = G_2$ and $N_{max} = 1$. Moreover $N = N_{max} = 1$ because there is indeed one solution of (11) which we denote by the singlet η. In contrast to the $\eta^I(y)$ of the round S^7, the components of η are actually constant in the orthonormal basis of (14). This yields an N=1, SO(5) × SU(2) vacuum. The massless sector consists of a 2-3/2 supermultiplet in the representation (1,1) and a 1-1/2 supermultiplet in the representation (10,1) + (1,3), together, possibly, with some 1/2-0 supermultiplets.

The squashed S^7, however, exhibits the following subtlety. If one reverses the orientation of the orthonormal basis of (14), by sending each e^a to $-e^a$ for example, one obtains a new S^7 solution of (6) hereafter referred to as right-squashed to distinguish it from the left-squashed S^7 discussed above. Although we still have $\mathcal{H} = G_2$, there are no longer any solutions of (11) and so this vacuum has $N = 0$. [The same effect could have been obtained by keeping the left-squashed S^7 but inserting a minus sign into (4)]. Because of triality, there are three inequivalent SO(5) × SU(2) subgroups of SO(8). Left squashing picks out $[SO(5) \times SU(2)]_c$ under which 8_c goes into (5,1) + (1,3), and both 8_s and 8_v go into (4,2), while right-squashing picks out $[SO(5) \times SU(2)]_s$ under which 8_s goes into (5,1) + (1,3) and both 8_c and 8_v go into (4,2). (The third subgroup $[SO(5) \times SU(2)]_v$ may also be obtained by squashing S^7 but in an inhomogeneous manner which does not yield an Einstein metric).

Let us now discuss the Higgs interpretation of the breaking $N=8/SO(8) \rightarrow N=1/[SO(5) \times SU(2)]_c$. First we observe that when SO(8) breaks to $[SO(5) \times SU(2)]_c$, the massless N=8 multiplet decomposes as follows $1 \rightarrow (1,1)$; $8_s \rightarrow (4,2)$; $28 \rightarrow (10,1) + (5,3) + (1,3)$; $56_s \rightarrow (16,2) + (4,4) + (4,2)$; $35_v \rightarrow (10,3) + (5,1)$; $35_c \rightarrow (14,1) + (5,3) + (1,5) + (1,1)$. These states by themselves can never form N=1 supermultiplets.In particular we note that <u>all</u> the eight gravitinos must acquire a mass. It follows that the single massless gravitino in the N=1 phase must come from the <u>massive</u> sector of the N=8 phase.

To proceed, we note that massless gravitinos in d=4 correspond to

$-7m/2$ modes of the Dirac operator $\not{D} \equiv \Gamma^m D_m$. That $\bar{D}_m \eta = 0$ implies $\not{D}\eta = -7m\,\eta/2$ is clear from (9). By consideration of the integral of $|\bar{D}_m\eta|^2$, the converse is also seen to be true after some straightforward algebra. In Ref [15], therefore, we calculated the Dirac operator in the squashed S^7 metric with arbitrary λ, and looked for the eigenmode ϕ which, in the case $\lambda^2 = 1/5$, is the solution η of $\bar{D}_m\eta = 0$. The result was that $\phi = \eta$ for all λ^2 and that the corresponding eigenvalue was

$$\frac{-3}{2} \cdot 7^{1/2} m \,(1+2\lambda^2)\,(1+8\lambda^2 - 2\lambda^4)^{-1/2}$$

Thus the eigenvalue $- 7m/2$ mode of $\not{D}$ on the $\lambda^2 = 1/5$ sphere has come from a $-9m/2$ mode on the $\lambda^2 = 1$ sphere. Now the Dirac eigenmodes on the round S^7 are $+ (n + \frac{7}{2})m$ and $-(n+\frac{7}{2})m$, $n \geqslant 0$, the modes being in the $(n,0,1,0)$ and $(n,0,0,1)$ representations respectively [15,28]. The singlet massless gravitino on the $\lambda^2 = 1/5$ squashed S^7 has therefore come from the next-to-lowest level of Dirac eigenmodes on the round S^7, which is a 56_c. It is, in fact, the singlet in the decomposition $56_c \rightarrow (10,1) + (10,3) + (5,3) + (1,1)$ of SO(8) breaking down to the $[SO(5) \times SU(2)_c]$ isometry group of the left squashed sphere. We refer to this mechanism, whereby states which are apparently very massive $M^2 \sim g^2/G$ in the N=8 phase zoom down from the Planckian sky to become massless in the N=1 phase as the 'Space Invaders Scenario'. See Table 3.

TABLE 3. THE 'SPACE INVADERS SCENARIO'

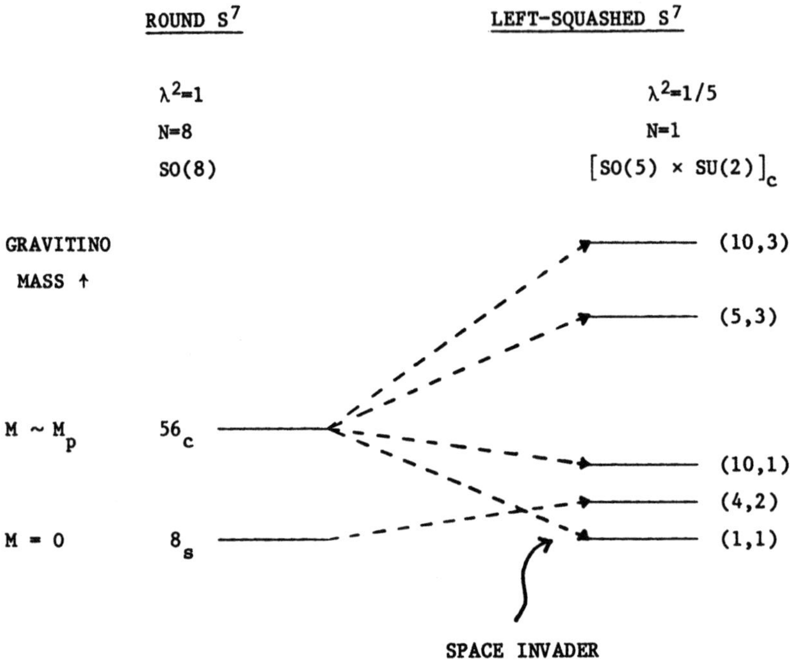

The Higgs and super-Higgs effect works as follows. Consider

$$g_{mn}(x,y) = \overset{o}{g}_{mn}(y) + h_{mn}(x,y)$$

where $\overset{o}{g}_{mn}(y)$ is the metric on the round S^7. Fourier expanding $h_{mn}(x,y)$ in harmonics on the round S^7

$$h_{mn}(x,y) = \sum_k S_k(x)\, Y_{mn}^{\;k}(y) \tag{17}$$

one obtains the 35_v scalars of the N=8 massless sector from the lowest mode together with an infinite tower of massive scalars, $S_k(x)$ where k denotes some representation of SO(8). In the symmetric phase

$$\langle g_{mn}(x,y)\rangle = \overset{o}{g}_{mn}(y) \quad \rightarrow \quad \text{ALL } \langle S_k(x)\rangle = 0, \tag{18}$$

but deformations of S^7 away from its maximally symmetric geometry correspond to nonzero vacuum expectation values for some of these scalars.

$$\langle g_{mn}(x,y)\rangle \neq \overset{o}{g}_{mn}(y) \quad \rightarrow \quad \text{SOME } \langle S_k(x)\rangle \neq 0. \tag{19}$$

In order to determine which are responsible for the breaking of SO(8) down to SO(5) × SU(2), we first note that in the orthonormal basis of (14), the metric deformation h_{ab} which takes the round S^7 into the squashed S^7 is of the form

$$h_{ab} \sim \text{diag}\ (0,0,0,0,1,1,1) \tag{20}$$

A straightforward calculation in the round S^7 background shows that this is a Killing tensor

$$\nabla_{(\ell}h_{mn)} = 0 \tag{21}$$

As discussed in [10], there are 336 symmetric rank-two Killing tensors on the round S^7, transforming as a 1, a 35_v, and a 300 under SO(8). The 35_v are associated with the massless scalars (in the n = o level of Table 2) discussed above, while the 1 and 300 describe massive scalars (in the n = 2 level of Table 2). The singlet Killing tensor is just a constant multiple of the metric and has constant trace; the trace of the 35_v gives a 35_v of eigenfunctions of the scalar Laplacian on S^7, while the 300 is tracefree. From (20) we see that h_{ab} has constant trace and so must be a linear combination of just the 1 and the 300. The singlet corresponds to an overall scaling of the metric and so the squashing of (14) corresponds to a nonzero VEV for the 300.
As a consistency check we note that under SO(8) → SO(5) × SU(2) we have 300 → (1,1) + (5,3) + (1,5) + (10,3) + (14,1) + (5,5) + (35,1) + (35,3) + (14,5) and we find as expected the singlet (1,1) together with the Goldstone bosons (5,3) which give a mass to the (5,3) spin 1 gauge bosons in the decomposition 28 → (10,1) + (1,3) + (5,3) .

iii) Vacuum III : The round S^7 with 'torsion'

If we substitute $\overset{o}{F}_{\mu\nu\rho\sigma} = 2m\ \varepsilon_{\mu\nu\rho\sigma}$ into (2) we find

$$\nabla_m \overset{o}{F}{}^{mnpq} = \frac{m}{6}\ \varepsilon^{npqrstu}\ \overset{o}{F}_{rstu} \tag{22}$$

and on the round S^7 non-trivial solutions may be found [3] which also solve (6). (We have rescaled $\overset{o}{F}_{\mu\nu\rho\sigma}$ so as to keep m^{-1} the radius of S^7). Moreover, the $\overset{o}{A}_{mnp}$ field of which $\overset{o}{F}_{mnpq}$ is the curl, $\overset{o}{F}_{mnpq} = 4\ \partial_{[m}\ \overset{o}{A}_{npq]}$, admits the interpretation of a parallelizing torsion i.e

$$R_{mnpq}\ (\Gamma^s{}_{tu} + S^s{}_{tu}) = 0 \tag{23}$$

where

$$S_{mnp} = S_{[mnp]} = 4m\ \overset{o}{A}_{mnp} \tag{24}$$

The group which leaves invariant $\overset{o}{g}_{mn}(y)$ and $\overset{o}{A}_{mnp}(y)$ is the SO(7) subgroup of SO(8) [40,20,21,22,23,42], denoted $[SO(7)]_c$, under which the massless N=8 supermultiplet decomposes $1 \to 1$; $8_s \to 8$; $28 \to 7 + 21$; $56_s \to 8 + 48$; $35_v \to 35$; $35_c \to 1 + 7 + 27$. Once again, all eight gravitinos must acquire a mass and all 8 supersymmetries are broken at a common scale, but this time there is no 'Invasion of the Killing Spinors' and N=0 [6,9].

The Higgs and super-Higgs now works as follows. Fourier expanding $A_{mnp}(x,y)$ in harmonics on the round S^7

$$A_{mnp}(x,y) = \sum_k P_k(x)\ Y^k_{mnp}(y) \tag{25}$$

one obtains the 35_c massless pseudoscalars of the N=8 massless sector from the lowest mode together with an infinite tower of massive pseudoscalars, $P_k(x)$ where k denotes some representation of SO(8). In the symmetric phase

$$\langle A_{mnp}(x,y)\rangle = 0 \quad \rightarrow \quad \text{ALL } \langle P_k(x)\rangle = 0$$

but non-zero 'torsion' means

$$\langle A_{mnp}(x,y)\rangle = \overset{\circ}{A}_{mnp}(y) \neq 0 \quad \rightarrow \text{SOME } \langle P_k(x)\rangle \neq 0$$

i.e. nonzero VEVs for some of these pseudoscalars, and hence a spontaneous breakdown of parity invariance. In order to determine which are responsible for the breaking of SO(8) down to $[SO(7)]_c$, we note that (23) and (24) imply that A_{mnp} is a 'Yano Killing Tensor' i.e.

$$\nabla_q A_{mnp} = \partial_{[q} A_{mnp]} \tag{26}$$

One can show that there are 70 rank-three totally antisymmetric Yano Killing tensors on the round S^7, transforming as a 35_s and a 35_c under SO(8). The 35_c are associated with the massless pseudoscalars discussed above (in the n=0 level of Table 2) while the 35_s describe massive pseudoscalars (in the n=2 level of Table 2). They obey

$$\nabla_m F^{mnpq} = \pm \frac{m}{6} \varepsilon^{npqrstu} F_{rstu} \tag{27}$$

where the + sign refers to 35_c and the - sign refers to 35_s. From (22), therefore we see that the non-vanishing 'torsion' corresponds to a nonzero VEV for the 35_c (in the n=0 level of Table 2). As a consistency check we note that under $SO(8) \rightarrow [SO(7)]_c$, $35_c \rightarrow 1 + 7 + 27$ and we find as expected the singlet together with the 7 Goldstone bosons which give a mass to the 7 spin 1 gauge bosons in the decomposition $28 \rightarrow 7 + 21$.

iv) Vacuum IV : The right-squashed S^7

As discussed in (ii) above, this vacuum breaks SO(8) to [SO(5) ×

$SU(2)]_s$ under which the massless N=8 supermultiplet decomposes $1 \to (1,1)$; $8_s \to (5,1) + (1,3)$; $28 \to (10,1) + (5,3) + (1,3)$; $56_s \to (10,1) + (10,3) + (5,3) + (1,1)$; $35 \to (10,3) + (5,1)$; $35_c \to (10,3) + (5,1)$. As in Vacuum (iii) all supersymmetries are broken but no longer at a common scale; the (5,1) gravitinos will have a different mass from the (1,3) See Table 4.

The Higgs effect works in the same way as Vacuum II with the 1 and 300 acquiring the non-zero VEV's. The mass spectrum for spins 2,1, and 0^+ (which have only a v subscript or no subscript) is insensitive to the handedness of squashing and will be the same for both Vacuum II and Vacuum IV, but the spectrum for spins 3/2, 1/2 and 0^- (which have s and c subscripts) will be different. In particular, there will be a single massless pseudoscalar on the right-squashed sphere since the relevant mass operator has a single zero eigenvalue mode. See the discussion in (v) below. The corresponds to a 0^- 'invasion' since it belongs to the massive 35_s (in the n=2 level of Table 2) of the N=8 phase.

TABLE 4. DIFFERENT SUPERSYMMETRY BREAKING SCALES

ROUND S^7	RIGHT-SQUASHED S^7
$\lambda^2=1$	$\lambda^2=1/5$
N=8	N=0
SO(8)	$[SO(5) \times SU(2)]_s$

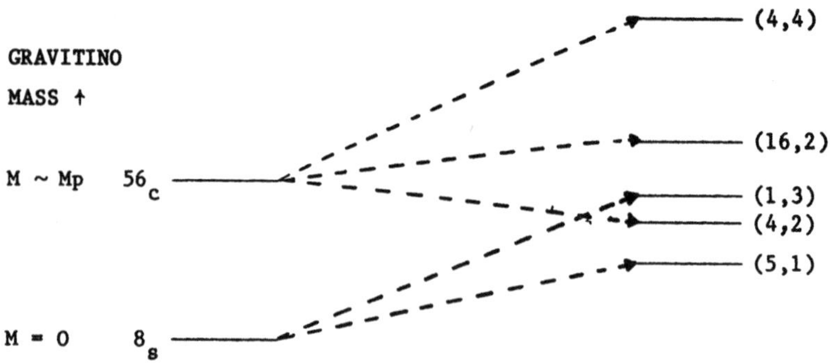

v) Vacuum V : The right-squashed S^7 with 'torsion'

Consider the constant three-form

$$\overset{o}{A}_{abc} = \pm \frac{1}{4} \bar{\eta}_{\ abc}\, \eta \quad ; \quad \bar{\eta}\eta = 1$$

where η is the constant Killing spinor of the left-squashed S^7. This satisfies the Einstein equations and

$$\nabla_m \overset{o}{F}{}^{mnpq} = - \frac{m}{6} \varepsilon^{npqrstu} \overset{o}{F}_{rstu} \tag{29}$$

However, comparison with (22) reveals a crucial sign difference and thus the left-squashed S^7 does not admit solutions of this type. Had we chosen $\overset{o}{F}_{\mu\nu\rho\sigma} = - 2m\, \varepsilon_{\mu\nu\rho\sigma}$ instead of $\overset{o}{F}_{\mu\nu\rho\sigma} = + 2m\, \varepsilon_{\mu\nu\rho\sigma}$, this would be a solution, but as discussed previously this sign change is equivalent to $\overset{o}{F}_{\mu\nu\rho\sigma} = +2m\, \varepsilon_{\mu\nu\rho\sigma}$ but with left-squashing replaced by right-squashing. So the right-squashed S^7 does admit solutions with 'torsion'.

In addition to the nonzero VEVs for the 1 and 300 of scalars (in the n=2 level of Table 2), the 35_s of pseudoscalars (also in the n=2 level) have acquired a nonzero VEV, as may be seen from (27). As a consistency check we note that under $SO(8) \rightarrow [SO(5) \times SU(2)]_s$, $35_s \rightarrow$ (14,1) + (5,3) + (1,5) + (1,1), and we find the singlet as expected. Since the non-zero torsion is a singlet under the vacuum symmetry of the zero-torsion vacuum, the unbroken gauge group remains $[SO(5) \times SU(2)]_s$ but there is, in addition, a spontaneous breakdown of parity. The pseudoscalar acquiring the non-zero VEV is the single pseudoscalar which was massless in the Vacuum IV above.

The $\overset{o}{A}_{mnp}$ field of the right-squashed sphere has, if anything, an even more interesting geometrical interpretation than that of the round sphere with torsion of Vacuum III. First of all, unlike that of the round sphere, it is constant in the orthonormal basis of (14) and is, in fact, the quantity apearing in the multiplication table of the octonions [12,13,17,18,21]. Moreover, with $S_{abc} = 4m\ \AA_{mnp}$

$$R_{mnpq}\,(\Gamma^s{}_{tu} + S^s{}_{tu}) = C_{mnpq}\,(\Gamma^s{}_{tu}) \tag{30}$$

so S_{abc} is not the parallizing torsion. Rather, it is the Ricci-flattening torsion [12], since from (30)

$$R_{mp}\,(\Gamma^s{}_{tu} + S^s{}_{tu}) = 0 \tag{31}$$

We shall return to this property in the next Section.

We emphasize once again that it is the right-squashed sphere of Vacuum IV and not the left-squashed sphere of Vacuum II which admits the 'torsion' leading to Vacuum V. In this respect, we find ourselves in disagreement with Refs. [12,20] which also discuss squashed seven-spheres with 'torsion' but do not distinguish between left and right.

vi) Stability

A remaining question is that of stability. One would expect Vacua I and II to be absolutely stable owing to the residual supersymmetry. Stability of both against dilations and squashing has recently been established at the classical level by Page [19]. See also [16]. This involves showing that the small scalar perturbations ϕ about the ground state satisfy the Breitenlohner-Freedman [41] criterion for positive energy in AdS, namely

$$-\Box\phi + \alpha\,\phi = 0 \qquad ; \qquad \alpha > +\frac{9R}{8.6} = -9m^2 \tag{32}$$

The peculiarities of AdS are such that the squashed sphere is stable against squashing even though, as a function of λ^2 the $\lambda^2 = 1/5$ sphere corresponds to a maximum of the effective potential and $\lambda^2 = 1$ to a minimum [66]. Unfortunately, this does not necessarily imply that the right-squashed sphere is also stable since its bosonic spectrum will differ in the 0^- sector. Indeed, there seems no apparent reason why Vacua III, IV and V should be stable but the analysis required to determine this has not yet been carried out. See, however, [23]. In the next Section we shall focus our attention on Vacuum II.

4. TOWARDS A REALISTIC THEORY

Since the discovery in 1976, Supergravity [44] has evolved along rather diverse lines; the N=8 route and the N=1 route. Those who do the maximal N=8 supersymmetry [35, 42] have several motivations. First, it is beautiful. Secondly, one has the feeling that if Nature cares at all about supersymmetry it would be crazy to stop at N=1. Thirdly, N=8 supergravity is the only truly unified field theory we have. It is the only known theory in which gravity and all the other particles of lower spin appear in one and the same multiplet. This is not true of N<8 supersymmetry and, in particular, it is not true of N=1. One also expects, therefore, that the N=8 theory will have the best behaviour from the point of view of ultraviolet divergences [1].

Those who do N=1 supersymmetry, on the other hand, do so because it is the only supersymmetry likely to be relevant for particle physics phenomenology at present energies. For a recent review of N=1 particle physics, see Zumino [45]. Consequently, the hope has sometimes been expressed that these two approaches could be linked if N=8 supergravity were to break spontaneously to N=1 at Planckian energies.

We have seen in the previous Section that this hope is indeed fulfilled: N=8 supersymmetry breaks spontaneously at the tree level to N=1 at a scale given by gM_P where g is the coupling constant of the Yang-Mills gauge group SO(8) and M_P is the Planck mass. At the same time SO(8) breaks spontaneously to SO(5) × SU(2) but parity remains unbroken. Note that of the three known N=8 supergravity theories, namely the ungauged theory of Cremmer and Julia [35], the gauged SO(8) theory of de Wit and Nicolai [42], and the theory proposed by Duff and Pope [1,2,10] obtained by spontaneous compactification of d=11 supergravity on S^7, it is only the last that posesses the above properties.

Indeed, spontaneous breaking of N=8 to 0<N<8 by a Higgs mechanism is impossible for conventional supergravity theories in four dimensions without a corresponding breakdown in parity. It is impossible all together for the Cremmer-Julia [35] theory because there is no effective potential for the spin-0 fields. One may break the supersymmetry from N=8 to N=6,4,2, or 0 via the Scherk-Schwarz [46] mechanism. This also relies on extra dimensions but the similarity to the mechanisms described in the present paper ends there. In particular, the surviving supersymmetries (massless gravitinos) are a subset of the original 8. Their mechanism is 'spontaneous' in the sense that the Lagrangian has a symmetry not shared by the vacuum. However, the symmetry breaking is not induced by nonzero VEVs for scalar fields but rather by assigning a specific dependence of the fields on the extra coordinates which are taken to parametrize a 7-torus T^7. This means that the symmetric phase and the broken phase are described by different Lagrangians with different symmetries, as opposed to the usual Higgs mechanism where the two phases are described by the same Lagrangian, as in the Weinberg-Salam model, for example. The gauged N=8 theory of deWit and Nicolai [42], on the other hand, does possess a non-trivial effective potential for which a conventional Higgs effect might take place. However, a surviving supersymmetry would necessarily imply parity violation. This is because the massive gravitinos would have to form spin 3/2 supermultiplets, but parity demands that massive spin -3/2 supermultiplets require both massive vectors and massive axial vectors[47]. Since one started only with 28 vectors, this is impossible without a nonzero VEV for some of the pseudoscalers and hence a spontaneous breakdown of parity. The reason that Kaluza-Klein supergravity circumvents this theorem is that the surviving supersymmetry is not one of the original eight. The 'Space Invaders Scenario' described in the previous Section provides an inverse super-Higgs effect whereby the massless gravitino of the N=1 phase comes from a massive N=8 supermultiplet, as do the Higgs scalars. The possibility of massive N=8 supermultiplets at the preon level is unique to Kaluza-Klein supergravity, and so the spontaneous breaking of N=8 to N=1 without P violation requires extra dimensions. Our conclusions about the breakdown of N=8 to N=1 thus differ from those reached by Ferrara and van Nieuwenhuizen [48] who claim that the Kaluza-Klein approach only describes a subset of all the breaking patterns discussed in their paper. Since they discussed only the case where the surviving massless gravitino of the N=1 phase is one of the eight massless gravitinos of the N=8 phase, they omitted the possibility of the 'Space Invaders Scenario'.

Although the hope of N=8 breaking to N=1 at the Planck scale has been fulfilled, it is far from clear that those who wished so fervently for it will be pleased with the result. To begin with, one of the main motivations of those who pursue the N=1 route is to use supersymmetry to solve the gauge hierarchy problem in grand unified theories (GUTS) based on simple groups like SU(5), but the SO(8) isometry group of S^7 not only fails to contain SU(5), it does not even contain the SU(3) × SU(2) × U(1) of the standard model. The N=1 symmetry of SO(5) × SU(2) is good enough for the electroweak SU(2) x U(1) but not the strong SU(3). Secondly, even leaving aside the SU(3) problem and even leaving aside the question of masses, none of the elementary spin 1/2 fermions in the N=1 theory appears in the right representations to be identified with any of the known quarks and leptons. The most obvious problem is that of chirality. Now N=1 supersymmetry is the only supersymmetry that can accommodate chiral fermion representations, so the fact that N=8 breaks to N=1 rather than N=2, say, might seem like a bonus. The trouble is, as was pointed out by Witten [49], that no classical compactification of d=11 supergravity could give rise to any asymmetry between the right and left handed spin 1/2 in the spectrum of d=4 elementary fermions. Thirdly, in common with all the other vacua of Table 1, the N=1 Vacuum II yields an enormous Planck-sized cosmological constant for spacetime which, unlike N=1 theories not obtained from N=8, cannot be fine-tuned to zero.

In the remainder of this section we pose the question of whether, in spite of these difficulties, the N=1 theory we have obtained from N=8 stands a chance of describing a realistic SU(3) × SU(2) × U(1) theory. The arguments will necessarily be highly speculative.

We shall argue that such a theory might emerge if, following Ellis, Gaillard, Maiani and Zumino [51] (EGMZ) and Ellis, Gaillard and Zumino [52] (EGZ), we postulate that some of the gauge bosons and some of the fermions of the standard model are composites formed the preons of supergravity. This was based on an earlier idea of Cremmer and

Julia [35]. It must be admitted at once that we have no better idea than these authors as to the precise dynamics by which such composites might form. However, we shall argue that the composite model based on the Kaluza-Klein theory described above has certain advantages over their composite model based on the four-dimensional Cremmer-Julia theory. Specifically,

i) Whereas EGZ merely supposed that N=8 breaks to N=1 through some unspecified mechanism, in our model this actually happens.

ii) The hoped-for bound states of EGMZ are based on an analogy with two and three dimensional σ-models, but it seems that gauge bosons are <u>not</u> generated dynamically in these models when the global invariance group is <u>non-compact</u>[53]. In contrast to the EGMZ model, Kaluza-Klein supergravity does not possess a non-compact $E_{7(7)}$. [The Cremmer-Julia E_7 may be broken by the gauging of the vectors fields as in the de Wit - Nicolai version of N=8. However, one recovers the troublesome E_7 in the limit of zero gauge coupling. The advantage of Kaluza-Klein, as explained in the previous section, is that the zero coupling limit does not exist.]

iii) EGMZ chose the massless SU(8) supermultiplet of bound states from which to construct a realistic GUT, but this supermultiplet contains many unwanted helicity states (see Table 5). These unwanted states cannot be made supermassive in a way consistent with an unbroken SU(3) × SU(2) × U(1). Since our N=1 theory contains the gauge bosons of SO(5) × SU(2) ⊃ SU(2) × U(1) already at the elementary level, we need look only for SU(3) at the composite level. As shown in Table 5 the unwanted helicity states of the EGMZ N=8 SU(8) supermultiplet may be made massive in a way consistent with an unbroken N=1 supersymmetry and G_2 gauge symmetry. G_2 contains SU(3) as a maximal subgroup.

The picture we have in mind differs from that of EGMZ in one other vital respect: it is <u>not</u> a GUT theory. Rather, the electroweak SU(2) × U(1) and the strong SU(3) have different origins and are not to be combined into a simple group.

TABLE 5. GIVING MASSES TO UNWANTED HELICITIES

N=8 SU(8) SUPERMULTIPLET

helicity	5/2	2	3/2	1	1/2	0	-1/2	-1	-3/2
SU(8) rep.	$\overline{8}$	$\overline{36}$	$\overline{168}$	$\overline{378}$	504	420	216	63	$\overline{8}$
		$\overline{28}$	$\overline{56}$	70	56	28	8	1	

Assume all spins 5/2,2, 3/2 and 1 get masses except 14 of 1 i.e. adjoint of G_2, and that the surviving spins 1/2 and 0 remain massless.

N=1 G_2 SUPERMULTIPLETS

$m \neq o$	1	7	14	27	64	77
(5/2,2,2,3/2)	1	1	0	0	0	0
(2,3/2,3/2,1)	0	1	1	1	0	0
(3/2,1,1,1/2)	1	2	1	1	1	0
(1,1/2,1/2,0)	1	2	0	2	0	1
m=0						
(1,1/2)	0	0	1	0	0	0
(1/2,0,0)	0	1	2	1	1	0

At this stage, there is one important aspect we should like to stress. The existence of two kinds of Yang-Mills gauge boson, 'elementary' and 'composite', is an automatic consequence of Kaluza-Klein theories and not to be regarded as an extra ad hoc ingredient in the theory. This is due to the fact, rarely stressed in the literature, that Kaluza-Klein theories unify gravity and Yang-Mills theories in two different ways. There are two gravitational bosonic symmetries in the d-dimensional theory: d-dimensional general covariance

$$x^M \to x'^M (x) \tag{33}$$

and d-dimensional local Lorentz invariance

$$\delta\, e_M{}^A\,(x) = \alpha^A{}_B\,(x)\, e_M{}^B\,(x)$$

$$\delta\, \omega_M{}^{AB}\,(x) = D_M\, \alpha^{AB}\,(x) \qquad (34)$$

which, in our d=11 case, is a local SO(1,10). Crudely speaking, the elfbeins $e_M{}^A(x)$ are the gauge bosons of general covarience and the spin connections $\omega_M{}^{AB}(x)$ are the gauge bosons of Lorentz invariance. However, whereas $e_M{}^A$ is an 'elementary' field the $\omega_M{}^{AB}$ are 'composite' fields. The spin connection has no kinetic term of its own and may therefore be expressed in terms of the elfbein and its derivatives. The Kaluza-Klein mechanism therefore gives rise to both elementary and compositie gauge bosons in d=4. The elementary fields $B_\mu(x)$ come from $e_\mu{}^a(x,y)$ and correspond to a gauge group given by the isometry group of the extra dimensions; SO(8) in the case of the round S^7. The composite gauge fields $A_\mu(x)$ come from $\omega_\mu{}^{ab}(x,y)$ and correspond to the tangent space group of the extra dimensions; SO(7) in the case of d=7. These latter fields have no kinetic energy term of their own and may therefore be expressed in terms of the scalar fields coming from $e_m{}^a(x,y)$ and their derivaties.

Thus, although we do not yet understand how the SO(5) × SU(2) might break to SU(2) × U(1) nor how the SO(7) might break to SU(3), the picture of Table 6 suggests itself as a possible breaking pattern. At any finite order of perturbation theory, of course, a non-vanishing VEV for $e_M{}^A$ means that both general covariance and SO(1,10) are spontaneously broken. Similarly a non-vanishing VEV for $e_m{}^a$ breaks both SO(8) and SO(7) leaving only a residual SO(8) vacuum symmetry. One must assume that this is an artefact of perturbation theory and that the full SO(8) × SO(7) (or at least some subgroup like SU(2) × U(1) × SU(3)) is realized dynamically in the non-perturbative regime.

From the work of Cremmer and Julia [35], we know that when the A_{MNP} field is taken into account, the hidden symmetry may be even bigger than SO(7) and might be enlargeable to SO(8) and then to SU(8) [The SO(7) generators Γ_{ab} are supplemented by Γ_a which together close on SO(8) and then by $\gamma_5\Gamma_{abc}$ to form the 63 generators of a chiral SU(8).] So the SU(8) supermultiplet of EGMZ and its breaking to G_2 may indeed be relevant for the theory obtained from S^7. The relation between this G_2 and the G_2 holonomy group of the N=1 phase remains somewhat obscure however.

TABLE 6 : HOW d=11 SUPERGRAVITY MIGHT GIVE SU(3) × SU(2) × U(1)

N=1	d=11 G.R	SO(1,10)
↓	↓	↓
N=8	d=4 G.R × SO(8)	SO(7) × SO(1,3)
↓	↓	↓
N=1	d=4 G.R × SO(5) × SU(2)	G_2 × SO(1,3)
↓	↓	↓
N=0	d=4 G.R × SU(2) × U(1)	SU(3) × SO(1,3)

Since we are assuming that the correct tree-level ground - state is that of Vacuum II with N=1 unbroken, it is unlikely that the breaking to N=0 will occur at any finite order of perturbation theory and we must again look to a non-perturbative mechanism. One possibility, which also has the advantage of providing the required parity breakdown would be spin -1/2 fermion condensates i.e non-zero VEVs for fermion bilinears like

$$\langle\bar{\chi}(x)\gamma_5\chi(x)\rangle = \text{constant} \neq 0 \qquad (35)$$

As it turns out such condensates are in any case required to solve one of the other outstanding problems: the cosmological constant.

Requiring that spacetime have vanishing cosmological constant imposes the following conditions on the curvatures and torsions of d=1 supergravity [7]

$$\mathring{R}_{\mu\nu\rho\sigma}(\omega) = 0$$

$$\mathring{R}_{mn}(\hat{\omega}) = 0 \qquad (36)$$

where $\hat{\omega}_{abc} = \omega_{abc} + S_{abc} - S_{acb} + S_{cab}$

$$S_{abc}(y) = \langle\bar{\psi}_a(x,y)\hat{\Gamma}_b\psi_c(x,y)\rangle \qquad (37)$$

$$S_{MNP} = 0 \ , \ \text{otherwise.}$$

In other words, there must exist a Ricci-flat connection in the extra dimensions but where the torsion $S_{abc}(y)$ is built of fermionic bilinears [The so-called 'torsion' of Vacua III and V built out of the bosonic A_{mnp} fields will not do the trick]. From (37) we see that the required fermion condensates are precisely of the spin -1/2 parity violating form (35) since the index of the gravitino fields ψ_M lies only in the extra dimensions and since $\hat{\Gamma}_a = \gamma_5\Gamma_a$.

One not very interesting solution is to have $S_{abc} = 0$ and $\mathring{R}_{mn}(\omega) = 0$ i.e Ricci flat solutions like T^7 [35] or $K3 \times T^3$ [54]. If on the other hand we require $\mathring{R}_{mn}(\omega) \neq 0$, as we must to get non-abelian elementary gauge fields, then for most geometries the problem has no solution. The remarkable property of S^7 topologies is that such a Ricci flat connection does exist. Indeed, we have already seen in (31) that the squashed S^7 admits a Ricci-flat torsion:

$$S_{abc} = \pm m\bar{\eta}\,{}_{abc}\,\eta \qquad (38)$$

where η is the Killing spinor of the N=1 left squashed sphere. [Since we are not obtaining the torsion from A_{mnp}, it does not matter if it satisfies (29) rather than (22)]. Thus if it were to turn out that in the true vacuum

$$\langle\bar{\psi}_a\hat{\Gamma}_b\psi_c\rangle = \pm m\bar{\eta}\Gamma_{abc}\,\eta \qquad (39)$$

then the true vacuum would have

$$\Lambda = 0 \tag{40}$$

[Another solution would be to take the parallelizing torsion on the round S^7 but this would be something of an overkill with $R_{mnpq}(\hat{\omega}) = 0$ whereas only $R_{mn}(\hat{\omega}) = 0$ is required for $\Lambda = 0$ We are, in any case, assuming Vacuum II at the tree level]. Of course we have succeeded only in swapping the problem of why (40) should be true for why (39) should be true. The points we are making are that a) in d=11 supergravity the cosmological constant of spacetime is calculable; a necessary condition for explaining $\Lambda = 0$. Theories where Λ may be 'tuned' can never explain this. (b) The unique properties of the seven-sphere allow the possibility that the answer is $\Lambda = 0$. Most geometries do not permit this possibility. (c) If $\bar{\phi}_a \hat{\Gamma}_b \phi_c$ has any geometrical significance, equation (39) seems the most natural one. Note, incidentally, that m^{-1}, the inverse radius of S^7, would now be a calculable multiple of the Planck mass. Thus, in common with the model suggested by Candelas and Weinberg [55], one would be able to calculate the Yang-Mills coupling constants as pure numbers.

Another calculable number is the CP violating angle θ. The term in the d=11 supergravity Lagrangian

$$\varepsilon^{M_1 M_2 \cdots M_{11}} F_{M_1 \cdots M_4} F_{M_5 \cdots M_8} A_{M_9 \cdots M_{11}}$$

yields a term in the d=4 Lagrangian of the form

$$f_{IJKL}(P) F_{\mu\nu}{}^{IJ} F_{\rho\sigma}{}^{KL} \varepsilon^{\mu\nu\rho\sigma}$$

where f is a function of the pseudoscalars. When $\langle P \rangle \neq 0$ as in Vacua III and V, one can obtain a

$$\Theta\, F_{\mu\nu}\, F_{\rho\sigma}\, \varepsilon^{\mu\nu\rho\sigma}$$

term. No such term is present in Vacuum II, though non-perturbative effects may introduce one. Note, however, that in the scheme we are advocating the $F_{\mu\nu}{}^{IJ}$ correspond only to the electroweak bosons and <u>not</u> to the gluons. Thus we have a rather drastic solution to the strong CP problem: not only are there no $F\tilde{F}$ terms for gluons there are no FF terms either! One must then ask whether supersymmetry provides any reason for generating the latter dynamically but not the former.

We have not addressed, nor at present are we able to address, the all important question of realistic fermion representations. And although N=8 breaks to the only $N > 0$ supersymmetry permitting chiral fermions, namely N=1, we are left with the question of whether a vectorlike preon theory can give rise to a flavour-chiral bound-state theory. Nor do we have anthing to say about families, except that SO(5) × SU(2) has room for another SO(3).

5. FURTHER QUESTIONS

It should be clear that the ideas set out in this paper, especially those of Section 4, are at a very preliminary stage and we do not wish to exclude other possible interpretations. Below we list a few random thoughts for future study.

1. We have argued that it is possible to give masses to the bound-states of unwanted helicities by the breaking of N=8/SU(8) to N=1/G_2. Yet another possibility is that there never were any N=8 bound states but that bound states form only in the N=1 phase. This picture would avoid having to answer embarrassing questions like: 'What does it mean to have massless spin-1 bosons not in the adjoint representation of the gauge group, massles spin -3/2 not in the vector representation, massless spin 2 which are not singlets and massless spin 5/2 of any kind?' However the amount of information obtainable from group theory alone would be accordingly diminished.

2. If our d=11 gravity Lagrangian had included terms quadratic in the Riemann tensor then the 'composite' gauge bosons coming from the spin connection would then be 'elementary' with their own Yang-Mills kinetic energy term. Aside from all the problems of higher derivatives, however, the crucial problem of symmetry restoration discussed in Section 4 would remain.

3. The chiral nature of the SU(8) was not exploited in the picture outlined in Section 4. Yet this may be the key to the question of chiral fermion representations in Kaluza-Klein. A comprehensive, though pessimistic, review of this crucial question has recently been given by Witten [60]. See also [61,62,63]. For the possible significance of SU(8) in the de Wit-Nicolai theory, see [42].

4. There was a certain amount of prejudice involved when in Section 2 we looked for spacetime vacua with maximal symmetry. Other solutions certainly exist. See, for example, [26]. Ideally, one should consider all possible solutions and exclude the undesirable candidates for the ground-state by stability considerations. Similar remarks apply to the alternative $F_{\mu\nu\rho\sigma}$ compactification of d=11 supergravity to a d=7 spacetime with 4 compact dimensions [36]. This has not yet been done. Indeed one can show that the ground-state solution of d=7 AdS $\times$ S^4 is stable and gives rise to an effective d=7 theory with local SO(5) invariance and N=4 supersymmetry. There is as yet no mathematical way to exclude this obviously unphysical vacuum state.

5. Indeed, although for reasons explained in this paper, we have focussed our attention exclusively on d=7 ground-state solutions topologically equivalent to S^7, there are an embarassingly large number (infinity) of other solutions with other topologies. The ones of obvious physical interest are those which fulfill the Witten [49] criteria of having a gauge group G $\supset$ SU(3) $\times$ SU(2) $\times$ U(1) and which therefore yield the correct gauge bosons already at the elementary level. Witten classified all such spaces in [49], and denoted them M^{pqr} where p,q and r are integers but did not enquire whether they provided solutions to the d=11 field equations. Duff & Toms [2]

pointed out the $M^{001} = CP^2 \times S^2 \times S^1$ did not admit an Einstein metric but that $M^{011} = CP^2 \times S^3$ and $M^{101} = S^5 \times S^2$ did, the latter admitting a spin structure so that fermions can be globally defined. More recently Castellani, D'Auria and Fre [42] have shown that all M^{pqr}, except M^{001}, admit an Einstein metric and, of particular interest, that the p = q spaces admit an N=2 supersymmetry. These will be stable whereas the rest are unlikely to be. The beauty of these solutions is that the SU(3), SU(2) and U(1) coupling constants will all be related by the geometry even though this is not a GUT theory [2]. Unfortunately, the problem for these solutions lies in the fermion spectrum. Aside from the absence of chirality anticipated by Witten [49,60], a harmonic expansion of fermion fields on M^{pqr} spaces reveals [64] that nowhere do the right quark and lepton representations appear either in the zero-modes or non-zero modes. One would then have to argue as in the case of S^7, that the quarks and leptons appear as bound states but in this case the whole idea of getting the right gauge bosons at the elementary level seems much less compelling. Thus it seems to us that by abandoning all the unique properties of S^7, by giving up the squashing = Higgs interpretation of Section 3, and giving up the spontaneous breakdown of N=8 to N=1 (the only supersymmetry compatible with chiral fermions) one has gained very little in return.

There are also solutions which are neither S^7 nor have SU(3) x SU(2) x U(1). Some of them do have an unbroken supersymmetry, however. Apart from the T^7 solution of Cremmer and Julia [35] for which $\mathcal{H} = \mathbb{1}$ and N=8, there is the $K3 \times T^3$ solution [54] for which $\mathcal{H}$ = SU(2) and N=4. If one likes hidden symmetries this one is quite rich with a hidden local SO(22) x SO(6) x U(1) with the 134 massless scalars belonging to the coset [SO(22,6)/SO(22) x SO(6)] x [SU(1,1)/U(1)]. (This observation is due to J. Schwarz and P.G.O. Freund, private communication.)

6. Just as we have conjectured [1,2,5,10] that the symmetric vacuum of the de Wit-Nicolai theory corresponds to the round S^7 solution of d=11 supergravity, so we have also conjectured that other asymmetric extrema of the de Wit-Nicolai effective potential might correspond to other solutions of the d=11 theory which deviate from the maximally symmetric

geometry but which still have the same topology, [5,10]. Of course, not all solutions with S^7 topology can correspond to de Wit-Nicolai extrema since they may involve a 'Space Invaders Scenario' as in Section 3. One possible equivalence is provided by the SO(7) invariant extremum with N=0 recently found by Warner [22] and the Englert solution of Vacuum III where in both cases only massless pseudoscalars acquire non-zero VEVs. Moreover, de Wit and Nicolai have shown the former extremum to be unstable [23]. As in the case of the round S^7 [27] the equality or otherwise of the cosmological constants would provide an interesting check. See [5,19,23,27]. Unfortunately this comparison is very difficult, not least because of the question of what to hold fixed in going from the symmetric phase to the broken one. Page [19] has suggested that the charge Q associated with A_{MNP} and defined in his paper is the relevant quantity. In this case he finds for the 5 vacua of Table 1.

$$\Lambda_I = -2^2.3^{4/3}|Q|^{-1/3}$$

$$\Lambda_{II} = -2^2.3^{11/3}.5^{-5/3}|Q|^{-1/3} = \Lambda_{IV}$$

$$\Lambda_{III} = -2^{1/3}.3^{1/3}.5^{4/3}|Q|^{-1/3}$$

$$\Lambda_V = -2^{1/3}.3^{8/3}.5^{-1/3}|Q|^{-1/3}$$

It is a strange empirical fact, not discussed by Page, that we find

$$\frac{\Lambda_{II}}{\Lambda_I} = \frac{\Lambda_V}{\Lambda_{III}}$$

the significance of which still eludes us.

Warner [22] has also found new extrema which appear to have no d=11 counterpart. In particular, he finds one for which N=8/SO(8) is broken down to N=1/G_2 with both scalars and pseudoscalars developing non-zero VEVs. This provides a concrete example of the necessity of parity breakdown in any spontaneous breaking of N=8 to N=1 without extra dimensions, as discussed in Section 4. The relation between this N=1 and this G_2 and the N=1 and G_2 discussed previously remains mysterious.

Finally there are the questions of quantization, Casimir energies, induced compactification, ultraviolet divergences and anomalies, not to mention string theories, d=10 supergravity, Kaluza-Klein cosmologies etc., etc. But this will be enough for now.

REFERENCES

[1] M.J. Duff in 'Supergravity 81', eds. S. Ferrara & J.G. Taylor (C.U.P. 1982), page 257.

[2] M.J. Duff & D.J. Toms in 'Unification of the Fundamental Interaction II', eds. J. Ellis & S. Ferrara (Plenum, 1982).

[3] F. Englert. Phys. Lett. 119B, 339 (1982).

[4] R. D'Auria & P. Fré. Phys. Lett. 121B, 141 (1983).

[5] M. J. Duff, Nucl. Phys. B219, 389 (1983), and to appear in proceedings of The Marcel Grossman Meeting, Shanghai, August 1982.

[6] R. D'Auria, P. Fré & P. van Nieuwenhuizen. Phys. Lett. 122B, 225 (1983).

[7] M.J. Duff & C. Orzalesi. Phys. Lett. 122B, 37 (1983).

[8] M.A. Awada, M.J. Duff & C.N. Pope. Phys. Rev. Lett. 50, 294 (1983).

[9] B. Biran, F. Englert, B. de Wit & H. Nicolai. Phys. Lett, 124B, 45 (1983).

[10] M.J. Duff & C.N. Pope, in 'Supersymmetry and Supergravity 82', eds. S. Ferrara, J.G. Taylor and P. Nieuwenhuizen. (World Scientific Publishing, 1983).

[11] P.G.O. Freund. Univ. of Chicago preprint EFI 82/83.

[12] F. Englert, M. Rooman & P. Spindel. Phys. Lett. 127B, 47 (1983).

[13] T. Dereli, M. Panahimoghaddam, A. Sudbery & R.W. Tucker. Phys. Lett. 126B, 33 (1983).

[14] B. Morel. Univ. of Geneva preprint UGVA-DPT 1983/02-378.

[15] M.J. Duff, B.E.W. Nilsson & C.N. Pope. Phys. Rev. Lett. 50, 2043 (1983) and Errata 51, 846 (1983).

[16] Y. Fujii, T. Inami, M. Kato & N. Ohta. Univ. of Tokyo preprint UT-Komaba 83-4. (Revised version).

[17] F. Gursey & C.H. Tze. Phys. Lett. 127B, 191 (1983).

[18] J. Lukierski & P. Minnaert. Univ. of Bordeaux preprint PTB-128.

[19] D.N. Page. Univ. of Texas preprint (to appear).

[20] F.A. Bais, H. Nicolai & P. van Nieuwenhuizen CERN preprint TH 3577 (1983).

[21] F. Englert, M. Rooman & P. Spindel. Univ. of Brussels preprint (1983).

[22] N.P. Warner. CALTECH preprints CALT-68-992 and CALT-68-1008 (1983).

[23] B. de Wit & H. Nicolai. NIKHEF preprint H/83-7 and H/83-8 (1983).

[24] P. Tataru-Mihai. Preprint (1983).

[25] R.G. Moorhouse & R.C. Warner. Univ. of Glasgow preprint (1983).

[26] P. van Baal, F.A. Bais & P. van Nieuwenhuizen. Univ. of Utrecht preprint (1983).

[27] M.J. Duff, C.N. Pope & N.P. Warner. Imperial College preprint ICTP/82-83/22.

[28] B.E.W. Nilsson & C.N. Pope. Imperial College preprint ICTP/82-83/27.

[29] Th. Kaluza, Sitzungsber. preuss. Akad. Wiss 966 (1921); O. Klein. Z. Phys. 37 895 (1926).

[30] W. Nahm. Nucl. Phys. B135, 149 (1979).

[31] E. Cremmer, B. Julia & J. Scherk. Phys. Lett. 76B, 469 (1978).

[32] P.G.O. Freund. Univ. of Chicago preprint (1981).

[33] A. Lichnerowicz. C.R. Acad. Sci. Paris, Ser A-B 257, 7 (1963).

[34] E. Cremmer & J. Scherk. Nucl. Phys. B108, 409 (1976); Nucl. Phys. B118, 61 (1977).

[35] E. Cremmer & B. Julia. Nucl. Phys. B159, 141 (1979).

[36] P.G.O. Freund & M.A. Rubin. Phys. Lett. 97B, 233 (1980).

[37] G. Jensen. J. Diff. Geom. 8, 599 (1973).

[38] E. Cartan & J.A. Schouten. Proc. K. Akad. Wet. Amsterdam 29, 933 (1926).

[39] R. Slansky. Phys. Reports C79, 1 (1981).

[40] L. Castellani & N.P. Warner. Caltech preprint CALT-63-1033.

[41] P. Breitenlohner & D.Z. Freedman. Phys. Lett. 115B, 197 (1982).

[42] B. de Wit & H. Nicolai, Phys. Lett. 108B, 285 (1982); Nucl. Phys. B208, 323 (1983).

[43] S. Weinberg. Phys. Lett. 125B, 265 (1983).

[44] D.Z. Freedman, P. van Nieuwenhuizen and S. Ferrara. Phys. Rev. D13, 3214 (1976); S. Deser & B. Zumino. Phys. Lett. 62B, 335 (1976).

[45] B. Zumino, Berkely preprint UCB-PTH-83/2; LBL-15819.

[46] J. Scherk & J.H. Schwarz. Phys. Lett. B153, 61 (1979); Nucl. Phys. B153, 61 (1979).

[47] P. Fayet & S. Ferrara. Phys. Rep. 32, 249 (1977).

[48] S. Ferrara & P. van Nieuwenhuizen. Phys. Lett. 127B, 70 (1983)

[49] E. Witten. Nucl. Phys. B186, 412 (1981).

[50] A. Salam & J. Strathdee. Ann. Phys. 141, 316 (1982).

[51] J. Ellis, M. Gaillard, L. Maiani and B. Zumino in 'Unification of the Fundamental Interactions' eds. J. Ellis, S. Ferrara & P. van Nieuwenhuizen (Plenum, 1980).

[52] J. Ellis, B. Zumino & M.K. Gaillard. Acta Physica Polonica. B13, 253 (1982).

[53] A.C. Davis, A.J. Macfarlane & J.W. Van Holten. Phys. Lett. 125B, 151 (1983)

[54] M.J. Duff, B.E.W. Nilsson & C.N. Pope. Univ. of Texas preprint (1983). To appear in Phys. Lett.

[55] P. Candelas & S. Weinberg. Univ. of Texas preprint UTTG-6-83.

[56] P.G.O. Freund. CERN preprint TH.3655.

[57] W. Heidenreich. Phys. Lett. 110B, 461 (1982).

[58] G.W. Gibbons, C.M. Hull & N.P. Warner. DAMTP preprint (1982);

[59] D.Z. Freedman & H. Nicolai (private communication).

[60] E. Witten. Lecture at 'Shelter Island II' conference.

[61] G. Chapline & R. Slansky. Nucl. Phys. B209, 461 (1982).

[62] C. Wetterich. Nucl. Phys. B223, 109 (1983).

[63] S. Randjbar-Daemi, J. Strathdee & A. Salam. Nucl. Phys. B (to be published).

[64] S. Randjbar-Daemi & J. Strathdee (unpublished).

[65] H. Nicolai, P.K. Townsend & P. van Nieuwenhuizen. Lett. Nuovo. Cimento. 30, 315 (1980).

[66] R. Coquereaux & A. Jadezyk. CERN preprint TH.3483 (1983).

CALCULATION OF FINE STRUCTURE CONSTANTS*

Steven Weinberg

Theory Group, Physics Department
University of Texas
Austin, Texas 78712

The title of my talk may seem a bit ambitious, but please note the plural "constants". To calculate the fine structure constant, 1/137, we would need a realistic model of just about everything, and this we do not have. In this talk I want to return to the old question of what it is that determines gauge couplings in general, and try to prepare the ground for a future realistic calculation.

As far as I know, the only theories that have a chance of predicting the gauge couplings are those which get Yang-Mills fields out of gravity in more than 4 dimensions. Everyone knows that in the original model of Kaluza and Klein, now over 60 years old, one begins with pure gravity in 5 spacetime dimensions. One of the spatial dimensions is assumed to be compact, a circle of circumference $2\pi\rho$, while the other 4 remain flat. Correspondingly, the metric is

$$\bar{g}_{LM}(x,y) = \begin{bmatrix} \eta_{\mu\nu} & 0 \\ 0 & \rho^2 \end{bmatrix} + \sum_{\ell=-\infty}^{\infty} \exp(i\,y\,\ell) \begin{bmatrix} g^{\ell}_{\mu\nu}(x) & A^{\ell}_{\mu}(x) \\ A^{\ell}_{\mu}(x) & \phi^{\ell}(x) \end{bmatrix} .$$

In the notation we will be using here, a bar denotes 4+N-dimensional quantities; indices L,M, etc. run over all 4+N coordinates; indices $\mu\nu$, etc., run over spacetime coordinates; and $\eta_{\mu\nu}$ is the Minkowski metric (diagonal, with elements +1, +1, +1, -1). Also, in the Kaluza-Klein model y is an angular variable running from 0 to 2π, and ℓ is an integer running over all positive and negative values. In 4 dimensions, the

---

 This incorporates the texts of talks given at the Fourth Workshop on Grand Unification and at the Shelter Island Conference on Quantum Field Theory and the Fundamental Problems of Physics, June 1, 1983. Research supported in part by the Robert A. Welch Foundation and NSF Contract No. PHY-82-15249.

metric excitations described by the fields $g^{\ell}_{\mu\nu}(x)$, $A^{\ell}_{\mu}(x)$, and $\phi^{\ell}(x)$ appear as particles of mass $|\ell|/\rho$ and spin 2, 1, and 0, respectively. Since ρ turns out very small, the only particles we observe experimentally are those with $\ell = 0$: a graviton described by $g^{0}_{\mu\nu}(x)$, a photon described by $A^{0}_{\mu}(x)$; and a scalar $\phi^{0}(x)$. Also, the electric charge of these particles is given by

$$q_{\ell} = \ell\sqrt{16\pi G}\,/\,\rho \ .$$

The Kaluza-Klein model thus provides an alternative to grand unification as an explanation of the quantization of electric charge. The fact that charge is quantized here was seen from the beginning as one of the most attractive features of the Kaluza-Klein model. It was this that especially interested Einstein, who helped Kaluza get a professorship after years as a lowly *privat dozent*. (Incidentally, I am told that the "ℓ" in "Kaluza" has the same Polish pronunciation as the "ℓ" in "Walesa", so whatever mispronounciation one uses for Walesa is equally appropriate for Kaluza.)

Unfortunately, the radius ρ of the fifth dimension is not dynamically fixed in the original Kaluza-Klein model, so it is not possible in this model to calculate the unit of electric charge from first principles. Instead, the experimentally determined value of the electronic charge e was historically used to calculate the size of the fifth dimension:

$$\rho = \sqrt{16\pi G}\,/\,e = 3.8\times 10^{-32}\ \text{cm} \ .$$

This is so small that one could well understand why the fifth dimension is unobservable, but the value of e remained mysterious.

In the last 20 years these ideas have been embodied in models of dimensionality 4+N higher than 5. I believe the first step was taken by Bryce de Witt in his lectures at the 1963 Les Houches Summer School (or perhaps I should say by de Witt and his students, since Bryce assigned the derivation as a take-home problem). In the early work the extra dimensions were taken to form the manifold of a compact Lie group, or (as in the review of Salam and Strathdee) an arbitrary compact homogenous manifold, but the key point does not even depend on homogeneity: The gauge group of the spin-one massless fields that are observed in 4 dimensions at low energy is identical with the group of symmetries of the compact manifold of extra dimensions; for each Killing vector of the manifold (i.e., each infinitesimal isometry) there is one massless gauge field. Thus for instance to get a low energy SU(3) × SU(2) × U(1) gauge group we need a compact manifold with an $SU(3) \times SU(2) \times U(1)$ isometry group, which as shown by Witten requires a spacetime of at least $4+7 = 11$ dimensions.

Recently I gave a simple prescription for calculating the gauge coupling constants in terms of the geometry of the compact manifold. Out of the Lie algebra of symmetries of the manifold, one can always choose a complete set of symmetry generators, each of which has the special property that if one starts at an arbitrary point of the manifold and follows the direction dictated by the symmetry transformation (i.e., follows the Killing vector) one comes back to the same point. The gauge coupling constant for the vector field associated with such a "Magellan curve" is given by

$$g = 2\pi\sqrt{16\pi G_0}/s \tag{1}$$

where s is the root-mean-square circumference of the manifold along these Magellan curves, the average being taken over starting points on the manifold, and G_0 is Newton's constant, apart from "induced gravity" corrections to which I will come back later. For instance, for the N-dimensional spherical surface S^N the isometry group is the group $O(N+1)$ of rotations in N+1-dimensions. Each of these rotations generates a family of Magellan curves, the "small circles", whose rms circumference is $2\pi\rho\sqrt{2/(N+1)}$, so the $O(N+1)$ gauge coupling constant is (for S^N radius ρ)

$$g = \sqrt{\frac{N+1}{2}}\,\frac{\sqrt{16\pi G_0}}{\rho} \; . \tag{1}$$

(Of course there are radiative corrections and as we shall see they can be very significant.) For the sphere all the gauge coupling constants are equal, as a result of the special symmetry of the sphere. In general, the gauge couplings within any one simple gauge group will automatically be predicted by Eq. (1) to have the ratios required by the group structure, while the ratios of couplings of different simple groups will be expressed in terms of dynamically determined circumferences. In any case, we cannot calculate the overall scale of the gauge couplings unless we know how to calculate the size of the compact manifold.

Starting with the work of Cremmer and Scherk, a number of authors have developed models in which one can calculate the metric of the compact manifold as a solution of Einstein's field equations in 4+N dimensions, with the energy-momentum tensor supplied by topologically non-trivial field configurations. (This is the case in particular for much of the work on higher-dimensional supergravity, which is covered by Duff's talk at this conference. I will not go into supergravity here.) These models lead in some cases to values for the ratios of various circumferences of the compact manifold, which can be used along with the general rules mentioned earlier to predict the ratios of various gauge coupling constants that are not related by any simple group-theoretic considerations. The ratios come out to be square-roots of rational numbers, just as in grand unified theories. Unfortunately, the strengths of the topological singularities in these models are free parameters (at least at the classical level), so the overall scale of the compact manifold and of the gauge coupling constants cannot be predicted, even if we know all parameters in the Lagrangian.

This is a serious problem, even apart from our natural ambition to be able to calculate fine structure constants. If the size of the compact manifold is not fixed by the underlying field equations, then we may expect it to evolve along with the cosmological expansion of ordinary spacetime. Then the ordinary fine structure constant α would have been rather different 10^{10} years ago from its present value. But we know experimentally that this is not the case: the spectrum of quasars shows that α was just about the same when the light was emitted as it is in our laboratories today. (I understand that this problem has been under consideration by Gross and Perry.) Hence we have some experimental evidence that the size of the compact manifold is actually locked by the field equations into a fixed value.

There is in fact a class of models in which the size ρ of the compact manifold is not only fixed but calculable. One may suppose that the energy-momentum tensor responsible for the compactification arises not from topologically non-trivial classical field configurations, but from the quantum fluctuations in a large number n of matter fields that in 4+N dimensions are free and massless. Now, this may seem like an unpromising approach, because you would think that if we include quantum matter fluctuations we also have to include quantum gravity fluctuations, and this gets us into all the unsolved difficulties of quantum gravitation. However for sufficiently large numbers of matter fields the quantum fluctuations (really, the Casimir pressure) of the matter can balance the classical action of gravitation, and the quantum effects of gravitation can simply be ignored. The quantum matter fluctuations of n massless matter fields give an energy density in 4+N dimensions of order $n\rho^{-4-N}$. The Einstein tensor $\bar{R}_{LM} - \frac{1}{2}\bar{g}_{LM}\bar{R}$ is of order ρ^{-2}, and the gravitational constant $\bar{G}$ in 4+N dimensions is given (at least in order of magnitude) by Newton's constant G times the volume of the compact manifold, and hence is of order $G\rho^N$. Einstein's field equations then give, in order of magnitude

$$\frac{1}{\rho^2} \approx 8\pi G\rho^N \times n\rho^{-4-N}$$

and therefore

$$\rho \approx \sqrt{8\pi Gn} \ . \tag{2}$$

The rms circumferences s are of order $2\pi\rho$, so (1) and (2) give gauge coupling constants roughly of order

$$g \approx 1 / \sqrt{n} \ . \tag{3}$$

This is highly satisfactory for a number of reasons. First, for a large number n of matter fields the gauge coupling constants turn out small, in agreement with observed values. Also, and not unrelatedly, the size of the compact manifold comes out larger by a factor $\sqrt{n}$ than the Planck length $\sqrt{8\pi G}$, and in consequence one can show that the energy-momentum tensor is dominated by the one-loop matter terms, all other terms being suppressed by powers of 1/n. Finally, as shown by Duff and Toms, these one-loop contributions to the energy-momentum tensor are actually finite in 4+N dimensions if N is odd, so at least for odd N we can calculate the ρ's and g's without having to know the coupling constants of counterterms (e.g., $R^{(4+N)/2}$) that might be needed to cancel infinities.

The need to add many matter fields here reminds me of the old childrens' story about stone soup. You probably remember the story: a poor peasant family hears a knock at the door of their hut one evening, and when they open it they find a starved-looking traveller. The peasants explain that they have no food at all, but the traveller tells them not to worry, he has a magic stone, which only has to be put into boiling water to make the most wonderful soup. Sure enough, when he puts the stone into boiling water, the traveller sniffs the steam, and exclaims that the soup will be really delicious. However, he says, it would be even better if a few potatoes were added. You know the rest: the peasants supply potatoes, then meat, and so on, until the soup is ready, and everyone has some, and they all agree that it is amazing how one can make such good soup from just a stone. These theories are like stone soup: we can get everything from just higher-dimensional gravity, but it would be even better if there were also some matter fields. (The problem raised by Witten, of getting low-mass fermions in non-real representations of the low-energy gauge group, may require that we add yet more ingredients to the soup.)

Philip Candelas and I have been working on the calculation of one-loop potentials and manifold sizes in this sort of model. The vacuum metric is taken to satisfy Poincare invariance in a 4-dimensional subspace, which constrains its form to be

$$\bar{g}^{VAC}_{MN}(x,y) = \begin{bmatrix} \eta_{\mu\nu} & 0 \\ 0 & \tilde{g}_{mn}(y) \end{bmatrix} \tag{4}$$

where $\tilde{g}_{mn}(y)$ is the metric of the compact N-dimensional manifold, with coordinates y^n. The total action then takes the form

$$I = -\int d^4x\, V_{eff} \tag{5}$$

$$V_{eff} = \int d^N y \sqrt{\tilde{g}}\, [\tilde{R}(y) + \bar{\Lambda}] / 16\pi\bar{G} + V[\tilde{g}] . \tag{6}$$

Here $V[\tilde{g}]$ is the "potential" (in the sense e.g. of Coleman and E. Weinberg) of the quantized matter fields in a classical background metric $\bar{g}^{VAC}_{MN}$. Also, $\bar{\Lambda}$ is a cosmological constant in 4+N dimensions, which we include here in order to be able to find a solution with a flat 4-dimensional spacetime. Of the Einstein field equations in 4+N dimensions, the nm components just say that V_{eff} is stationary with respect to $\tilde{g}_{nm}$; the $n\mu$ and μn components are automatically satisfied; and the $\mu\nu$ components require that $\bar{\Lambda}$ be fine-tuned to make V_{eff} vanish at its stationary point. (Of course it is very unpleasant to have to adjust $\bar{\Lambda}$ in this way, but I don't know of any remotely realistic model that does not have this problem.)

It is especially easy to solve these field equations for a class of one-parameter homogeneous manifolds, for which the symmetries of the manifold dictate that the metric takes the form

$$\tilde{g}_{nm}(y) = \rho^2 \gamma_{nm}(y) \tag{7}$$

where ρ is a free radius parameter, and γ is fixed (up to a choice of coordinates) by the symmetries of the manifold. This is the case for instance for spheres S^N, complex projective spaces CP^N, and the manifolds of simple compact Lie groups (and more generally, for homogeneous spaces G/H for which the generators of G that are not in H form a representation of H that is not reducible into real representations). In general, the relative normalization of ρ and Υ can be fixed by specifying that the curvature scalar $\tilde{R}$ of the compact manifold is

$$\tilde{R} = -N(N-1)\rho^{-2} \ . \tag{8}$$

(The factor N(N-1) is inserted so that for spheres ρ will be the usual radius. Note that $\tilde{R}$ is constant because by assumption the manifold is homogeneous; that is, any point can be carried into some other point by a symmetry of the manifold.) The potential (6) then takes the form

$$V_{eff} = \frac{1}{16\pi G_0}\left(\frac{-N(N-1)}{\rho^2} + \bar{\Lambda}\right) + V(\rho) \ . \tag{9}$$

We have here introduced a constant G_0

$$G_0 \equiv \bar{G} \Big/ \int d^N y \sqrt{\tilde{g}(y)} \ . \tag{10}$$

This equals the observed Newton constant, apart from "induced gravity" corrections I'll come to later. Note however that it is $\bar{G}$ that must be regarded as a fixed constant, with G_0 varying like ρ^{-N}. Therefore the condition that V_{eff} be stationary in ρ yields

$$0 = \frac{1}{16\pi G_0}\left(\frac{-N(N-1)(N-2)}{\rho^2} + N\bar{\Lambda}\right) + \frac{\rho dV(\rho)}{d\rho} \ . \tag{11}$$

Eliminating $\bar{\Lambda}$ by requiring also that $V_{eff} = 0$, we find

$$NV(\rho) - \frac{\rho dV(\rho)}{d\rho} = \frac{N(N-1)}{8\pi G_0 \rho^2} \ . \tag{12}$$

We can calculate the size of the compact manifold for any set of matter fields by simply calculating the matter potential $V(\rho)$ in one-loop order and solving Eq. (12) for ρ.

This is especially simple if the matter fields are massless and N is odd. There is no one-loop conformal anomaly for odd N, so in this case we can use ordinary dimensional analysis to see that $V(\rho)$ is proportional to $1/\rho^4$:

$$V(\rho) = C_N/\rho^4 . \tag{13}$$

The constant C_N depends on N, on the number of various types of massless matter fields, and on the symmetries of the manifold, but need not depend on any continuous parameter. Using (13) in (12) then yields our result for ρ:

$$\rho^2 = \frac{(N+4)8\pi G_0 C_N}{N(N-1)} . \tag{14}$$

Note that this solution breaks down for the case of the original Kaluza-Klein model, with $N=1$. This is because in this case the curvature $\tilde{R}$ automatically vanishes, so in order to allow $V_{eff} = 0$ the sign of $\bar{\Lambda}$ must be opposite to that of C_N, but in this case V_{eff} would have no stationary point for any finite ρ.

The constant C_N is roughly (as it turns out, very roughly) of the order of the number n of matter fields, so for large n, the radius ρ is indeed larger than the Planck length by a factor $\sqrt{n}$, as anticipated earlier. However, we can have a valid solution only for C_N positive; otherwise $\bar{\Lambda}$ cannot be adjusted to make V_{eff} vanish at its stationary point.

Now, to the calculation of C_N and ρ. Each of the 4+N-dimensional matter fields is manifested in 4 dimensions as an infinite tower of particles with masses proportional to $1/\rho$. To calculate the contribution of each of these particles to the potential, we use dimensional regularization, replacing the dimensionality 4 of spacetime by a complex number d. The one-loop potential of a particle of mass M in d dimensions is

$$\frac{-ih}{2(2\pi)^d}\int d^d k\, \ell n(k^2 + M^2 - i\varepsilon) = \frac{-h}{2}(4\pi)^{-d/2}\,\Gamma\left(\frac{-d}{2}\right)M^d \tag{15}$$

where h is the number of helicity states, with an extra minus sign for fermions. The total potential is thus proportional to a sum over particle states of M^d (weighted by their degeneracies) which diverges for all $\mathrm{Re}\,d > -N$, times a factor $\Gamma(-d/2)$, which diverges for $d \to 4$. Fortunately the sum can be analytically continued from $\mathrm{Re}\,d < -N-1$, where it converges, to $d=4$, where its analytic continuation for odd N has a simple zero that cancels the poles in $\Gamma(-d/2)$. With b massless minimally coupled scalar bosons and f massless Dirac fermions in 4+N dimensions, the result is

$$C_N = b\, C_N^{(0)} + f\, C_N^{(\frac{1}{2})} \tag{16}$$

where $C_N^{(j)}$ are ρ-independent numerical constants. In the special case of a spherical compact manifold, we find

$$C_N^{(0)} = \frac{1}{4\pi\nu}\int_0^{\pi/2} \frac{d\theta}{[2\cosh(\frac{\pi}{2}\tan\theta)]^{2\nu}}$$

$$\times \left\{\left(\frac{\nu^3}{\pi^3}\cos\theta\sin 3\theta - \frac{15\nu}{\pi^5}\cos^3\theta\sin 5\theta\right)\sinh(\nu\pi\tan\theta)\right.$$

$$\left. + \left(\frac{6\nu^2}{\pi^4}\cos^2\theta\cos 4\theta - \frac{15}{\pi^6}\cos^4\theta\cos 6\theta\right)\cosh(\nu\pi\tan\theta)\right\} \qquad (17)$$

$$C_N^{(\frac{1}{2})} = \frac{(-)^{\nu+1}\, 3\cdot 2^{\nu+1}}{\pi^6}\int_0^{\pi/2} \frac{\cos^3\theta\cos 5\theta\, d\theta}{[2\cosh(\frac{\pi}{2}\tan\theta)]^{2\nu+1}} \qquad (18)$$

where $\nu \equiv (N-1)/2$. Here are some numerical results:

N	$C_N^{(0)}$	$C_N^{(\frac{1}{2})}$
1	-5.05576×10^{-5}	$+2.022304\times 10^{-4}$
3	$+7.56870\times 10^{-5}$	$+1.945058\times 10^{-4}$
5	$+4.28304\times 10^{-4}$	-1.140405×10^{-4}
7	$+8.15883\times 10^{-4}$	$+5.958744\times 10^{-4}$
9	$+1.13389\times 10^{-3}$	-2.992172×10^{-5}
11	$+1.32932\times 10^{-3}$	$+1.477709\times 10^{-5}$
13	$+1.37403\times 10^{-3}$	-7.242740×10^{-6}
15	$+1.25249\times 10^{-3}$	$+3.537614\times 10^{-6}$
17	$+9.55916\times 10^{-4}$	-1.725405×10^{-6}
19	$+4.79352\times 10^{-4}$	$+8.412070\times 10^{-7}$
21	-1.79909×10^{-4}	-4.101970×10^{-7}
≥ 23	negative	alternates

The value of $C_N^{(0)}$ for $N = 1$ is the one found (for untwisted scalars) in the original Kaluza-Klein model by Appelquist and Chodos. By itself, it might lead to pessimism about our chances of satisfying the condition $C_N > 0$, necessary for a satisfactory solution of Einstein's equations. Fortunately, $C_N^{(0)}$ changes sign between $N = 1$ and $N = 3$, and remains positive for 9 cases, from $N = 3$ to $N = 19$, after which it becomes and remains negative for all higher N. Also, $C_N^{(\frac{1}{2})}$ alternates sign between $N = 3 \pmod 4$ and $N = 5 \pmod 4$. We conclude that the necessary condition $C_N > 0$ is satisfied for any mix of bosons and fermions when the total dimensionality N+4 is equal to 3, 7, 11, 15, or 19; for suitable boson-fermion mixes when N+4 = 5, 9, 13, 17, 23, 27, 31, ..., and not at all when N+4 = 21, 25, 29, ...

For spheres S^N the gauge group is O(N+1). The formula I gave earlier for the O(N+1) coupling constant can be derived by noting that the Einstein-Hilbert action in 4+N dimension when expanded in eigenmodes of definite 4-dimensional mass contains the term

$$-\frac{1}{4}\frac{g^2}{16\pi G_0}\left(\frac{s}{2\pi}\right)^2 F_{\mu\nu}F^{\mu\nu}$$

so (1) is obtained as the requirement that the Yang-Mills curl $F_{\mu\nu}$ be canonically normalized. But one must not forget (though at first we did) that with a large number of matter fields there is also an "induced Maxwell" term in the effective 4-dimensional action of the form

$$-\frac{1}{4}g^2 D_N F_{\mu\nu}F^{\mu\nu}$$

where D_N is a numerical coefficient like C_N, proportional to the number of matter fields. Since s^2/G_0 is of the order of the number of matter fields, these two terms are comparable, and we must write the normalization condition for the gauge fields as

$$g^2 = \left[\frac{(s/2\pi)^2}{16\pi G_0} + D_N\right]^{-1} . \qquad (19)$$

For spheres $s/2\pi = \rho\sqrt{2/(N+1)}$, and ρ is given by (14), so the O(N+1) coupling is

$$g_N^2 = \left[\frac{(N+4)}{N(N^2-1)}C_N + D_N\right]^{-1} . \qquad (20)$$

This is what we wanted - a formula for a gauge coupling with no unknowns, except for the dimensionality N and the numbers of various types of matter fields that enter in C_N and D_N. As anticipated, we can make g_N^2 as small as we like by having enough matter fields, but one now has a new consistency condition, that (20) should be positive. We have not had time yet to calculate D_N for $N > 1$. (D_N is calculated for $N = 1$ in a recent paper of Toms.) Unfortunately the coefficients $C_N^{(0)}$ and $C_N^{(\frac{1}{2})}$ are very small, and if the same is true of D_N then we will need a very large number of matter fields to get a reasonable gauge coupling. For instance, in 11 total dimensions (and neglecting D_N), we need 10^4 massless scalars to get $g_N^2/8\pi$ down to 0.23. It may be that the extreme smallness of $C_N^{(j)}$ which produces this problem is due to the great degree of symmetry of a sphere, and will not persist when we consider other less symmetric compact manifolds. At any rate, it seems to me that the important thing here is not to get the right value of the gauge coupling - the models studied so far are far from realistic - but to be able to calculate gauge couplings at all.

As everyone today knows, in specifying the value of a gauge coupling constant it is necessary to say not only what its value is but where - that is at what renormalization scale - it has that value. For the models discussed here, the answer is easy. In using the one-loop approximation, we were really assuming not only that there are many matter fields, but also that there are no large logarithms that could compensate for the factors $1/n$. (n is b or f or some linear combination of them.) If we set out to calculate the gauge coupling constants at a renormalization scale of say 1 eV, then the higher loop contributions to the potential could be characterized by factors not only of $1/n$, but also of $\ell n(\rho \times 1 \text{ eV})$, and our one-loop calculations would therefore be invalid. Our calculations are only valid if we interpret them as giving the gauge couplings at energies roughly of order $1/\rho$, about 10^{17} to 10^{18} GeV. In order to calculate the gauge couplings observed at accessible energies, we would have to use the results of our one-loop calculation as the initial condition for a set of Gell-Mann - Low differential equations, and integrate down from 10^{17} GeV to ordinary energies. Even if all this were feasible, in order to calculate the fine structure constant 1/37 from first principles we would also know how to calculate the electron mass (in units of $1/\rho$), and for this as yet we have no theory at all.

Up to now I have emphasized the calculation of the manifold size ρ in units of $\sqrt{8\pi G_0}$, because this is what we need to know in calculating gauge couplings. However it is also of some interest to know what ρ is in terms of the experimentally determined Newton constant G. Matter loops induce a term in the effective 4-dimensional action of the form

$$-\frac{E_N}{\rho^2}\,[\sqrt{g}\,R]_{4\ \mathrm{dim}}$$

where E_N is another numerical constant, proportional like C_N and D_N to the number of species of matter fields. (This is the "induced gravity" term used by Sakharov, Adler, Zee, etc. in calculations of the Newton constant.) Putting this together with the familiar term $-\sqrt{g}\,R/16\pi G_0$ from the classical action, we see that the true Newton constant G is given by

$$\frac{1}{16\pi G} = \frac{1}{16\pi G_0} + \frac{E_N}{\rho^2} \; . \tag{21}$$

Using Eq. (14) to eliminate G_0, we find

$$\rho^2 = 8\pi G \left[\frac{(N+4)\,C_N}{2N(N-1)} + E_N\right] \; . \tag{22}$$

E_N has been calculated for $N = 1$ by Toms. We have not yet had time to calculate E_N for $N > 1$, but we expect it to be comparable in magnitude with C_N. We see that one more consistency condition that will need to be satisfied is that

$$\frac{(N+4)\,C_N}{2N(N-1)} + E_N > 0 \; .$$

I should also say a word about the stability of the compact manifolds. The calculation of ρ that we did earlier should make it clear that these manifolds for $C_N > 0$ are stable against a uniform dilation or compression. The potential (9) is the sum of three terms: the matter potential is positive and varies like ρ^{-4}, the cosmological constant term is positive and varies like ρ^{+N} (recall that $G \propto \bar{G}/\rho^N$), and the the curvature term is negative and varies like ρ^{N-2}. Thus the manifold is stabilized against implosion by the Casimir pressure of the matter quantum fluctuations and stabilized against explosion by the cosmological constant. This does not say however that a particular manifold like the sphere is necessarily stable against deformation. Page has recently studied the stability of the sphere (with Casimir pressure supplied by massless scalars) against deformation into a simple kind of homogeneous 2-parameter manifold. He finds stability for $N = 5$, 7, 9, 11, 13, or 15, but instability for $N = 3$, 17, 19. Of course, instability here is not necessarily a bad thing - we do not know that the compact manifold or the physical vacuum is a sphere, or in other words, that the gauge group at 10^{17} GeV is an O(N+1).

There is also an instability of a rather different sort. So far, everything here has been for the case of zero temperature. Now, at finite temperature the four-dimensional spacetime has a Robertson-Walker rather than a Minkowski metric, and the time-dependent treatment given earlier is inappropriate. However, if we shut our eyes to this problem, we can try to judge the stability of the compact manifold at finite temperature by simply including temperature-dependent effects in the static matter potential. These add a negative term to V_{eff}, of order $-T^{4+N}\rho^N$. Hence for small T the cosmological term still dominates for large ρ, but for T above a certain critical value the thermal term wins and V_{eff} decreases without limit as $\rho \to \infty$. This suggests (though it does not prove) that there is a critical temperature at which the compact manifold explodes.

In closing I want to return to the problem of ultraviolet divergences. I don't think that there is any profound significance to the fact that there are no one-loop divergences in odd dimensions. This just provides us with a chance (if there are many matter fields and N is odd) to calculate the gauge couplings without worrying about the difficult problems of quantum gravity. In this respect, the situation here is like the use of general relativity in astronomy. Even when we study a compact object like a neutron star, where the gravitational field is so strong that we have to take the nonlinearities in Einstein's equations into account, we are comfortable in ignoring effects of quantum gravitation. Why is this? It is not that G is small; G is not dimensionless, and the dimensionless quantity GM/r for neutron stars is of order unity. Rather, it is that $\sqrt{G}M$ is large; that is, there are a large number (about 10^{38}) of Planck masses in the star. When we ignore quantum effects in our calculations of compact manifold sizes, we are neglecting the same sorts of 1/n corrections as when we ignore quantum gravity in astronomy - only for us n is presumably a few hundred, not 10^{38}. But the ultraviolet divergences are still there, and will eventually have to be dealt with.

My own guess is that there is no renormalizable theory of gravitation, and that the final theory (whether in 4 or 4+N dimensions) will be infinitely complicated, with a Lagrangian (or something like it) containing all possible terms allowed by symmetry principles. There is no problem with infinities in such a theory; for every ultraviolet divergence there is a counterterm ready to absorb it. The real problem is rather to understand why the infinite number of coupling constants have any specific values.

There is a possible answer to this problem, one that I have been advocating at every opportunity for some years past. The renormalized coupling constants depend of course on renormalization scale μ, and trace out trajectories in coupling-constant space as μ is varied. It seems to me very likely that the generic trajectory runs off to infinity as μ increases, and that in consequence the theory develops diseases of one sort or another: tachyons, Landau ghosts, etc. One way, and perhaps the only way, to save the consistency of relativity and quantum mechanics is for the coupling constants (scaled by powers of μ to make them dimensionless) to lie on a trajectory that hits a fixed point of the renormalization group equations for $\mu \to \infty$. The trajectories that hit a given fixed point will form an "ultraviolet critical surface" in coupling constant space, and the requirement that the couplings lie on this surface will leave us with a smaller number of free dimensionless parameters, equal to the dimensionality of the critical surface minus one, plus one scale parameter.

This is very speculative, but there are solid reasons to believe that complicated field theories have ultraviolet critical surfaces of finite (and in fact small) dimensionality. Consider, for instance, the effective field theory that is used to study critical phenomena in water. This theory has an infinite number of coupling parameters, that depend on the temperature and pressure and all the microscopic properties of water molecules. Nevertheless, in order to bring about a second-order phase transition it is only necessary to adjust 2 parameters; say the temperature and pressure. If there were some sort of external field that would allow us to change a microscopic parameter like the mass of the water molecules, we could instead adjust that parameter and the temperature <u>or</u> pressure. The important thing is that just two quantities need to be adjusted. It is now understood that this means that there is a fixed point in coupling parameter space, and that the surface of trajectories attracted to this fixed point as $\mu \to 0$ has dimensionality ∞ minus 2; that is, just 2 parameters need to be adjusted to place us on the <u>infrared</u> critical surfaces. But a fixed point is a fixed point, and if its infrared critical surface has dimensionality (in the above sense) of ∞ minus 2, then its ultraviolet critical surface has dimensionality 2. In other cases the dimensionality is even smaller; for instance for ferromagnets it is 1. (Even though two parameters, T and H, are at our disposal, only one needs to be adjusted to produce a second-order phase transition.) But in any case, the ultraviolet critical surface seems always to be finite dimensional. Thus there are grounds to hope that in this way we will wind up with only a small number of free dimensionless parameters - perhaps none at all.

Of course, if one studies short-range or high energy phenomena in real water or real ferromagnets, the description of the system in terms of an effective field theory for pressure or magnetization fluctuations eventually breaks down as μ increases, and we must go over to a description in terms of water molecules or iron atoms, or even their constituents. The requirement that the true theory must lie on an ultraviolet critical surface makes sense only when we describe nature in terms of the final short-range degrees of freedom, whatever they are. In the end I think that all physical constants will be determined in this way, but if we are lucky - say if there are many matter fields in higher odd dimensions - we may be able to calculate some of them long before we get to the fixed point.

Added Note: Candelas and I have now calculated the coefficients D_N and E_N for spinors and minimally coupled scalars. Our results show that the necessary conditions for g^2 and G/ρ^2 as well as $V(\rho)$ to be positive are satisfied in N = 5,9,13,17, and 21 compact dimensions, for a variety of suitable mixes of fermions and scalars.

D=10 SUPERSTRING THEORY

Edward Witten*
Joseph Henry Laboratories, Princeton University
Princeton, New Jersey 08544

*Manuscript prepared by Yutaka Hosotani

Introduction

The supersymmetric string of Schwarz and Green [1] is a variant of the old fermionic string theory of Ramond-Neveu-Schwarz [2]. It is consistent only in ten dimensions and apparently must be interpreted in the sense of a Kaluza-Klein theory, the ten dimensions consisting of four Minkowskian ones and six compact ones.

Naively, the ten dimensional superstring theory reduces at low energies to ten dimensional supergravity. But unlike that theory it is apparently finite or renormalizable. In fact the superstring theory would appear at the moment to be the one real contender as a renormalizable, physically sensible quantum theory of gravity.

To see why a string theory describes gravity, note first that it contains a rich spectrum of particles. String oscillations include oscillations of the extra dimensions. It turns out that the lightest states of this spectrum are massless particles of spins 0, ½, ..., 2 with just the long wavelength couplings of supergravity.

The phenomenology of the superstring theory includes whatever is possible in N=8 supergravity in four dimensions or in D=10 supergravity. It could describe many other possibilities as well, if the compactification inherently depends on properties of strings.

The status of these theories is summaried in the following table.

	N=8 Supergravity	10 dim. Supergravity	10 dim. Superstring
Finite renormalizable?	Probably not	No	Probably
Uniqueness	Far from it	Almost and maybe	Almost and maybe
Phenomenological prospects	Poor	Problematic	Problematic
Aesthetics	O.K. except for nonuniqueness	Beautiful	Unsatisfactory in present formulation

What is really unsatisfactory about the string theory at the moment is that it isn't yet a theory. It is a (not entirely complete) set of Feynman rules for three string vertices, four string verticies, etc. Suppose that general relativity had never been invented and someone was trying to construct a Lorentz invariant theory of a massless spin two particle. After constructing the free field theory, one tries to find an acceptable three body vertex:

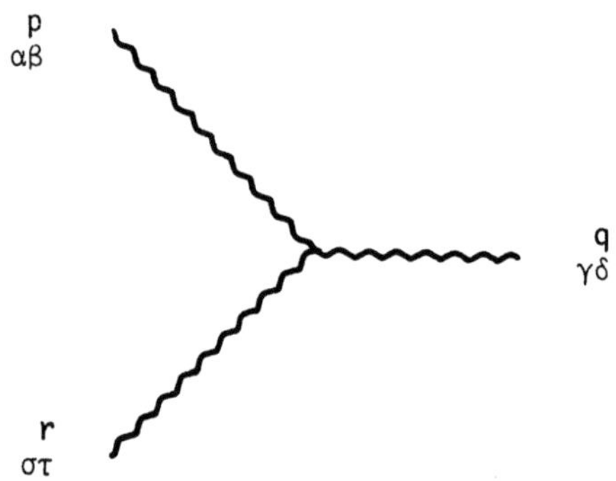

A priori it would be a tensor $\Gamma_{\alpha\beta\gamma\delta\sigma\tau}(pqr)$. But there is a unique acceptable choice. Even more complex, if one has never heard of Riemannian geometry, is the four body vertex:

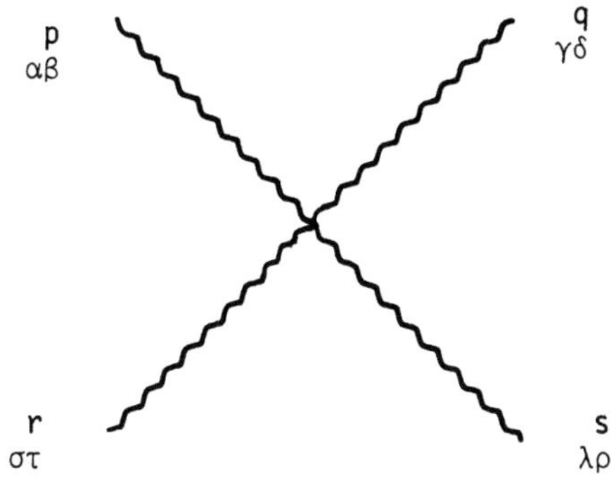

The existence of a physically acceptable four body vertex is a bizarre miracle until one discovers Riemannian geometry and writes

$$\sqrt{g}\ R. \tag{1}$$

The string theory is in such a state. There seem to be consistent string interactions

and a finite, physically sensible quantum theory of gravity which contains no ghosts and tachyons. But the consistency is a miracle; the vertices are laboriously constructed and proved consistent.

We don't have an analogue of Riemannian geometry underlying string theory and we cannot sum the vertices in a form analogous to $\sqrt{g}\ R$. This is a crucial problem on the esthetic plane. If there is not an analog of Riemannian geometry and $\sqrt{g}\ R$ in string theory, I doubt that string theory is attractive as a fundamental theory of nature.

Also on the practical level, it is a crucial problem. It is very unreasonable to describe the earth going around the sun by exchange of 10^{80} or 10^{90} strings. We should work with a nontrivial solution of something. Even worse, to do any reasonable phenomenology in string theory will require a Kaluza-Klein approach and probably not with flat extra dimensions (a case already considered by Schwarz and Green). To discuss non-trivial compactification one needs the string analogue of Riemannian geometry, i.e. what equations should one solve and what is a "non-singular" solution to those?

Supersymmetry from String Theory

How do massless particles and supergravity arise in string theory? I will consider the slightly simpler case of open strings leading to massless particles of spins $(\frac{1}{2}, 1)$ and super Yang-Mills theory.

The ordinary (D=26) bosonic string theory describes the motion of a string, parametrized by (σ, τ), in a spacetime of coordinates X^μ. In some gauge

$$L = \int \frac{1}{2} \left(\partial\alpha X^\mu \partial\alpha X_\mu\right) d\sigma d\tau$$

$$\alpha = \sigma, \tau, \qquad 0 \leq \sigma \leq \pi \tag{2}$$

The X^μ, although spacetime coordinates, appear as free scalar fields in a two dimensional σ-τ world.

In the Ramond-Neveu-Schwarz fermionic string theory [2], the string also carries anti-commuting degrees of freedom $\psi^\mu(\sigma,\tau)$. The ψ^μ field is rather odd. It anticommutes and is a spinor under transformations of (σ, τ); however, under Lorentz rotations of X^μ the ψ^μ transform as a vector. Hence ψ^μ maps bosons into bosons and fermions into fermions.

$$\psi^\mu |B\rangle = |B'\rangle \qquad \Delta J = 0, 1 \qquad \psi^\mu |F\rangle = |F'\rangle \tag{3}$$

In fact, as we will see, the model can be constructed so that all states are bosons or all are fermions.

The Lagrangian is

$$L = \int d\sigma d\tau \, [\frac{1}{2} \partial^{\alpha} X^{\mu} \partial_{\alpha} X_{\mu} + \frac{1}{2} i\bar{\psi}_{\mu} \gamma^{\alpha} \partial_{\alpha} \psi^{\mu}] \tag{4}$$

It has very peculiar symmetry:

$$\begin{aligned} \delta X^{\mu} &= i\varepsilon\psi^{\mu} \\ \delta\psi^{\mu} &= \partial X^{\mu} \cdot \varepsilon \end{aligned} \tag{5}$$

where ε^{α} is a two component anti-commuting spinor in σ-τ space but a Lorentz scalar. The corresponding conserved charge Q_{α} commutes with angular momentum

$$Q_{\alpha} \, |J, J_z; \eta> = |J, J_z; \tilde{\eta}> \tag{6}$$

This peculiar symmetry was the genesis of supersymmetry. Wess and Zumino erased the μ index from X^{μ}, interpreted X^{μ} as a scalar field ϕ, and generalized the string parameters (σ, τ) to four spacetime coordinates. This what supersymmetry developed from [3].

The Lagrangian (4) does not have spacetime supersymmetry, i.e., it does not have Bose-Fermi symmetry. But a slight variant of it does.

First of all, I have written the Lagrangian (4) in a particular gauge. It is not necessary to write the complicated gauge invariant form. But we must impose the constraint equations (analogous to Gauss's law) that are conjugate to the gauge conditions. The constraints are the vanishing of the symmetry generators (reparametrizations of strings and "supersymmetry") or of the string energy-momentum tensor and supercurrent.

$$\begin{aligned} \partial\alpha X^{\mu} \partial_{\beta} X_{\mu} - \frac{1}{2} \eta_{\alpha\beta} (\partial\alpha X^{\mu})^2 &= 0 \\ (\partial\alpha X^{\mu}) \, \gamma^{\alpha} \gamma^{\beta} \psi_{\mu} &= 0 \end{aligned} \tag{7}$$

The constraints can be used to eliminate two components of X^μ (say X^o, X^9) and the corresponding two components of $\psi^\mu(\psi^o, \psi^9)$. We are then left with eight free Bose and Fermi fields.

$$L = \frac{1}{2} \sum_{i=1}^{8} \int d\sigma d\tau \, [(\partial_\alpha X^i)^2 + i\bar{\psi}^i \gamma^\alpha \partial_\alpha \psi^i] \tag{8}$$

As I have mentioned, this Lagrangian can be quantized so that the states are all bosons or all fermions.

For open strings the field ψ has a Fourier expansion

$$\psi^i(\sigma) = \Sigma \, \psi_n^i \, e^{in\sigma} \quad (0 \leq \sigma \leq \pi) \tag{9}$$

where the sum runs over integer n or half odd integer n corresponding to fermions or bosons. Indeed canonical quantization gives

$$\{\psi_{-n}^i, \psi_m^j\} = \delta^{ij} \, \delta_{nm}. \tag{10}$$

One may regard ψ_m^j $(m > 0)$ as "annihilation operators" and ψ_{-m}^j as "creation operators". If n runs over half odd integers $(n \varepsilon \, z + \frac{1}{2})$, the Hilbert space is as follows: There is a unique ground state of the string, $|\Omega\rangle$, which is a bosonic state with J=0. Excitations are given by

$$\psi_{-m1}^{i1} \, \psi_{-m2}^{i2} \cdots \psi_{-mk}^{ik} \, |\Omega\rangle \tag{11}$$

All are bosons since ψ^i is a vector. On the other hand, if n is integer, the ψ_n^i for $n > 0$ $(n < 0)$ are annihilation (creation) operators. But ψ_o^i is left over, being self conjugate. The ψ_o^i obey

$$\{\psi_o^i, \psi_o^j\} = 2\delta^{ij}, \; i, j = 1, \ldots, 8, \tag{12}$$

namely an 8 dimensional Clifford algebra (γ matrices of O(8)). Its irreducible representation is unique and sixteen dimensional. It is the spinor of O(8). So the ground state is a spinor $|\Omega^{\alpha}\rangle$ (α=1, ..., 16) and the excitations

$$\psi^{i1}_{-m1}\ \psi^{i2}_{-m2}\ \cdots\ \psi^{ik}_{-mk}\ |\Omega^{\alpha}\rangle$$

are fermions

The low-lying states are

Boson		Fermion		
$\lvert\Omega\rangle$	1 state			
		$\lvert\Omega^{\alpha}\rangle$	16 states	(13)
$\psi^{i}_{-1}\lvert\Omega\rangle = \lvert\Omega^{i}\rangle$	8 states			

It is obvious that the theory cannot have spacetime supersymmetry, since the counting of states is different. Actually the first bose excitation $\psi^{i}_{-1}\ |\Omega\rangle$ is massless; it is a vector but a massive vector needs nine components. The ground state $|\Omega\rangle$ of the bose spectrum is thus a tachyon.

However, Gliozzi, Scherk, and Olive [4] made the observation that half the bosons are in correspondence with half the fermions. For bosons we keep only "even G" states:

$$\psi^{i1}_{m1}\ \psi^{i2}_{m2}\ \cdots\ \psi^{ik}_{mk}\ |\Omega\rangle \qquad \text{(odd k)} \tag{14}$$

This eliminates the tachyon so that the ground state of the string is now a massless vector. For fermions we define an operator $\tilde{\psi}$ = $\psi^1_o\psi^2_o\ \ldots\psi^8_o$, an analogue of γ_5, and keep only states of $\tilde{\psi}$ = + 1. The ground state of the Fermi spectrum has now 8 components and must be m=0 since an m≠0 fermion in 10 dimensions needs 16 components. So the lowest states are

Boson	Fermion	
m=0 vector	m=0 spin $\frac{1}{2}$	
		(15)
	with definite chirality	

This is the $(\frac{1}{2}, 1)$ multiplet of D=10 supersymmetry.

Gliozzi, Scherk, and Olive showed, by counting, that also the excitations might be supersymmetric; and from the recent work of Schwarz and Green we know that restricted to half the bosons and half the fermions, the interactions too are supersymmetric [5].

Our Lagrangian was obtained by fixing a gauge from one that was mainfestly Lorentz invariant, but had no 10 dimensional supersymmetry. It can be converted to a form that has supersymmetry but has no manifest Lorentz invariance by using a transformation introduced by Shankar in studying the O (8) $(\bar{\psi}\psi)^2$ model [6]. This procedure automatically drops half the states.

First we bosonize the fermions ψ^i (i=1, ..., 8) by introducing real scalars ϕ^a (a=1, ..., 4):

$$
\begin{aligned}
\frac{1}{\sqrt{\pi}} \varepsilon_{\mu\nu}\partial_\nu\phi_1 &= \bar{\psi}_1\gamma_\mu\psi_2 \\
\frac{1}{\sqrt{\pi}} \varepsilon_{\mu\nu}\partial_\nu\phi_2 &= \bar{\psi}_3\gamma_\mu\psi_4 \\
\frac{1}{\sqrt{\pi}} \varepsilon_{\mu\nu}\partial_\nu\phi_3 &= \bar{\psi}_5\gamma_\mu\psi_6 \\
\frac{1}{\sqrt{\pi}} \varepsilon_{\mu\nu}\partial_\nu\phi_4 &= \bar{\psi}_7\gamma_\mu\psi_8
\end{aligned}
\tag{16}
$$

Next we shuffle the scalars

$$
\begin{aligned}
\sigma_1 &= \frac{1}{2}(\phi_1 + \phi_2 + \phi_3 + \phi_4) \\
\sigma_2 &= \frac{1}{2}(\phi_1 + \phi_2 - \phi_3 - \phi_4) \\
\sigma_3 &= \frac{1}{2}(\phi_1 - \phi_2 + \phi_3 - \phi_4) \\
\sigma_4 &= \frac{1}{2}(\phi_1 - \phi_2 - \phi_3 + \phi_4)
\end{aligned}
\tag{17}
$$

and re-fermionize them, introducing eight new majorana fermions λ_α with

$$
\begin{aligned}
\frac{1}{\sqrt{\pi}} \varepsilon_{\mu\nu}\partial_\nu\sigma_1 &= \bar{\lambda}_1\gamma_\mu\lambda_2 \\
\frac{1}{\sqrt{\pi}} \varepsilon_{\mu\nu}\partial_\nu\sigma_2 &= \bar{\lambda}_3\gamma_\mu\lambda_4 . \\
\frac{1}{\sqrt{\pi}} \varepsilon_{\mu\nu}\partial_\nu\sigma_3 &= \bar{\lambda}_5\gamma_\mu\lambda_6 \\
\frac{1}{\sqrt{\pi}} \varepsilon_{\mu\nu}\partial_\nu\sigma_4 &= \bar{\lambda}_7\gamma_\mu\lambda_8
\end{aligned}
\tag{18}
$$

Shankar showed that the λ_α transform as a spinor of 0 (8).

The Lagrangian is now

$$\bar{L} = \frac{1}{2} \int d\sigma d\tau \; \{(\partial_\alpha X^i)^2 + \bar{\lambda}^\alpha i \not{\partial} \lambda_\alpha\} \tag{19}$$

But it is not quite equivalent to the old one. Bosonization of fermions is exact on the open line. Here, on $0 \le \sigma \le \pi$, it is not quite exact. $\bar{L}$ differs from L exactly as desired:

(1) It automatically describes both fermions and bosons, unlike L which can be quantized with only fermions or only bosons. After all λ^α has $J = \frac{1}{2}$.

(2) Quantization of $\bar{L}$ gives half the bose sector of L and half the fermi sector.

(3) $\bar{L}$ is supersymmetric, the conserved currents being

$$S_\alpha^\varepsilon = \gamma_\alpha \lambda^\varepsilon$$

$$\tilde{S}_\alpha^\varepsilon = (\gamma^\beta \partial_\beta X^i \gamma_i \lambda_\alpha)^\varepsilon \tag{20}$$

$\bar{L}$ is not manifestly Lorentz invariant and this is not obvious since $\bar{L}$ is not really equivalent to L. $\bar{L}$ (to our knowledge) cannot be obtained by gauge fixing in a manifestly Lorentz invariant Lagrangian. $\bar{L}$ can be laboriously shown to be Lorentz invariant (for instance, S^ε and $\tilde{S}^\varepsilon$ above combine into one Lorentz multiplet) and interactions can be laboriously constructed.

To find a better approach one must solve the outstanding problem explained at the beginning of the talk: to find the analogue of general covariance and $\sqrt{g}\, R$.

Uniqueness and Anomalies

I now wish to explain why the superstring theory and ten-dimensional supergravity are "almost and maybe" unique.

In four dimensions the triangle diagram is anomalous in certain gauge theories.

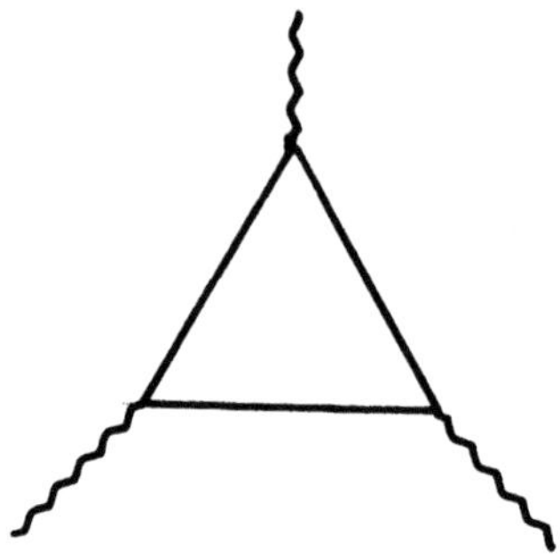

In ten dimensions the hexagon diagram is anomalous [7]. There are various anomalies:

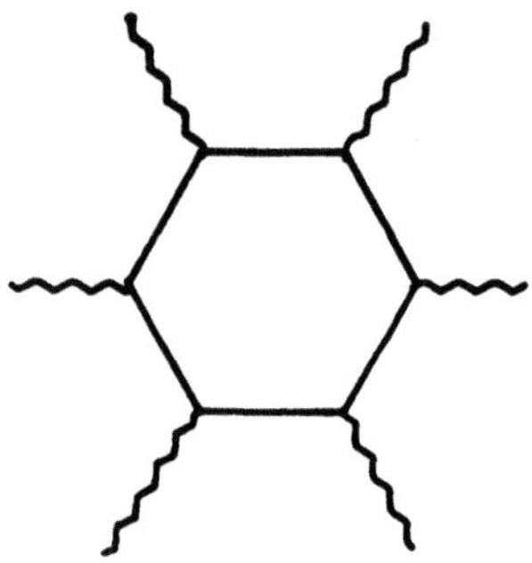

6 gluons
4 gluons + 2 gravitons
2 gluons + 4 gravitons
6 gravitons

Cancellation of anomalies, if there are only spin $\frac{1}{2}$ charged fermions, requires

$$(TrT^6)_L = (TrT^6)_R$$

$$(TrT^4)_L = (TrT^4)_R \qquad (21)$$

$$(TrT^2)_L = (TrT^2)R$$

and an extra condition from the six graviton graph.

These constraints rule out all open string theories and some closed string theories. Only the parity conserving closed string theory with two gravitinos is manifestly anomaly free. These constraints must hold in the string theory, because, as 't Hooft taught us, anomalies can be extracted from long wave length physics.

Let me prove that the six graviton graph has an anomaly. First, we note the anomaly in two dimensions for Weyl fermions in external gravitational fields. In light cone variables

$$X^{\pm} = \frac{1}{\sqrt{2}} (X^0 \pm X^1)$$

we have an energy-momentum tensor for Weyl fermions

$$T_{--} = T_{+-} = 0$$
$$T_{++} = \bar{\psi} \frac{\partial}{\partial X^+} \psi \tag{22}$$

The naive conservation law is $\partial_- T_{++}=0$, but it is nonsense. Applied to $\langle T_{++}(p)T_{++}(-p)\rangle$, it says

$$\partial_- T_{++} = 0,$$

$$P_- \langle T_{++}(p)T_{++}(-p)\rangle = 0,$$

so

$$\langle T_{++}(p)T_{++}(-p)\rangle = 0, \tag{23}$$

which is impossible. In fact

$$\langle T_{++}(p)T_{++}(-p)\rangle = \frac{1}{2\pi} \frac{P_+^{\,3}}{P_-} \tag{24}$$

Now let us look at 4k + 2 dimensional Kaluza-Klein theory on $M^2 \times B$, where B is a compact space of dimensions 4k with nonzero index of Dirac operator. As

$$\not{D} = \not{D}^{(2)} + \not{D}^{(B)}, \tag{25}$$

at long wavelengths the zero modes of $\not{D}^{(B)}$ are observed as massless fermions in two dimensions. Furthermore

$$(4k + 2 \text{ dim. chirality}) = \gamma_1\gamma_2 \cdots \gamma_{4k+2}$$

$$= \gamma_1\gamma_2\gamma_3\gamma_4 \cdots \gamma_{4k+2}$$

$$= (2 \text{ dim. chirality}) \cdot (4k \text{ dim. chirality}) \qquad (26)$$

If $\not{D}^B$ has nonzero index, the massless particles have preferentially one chirality in 4k dimensions; hence (if we start with Weyl fermions in the original 4k + 2 dimensional theory) they have preferentially one chirality in 2 dimensions. Thus a 4k + 2 dimensional theory with Weyl spinors can reduce to an anomalous theory in D=2.

But there is a simple fact: an anomaly free theory always remains anomaly free after compactification. The anomaly is a non-gauge invariance or non-general coordinate invariance of the fermion one loop effective action.

$$\frac{\delta}{\delta A^a_\mu} \ln \det i \not{D} = \langle\psi\gamma^\mu\lambda^a\psi\rangle_A$$

$$D_\mu \frac{\delta}{\delta A_\mu} \ln \det i\not{D} = D_\mu\langle\psi\gamma^\mu\lambda^a\psi\rangle_A$$

$$= \frac{d^{abc}}{16\pi^2} F^b\tilde{F}^c \qquad (27)$$

In our example, a Weyl theory in 4k + 2 dimensions is compactified to a Weyl theory in 2 dimensions which is known to be anomalous. Hence the original Weyl theory in D=4k + 2 was anomalous. Q.E.D. (However, an explicit calculation of the hexagon anomaly has not been done.)

Finally I want to mention the only application I know of gravitational anomalies to four dimensional physics.

Experimentally, for the hyper-charge Y of SU(2) x U(1),

$$\sum_{\substack{\text{left-handed}\\ \text{fermions}}} Y_i = 0, \qquad (28)$$

which is a necessary condition for grand unification and often regarded as evidence for grand unification. But it has another interpretation. The triangle with two gravitons and an axial current is anomalous [8]. In the Weinberg-Salam model coupled to gravity, the U(1)-graviton - graviton anomaly spoils gauge invariance or general coordinate invariance unless $\sum_i y_i=0$. No short distance effect can remove the anomaly (again, this is crucial in 't Hooft's preon argument) so the non-renormalizability of gravity is irrelevant.

By the same token, in preon models coupled to gravity there is an extra condition

$$\sum_{\substack{\text{left-handed} \\ \text{preons}}} q_i = \sum_{\substack{\text{left handed} \\ \text{physical} \\ m=0 \text{ fermions}}} q_i \tag{29}$$

for an unbroken U(1) charge Q, that is analogous to 't Hooft's constraint

$$\sum_{\text{preons}} q_i^3 = \sum_{\text{composites}} q_i^3 \tag{30}$$

Conclusion

If it were just a question of asthetic judgment, I doubt seriously if I would have chosen to talk about strings. However, it seems to me that the fact that there is a string theory whose low energy limit gives some of the most beautiful of the known field theories, namely supergravity and supersymmetric Yang-Mills theory, is not something that should be dismissed lightly. The fact that one of the string theories seems to be a physically consistent and finite theory of quantum gravity certainly makes it worthy of serious attention.

I would like to thank Y. Hosotani for preparing the written version of these notes.

References

[1] For review, see, J. Schwarz, Phys. Rep. 89, 223 (1982).

[2] P. Ramond, Phys. Rev. D3, 2415 (1971)
A. Neveu and J. Schwarz Nucl. Phys. B31, 86 (1971); Phys. Rev. D4, 1109 (1971).

[3] J. Wess and B. Zumino, Nucl. Phys. B70, 39 (1974).

[4] F. Gliozzi, J. Scherk and D. Olive, Phys. Lett. 65B, 282 (1976); Nucl. Phys. B122, 253 (1977).

[5] M. Green and J. Schwarz, CALT-68-956 (Oct. 1982).

[6] R. Shankar, Phys. Lett. 92B, 333 (1980).

[7] P. Frampton and T. Kephart, Phys. Rev. Lett. 50, 1343 (1983); ibid. 50, 1347 (1983);
P.K. Townsend and G. Sierra, LPTENS 83/12 (Feb. 1983).

[8] R. Delbourgo and A. Salam, Phys. Lett. 40B, 381 (1972).

FOURTH WORKSHOP ON GRAND UNIFICATION

PROGRAM

Thursday (April 21, 1983)

Morning	B. Kayser, Chairman
T. Ehrlich	Welcome
H. Georgi	Opening Remarks
W. Marciano	Proton Decay Theory
B. Sreekantan	The K.G.F. Nucleon Decay Experiment
T. Gaisser	The Angular Distribution and Flux of Atmospheric Neutrinos
A. Dar	Atmospheric Neutrinos and Astrophysical Neutrinos in Proton Decay Experiments
E. Peterson	New Results from the Soudan 1 Detector
Afternoon	R. Arnowitt, Chairman
J. Polchinski	Low Energy Supergravity: The Minimal Model
F. Boehm	Neutrino Mass and Neutrino Oscillations
G. Kane	Experimental Searches for Supersymmetric Particles
F. Avignone	Double Beta Decay: Recent Developments and Projections
M. Duff	Superunification from Eleven Dimensions

Friday (April 22, 1983)

Morning	L. Sulak, Chairman
B. Cortez	Results from the IMB Detector
E. Fiorini	Results from Mont Blanc
S. Weinberg	Calculation of Fine Structure Constants
A. Grant	Review of Future Nucleon Decay Experiments

Afternoon	M. Goldhaber and M. Dresden, Chairmen
M. Turner	Inflation Circa 1983
J. Primack	Dark Matter, Galaxies, Superclusters and Voids
G. Fidecaro	$n\bar{n}$ Oscillation Experiments
A. Szalay	The Late Evolution of Density Perturbations
P. Sikivie	Invisible Axions

Saturday (April 23, 1983)

Morning	R. Slansky, Chairman
C. Callan	Monopole Catalysis of Baryon Decay
A. Goldhaber	Monopoles, Gauge Fields, and Anomalies
P. Bosetti	Searches for Magnetic Monopoles
C. Tesche	The IBM Monopole Experiments
E. Witten	D=10 Superstring Theory

FOURTH WORKSHOP ON GRAND UNIFICATION

ORGANIZING COMMITTEE

Gino Segre (Chairman)

Paul Langacker (Chairman)

Ettore Fiorini

Willy Fischler

Paul Frampton

Mary K. Gaillard

Sheldon Glashow

Kenneth Lande

Alfred Mann

Pierre Ramond

Paul Steinhardt

Jack Vander Velde

Arthur Weldon

LIST OF PARTICIPANTS

1.	ALBRECHT, Andreas	University of Pennsylvania
2.	ALBRIGHT, Carl H.	Northern Illinois University
3.	AMADO, Ralph	University of Pennsylvania
4.	ARNOWITT, R.	Northeastern University
5.	ATKINSON, Gare	University of Maryland
6.	AULAKH, S. C.	City College of CUNY
7.	AVIGNONE, Frank T.	University of South Carolina
8.	AXELROD, Alan	Argonne National Laboratory
9.	BAGGER, Jonathan A.	SLAC
10.	BARAD, Karen	SUNY, Stony Brook
11.	BARBER, James S.	SUNY, Stony Brook
12.	BARNHILL, Maurice V. III	University of Delaware
13.	BARR, Stephen	University of Washington
14.	BARS, Itzhak	Yale University
15.	BARTELT, John E.	University of Minnesota
16.	BEIER, Eugene W.	University of Pennsylvania
17.	BLUDMAN, Sidney	University of Pennsylvania
18.	BOEHM, Felix	California Institute of Technology
19.	BONATSOS, Dennis	University of Pennsylvania
20.	BOSETTI, Peter C.	III Physikal. Institut A, Aachen
21.	BRACKEN, Paul	University of Michigan
22.	BRANCO G. C.	Inst. Nac. de Univ. Cient., Lisboa
23.	BRATTON Clyde B.	Cleveland State University
24.	BRODY, Howard	University of Pennsylvania
25.	BYERS, Nina	UCLA
26.	CAHILL, Kevin	University of New Mexico
27.	CALLAN, Curt	Princeton University
28.	CALLEN, Bruce	University of Pennsylvania
29.	CARONMBOLIS, Dionysios E.	Columbia University
30.	CHANG, Lay Nam	Virginia Polytechnic Inst. & S. U.
31.	CHANG, Ngee Pong	City College of CUNY
32.	CHAO, Yu-Chiu	University of Michigan
33.	CHATTOPADHYAY, Utpal	SUNY, Stony Brook
34.	CHAUDHARI, P.	IBM
35.	CHEN, Chia Chu	Northeastern University
36.	CHERRY, Michael	University of Pennsylvania
37.	CHODOS, Alan	Yale University
38.	CLARK, Thomas E.	Purdue University
39.	CLEVELAND, Bruce	Brookhaven National Laboratory
40.	CONNOLLY, Philip L.	Brookhaven National Laboratory
41.	CONNORS, David	University of Pennsylvania
42.	COOPER, John	University of Pennsylvania
43.	CORBATO, Steve	University of Pennsylvania
44.	CORTEZ, Bruce	University of Michigan
45.	CUDELL, Jean	University of Wisconsin
46.	CVETIC, Mirjam	University of Maryland
47.	DAR, Arnon	Technion Inst./Univ. of Pennsylvania
48.	DAWSON, S.	Lawrence Berkeley Laboratory
49.	DE POMMIER, Pierre	University of Montreal
50.	DESHPANDE, N. G.	University of Oregon
51.	DINE, Michael	Institute for Advanced Study
52.	DRESDEN, M.	SUNY, Stony Brook
53.	DUFF, Michael	University of Texas, Austin
54.	DURKIN, L. Stanley	University of Pennsylvania
55.	EASTAUGH, Alex	Yale University
56.	EDER, Monica	ETH, Zurich
57.	ELIASSON, Ernst	Northeastern University
58.	FARRAR, Glennys	Rutgers University

59.	FERNANDEZ-LABASTIDA, Jose	SUNY, Stony Brook
60.	FIDECARO, Giuseppe	CERN
61.	FIORINI, Ettore	Universita di Milano
62.	FISCHBACH, Ephraim	Purdue University
63.	FISCHLER, Willy	University of Pennsylvania
64.	FLEISCHMAN, Jack	University of Pennsylvania
65.	FOGELMAN, Guy	TRIUMF
66.	FRAMPTON, Paul	University of North Carolina
67.	FRANKEL, Sherman	University of Pennsylvania
68.	FRY, James N.	University of Chicago
69.	GAISSER, Thomas K.	Bartol Research Foundation
70.	GALLATIN, Gregg	Fairfield University
71.	GEORGI, Howard	Harvard University
72.	GHOSH, Ranjan K.	SUNY, Stony Brook
73.	GIPSON, John M.	Virginia Polytechnic Inst. & S. U.
74.	GOLDBERG, Hyman	Northeastern University
75.	GOLDHABER, A. S.	SUNY, Stony Brook
76.	GOLDHABER, Maurice	Brookhaven National Laboratory
77.	GOLOWICH, Gene	University of Massachusetts
78.	GONZALES, Daniel	Massachusetts Institute of Technology
79.	GRANT, Alan L.	CERN
80.	GREENWOOD, Zeno	UC, Irvine/Savannah River Plant
81.	GROSSMAN, Bernard	Rockefeller University
82.	HA, Yuan K.	New York University
83.	HABER, Howard	University of California, Santa Cruz
84.	HALPRIN, Arthur	University of Delaware
85.	HARVEY, Jeff	Princeton University
86.	HAXTON, Wick	Los Alamos National Laboratory
87.	HEAGY, Stuart M.	University of Pennsylvania
88.	HILL, Alfred	University of Michigan
89.	HOSOTANI, Yutaka	University of Pennsylvania
90.	HUNG, Pham Quang	University of Virginia
91.	JENSEN, Lars	University of Pennsylvania
92.	JONES, Keith	Nordita
93.	KALARA, Sunny	University of Rochester
94.	KALYNIAK, Pat	TRIUMF
95.	KANE, Gordon	University of Michigan
96.	KANG, K.	Brown University
97.	KAPLAN, Norman C.	Case Western Reserve University
98.	KARL, G.	University of Guelph
99.	KARLHEDE, Anders	SUNY, Stony Brook
100.	KAUFMAN, William	University of Michigan
101.	KAYMAKCALAN, Omer	Syracuse University
102.	KAYSER, Boris	National Science Foundation
103.	KIEDA, David	University of Pennsylvania
104.	KIM, C. W.	Johns Hopkins University
105.	KIM, Jai Sam	California Institute of Technology
106.	KROPP, William R.	University of California, Irvine
107.	LANDE, Kenneth	University of Pennsylvania
108.	LANGACKER, Paul	University of Pennsylvania
109.	LASSILA, K. E.	Iowa State University
110.	LEARNED, John	University of Hawaii
111.	LEE, C. K.	University of Pennsylvania
112.	LEE, H. C.	Chalk River Nuclear Laboratory
113.	LEE, Haeshim	University of Pennsylvania
114.	LESSURE, Harold	University of Michigan
115.	LI, Da-Xi	City College of CUNY
116.	LINDBLOOM, Peter	University of Pennsylvania
117.	LING, T. Y.	Ohio State University
118.	LIPTON, Gary	University of North Carolina

119.	LIZZI, Fedeie	Syracuse University
120.	LOVE, Sherwin	Purdue University
121.	LU, Hui	University of Michigan
122.	LUBKIN, Gloria B.	Physics Today
123.	MACHACEK, Marie E.	Northeastern/Harvard University
124.	MAHANTHAPPA, K. T.	University of Colorado
125.	MANDELBAUM, G.	Harvard University
126.	MANN, Alfred K.	University of Pennsylvania
127.	MANN, W. Anthony	Tufts University
128.	MARCIANO, William J.	Brookhaven National Laboratory
129.	MARSHAK, Robert E.	Virginia Polytechnic Inst. & S. U.
130.	MATSUKI, Takayuki	Ohio State University
131.	MC CABE, John	Ohio State University
132.	MILLER, Marshall	University of Pennsylvania
133.	MIMOUNI, Jamal	University of Pennsylvania
134.	MOHANTY, Ajaya K.	University of Maryland
135.	MOHAPATRA, R. N.	University of Maryland
136.	NAIR, V. Parameswaran	Syracuse University
137.	NEMESCHANSKY, Dennis	Princeton University
138.	NIEVES, Jose F.	University of Puerto Rico
139.	O'DONNELL, Patrick	University of Toronto
140.	OAKES, Robert J.	Northwestern University
141.	OUVRY, Stephane	City College of CUNY
142.	OVRUT, Burt	Rockefeller University
143.	PAL, Palash B.	Carnegie-Mellon University
144.	PARANJAPE, Manu	Massachusetts Institute of Technology
145.	PARSA, Zohreh	New Jersey Institute of Technology
146.	PASUPATHY, J.	Indian Institute of Science, Bangalore
147.	PATI, J.	University of Maryland
148.	PEREZ-MERCADER, Juan A.	Louisiana State University
149.	PERNICI, Mario	SUNY, Stony Brook
150.	PETERSON, Earl A.	University of Minnesota
151.	PETSCHEK, Albert G.	Los Alamos National Laboratory
152.	PHELPS, Rick	University of Michigan
153.	POLCHINSKI, Joseph	Harvard University
154.	PRIMACK, Joel	University of California, Santa Cruz
155.	PRIMAKOFF, Henry	University of Pennsylvania
156.	RAMOND, Pierre	University of Florida
157.	RAO, Sumathi	SUNY, Stony Brook
158.	REISS, David	University of Washington
159.	RIM, Chaiho	SUNY, Stony Brook
160.	ROBINETT, R. W.	University of Wisconsin
161.	ROBINSON, Barry	University of Pennsylvania
162.	ROCEK, M.	SUNY, Stony Brook
163.	ROHM, Ryan M.	Princeton University
164.	ROSS, Douglas	University of Southampton
165.	RUDAZ, Serge	University of Minnesota
166.	SAHDEV, Deshdeep	University of Pennsylvania
167.	SAKAI, Norisuke	Tokyo Institute of Technology
168.	SCANIO, Joe	University of Cincinnati
169.	SCHELLEKENS, A. H.	Fermilab
170.	SEGRE, Gino	University of Pennsylvania
171.	SELOVE, Walter	University of Pennsylvania
172.	SEN, Ashore	Fermilab
173.	SENJANOVIC, Goran	Brookhaven National Laboratory
174.	SHAYEB, A. G.	Princeton University
175.	SHROCK, Robert	SUNY, Stony Brook
176.	SIKIVIE, Pierre	University of Florida
177.	SINGK, Rajiv Ranjan	SUNY, Stony Brook
178.	SIRLIN, A.	Rockefeller University

179.	SLANSKY, Richard	Los Alamos National Laboratory
180.	SMITH, J.	SUNY, Stony Brook
181.	SOBEL, Henry W.	University of California, Irvine
182.	SONI, Sanjeev K.	University of Pennsylvania
183.	SPARROW, David	University of Pennsylvania
184.	SREEKANTAN, B. V.	Tata Institute of Fundamental Research
185.	STANEV, Todor S.	Bartol Research Foundation
186.	STECKER, Floyd W.	NASA, Goddard Space Flight Center
187.	STEINBERG, Richard	University of Pennsylvania
188.	STEINHARDT, Paul	University of Pennsylvania
189.	STERMAN, George	SUNY, Stony Brook
190.	SULAK, Lawrence R.	University of Michigan
191.	SURANYI, Peter	University of Cincinnati
192.	SZALAY, Alexander	Eotvos University
193.	TANAKA, Katsumi	Ohio State University
194.	TAYLOR, Cyrus C.	Massachusetts Institute of Technology
195.	TESCHE, Claudia	IBM, T. J. Watson Research Center
196.	THEWS, Robert	DOE
197.	TOSA, Yasunari	Virginia Polytechnic Inst. & S. U.
198.	TURNER, Michael	University of Chicago
199.	UNGER, David	Carnegie-Mellon University
200.	VAFA, Cumrun	Princeton University
201.	VAN DE VEN, Anton E.	SUNY, Stony Brook
202.	VAN DER VELDE, Jack	University of Michigan
203.	VAUGHN, Michael T.	Northeastern University
204.	VELASCO, Eduardo Sanchez	SUNY, Stony Brook
205.	WADA, Walter	Ohio State University
206.	WALI, K. C.	Syracuse University
207.	WEINBERG, Erick	Columbia University
208.	WEINBERG, Steven	University of Texas, Austin
209.	WEISBERGER, William I.	SUNY, Stony Brook
210.	WELDON, Arthur	University of Pennsylvania
211.	WIJEWARDHANA, L. C. R.	Massachusetts Institute of Technology
212.	WILLIAMS, H. H.	University of Pennsylvania
213.	WITTEN, Edward	Princeton University
214.	WU, Jin-Chu	University of Pittsburgh
215.	WU, Xizeng	City College of CUNY
216.	YAMAWAKI, Mieko	University of Rochester
217.	YOUNG, Roberta E.	Princeton University
218.	YU, Rong-Qing	University of Pennsylvania
219.	YUAN, Tze-Chiang	Northeastern University
220.	ZAJC, William	University of Pennsylvania
221.	ZEPEDA, Arnulfo	Centro de Investigacion IPN, Mexico

Progress in Mathematics

Edited by J. Coates and S. Helgason

Progress in Physics

Edited by A. Jaffe and D. Ruelle

- A collection of research-oriented monographs, reports, notes arising from lectures or seminars
- Quickly published concurrent with research
- Easily accessible through international distribution facilities
- Reasonably priced
- Reporting research developments combining original results with an expository treatment of the particular subject area
- A contribution to the international scientific community: for colleagues and for graduate students who are seeking current information and directions in their graduate and post-graduate work.

Manuscripts

Manuscripts should be no less than 100 and preferably no more than 500 pages in length.

They are reproduced by a photographic process and therefore must be typed with extreme care. Symbols not on the typewriter should be inserted by hand in indelible black ink. Corrections to the typescript should be made by pasting in the new text or painting out errors with white correction fluid.

The typescript is reduced slightly (75%) in size during reproduction; best results will not be obtained unless the text on any one page is kept within the overall limit of 6x9½ in (16x24 cm). On request, the publisher will supply special paper with the typing area outlined.

Manuscripts should be sent to the editors or directly to:
Birkhäuser Boston, Inc., P.O. Box 2007, Cambridge, Massachusetts 02139

PROGRESS IN MATHEMATICS
Already published

PM 1 Quadratic Forms in Infinite-Dimensional Vector Spaces
Herbert Gross
ISBN 3-7643-1111-8, 432 pages, paperback

PM 2 Singularités des systèmes différentiels de Gauss-Manin
Frédéric Pham
ISBN 3-7643-3002-3, 346 pages, paperback

PM 3 Vector Bundles on Complex Projective Spaces
C. Okonek, M. Schneider, H. Spindler
ISBN 3-7643-3000-7, 396 pages, paperback

PM 4 Complex Approximation, Proceedings, Quebec, Canada, July 3-8, 1978
Edited by Bernard Aupetit
ISBN 3-7643-3004-X, 128 pages, paperback

PM 5 The Radon Transform
Sigurdur Helgason
ISBN 3-7643-3006-6, 207 pages, hardcover

PM 6 The Weil Representation, Maslov Index and Theta Series
Gérard Lion, Michèle Vergne
ISBN 3-7643-3007-4, 348 pages, paperback

PM 7 Vector Bundles and Differential Equations
Proceedings, Nice, France, June 12-17, 1979
Edited by André Hirschowitz
ISBN 3-7643-3022-8, 256 pages, paperback

PM 8 Dynamical Systems, C.I.M.E. Lectures, Bressanone, Italy, June 1978
John Guckenheimer, Jürgen Moser, Sheldon E. Newhouse
ISBN 3-7643-3024-4, 305 pages, hardcover

PM 9 Linear Algebraic Groups
T. A. Springer
ISBN 3-7643-3029-5, 314 pages, hardcover

PM 10 Ergodic Theory and Dynamical Systems I
A. Katok
ISBN 3-7643-3036-8, 346 pages, hardcover

PM 11 18th Scandinavian Congress of Mathematicians, Aarhus, Denmark, 1980
Edited by Erik Balslev
ISBN 3-7643-3040-6, 526 pages, hardcover

PM 12 Séminaire de Théorie des Nombres, Paris 1979-80
Edited by Marie-José Bertin
ISBN 3-7643-3035-X, 404 pages, hardcover

PM 13 Topics in Harmonic Analysis on Homogeneous Spaces
Sigurdur Helgason
ISBN 3-7643-3051-1, 152 pages, hardcover

PM 14 Manifolds and Lie Groups, Papers in Honor of Yozô Matsushima
Edited by J. Hano, A. Marimoto, S. Murakami, K. Okamoto, and H. Ozeki
ISBN 3-7643-3053-8, 476 pages, hardcover

PM 15 Representations of Real Reductive Lie Groups
David A. Vogan, Jr.
ISBN 3-7643-3037-6, 776 pages, hardcover

PM 16 Rational Homotopy Theory and Differential Forms
Phillip A. Griffiths, John W. Morgan
ISBN 3-7643-3041-4, 258 pages, hardcover

PM 17 Triangular Products of Group Representations and their Applications
S. M. Vovsi
ISBN 3-7643-3062-7, 142 pages, hardcover

PM 18 Géométrie Analytique Rigide et Applications
Jean Fresnel, Marius van der Put
ISBN 3-7643-3069-4, 232 pages, hardcover

PM 19 Periods of Hilbert Modular Surfaces
Takayuki Oda
ISBN 3-7643-3084-8, 144 pages, hardcover

PM 20 Arithmetic on Modular Curves
Glenn Stevens
ISBN 3-7643-3088-0, 236 pagers, hardcover

PM 21 Ergodic Theory and Dynamical Systems II
A. Katok, editor
ISBN 3-7643-3096-1, 226 pages, hardcover

PM 22 Séminaire de Théorie des Nombres, Paris 1980-81
Marie-José Bertin, editor
ISBN 3-7643-3066-X, 374 pages, hardcover

PM 23 Adeles and Algebraic Groups
A. Weil
ISBN 3-7643-3092-9, 138 pages, hardcover

PM 24 Enumerative Geometry and Classical Algebraic Geometry
Patrick Le Barz, Yves Hervier, editors
ISBN 3-7643-3106-2, 260 pages, hardcover

PM 25 Exterior Differential Systems and the Calculus of Variations
Phillip A. Griffiths
ISBN 3-7643-3103-8, 349 pages, hardcover

PM 26 Number Theory Related to Fermat's Last Theorem
Neal Koblitz, editor
ISBN 3-7643-3104-6, 376 pages, hardcover

PM 27 Differential Geometric Control Theory
Roger W. Brockett, Richard S. Millman, Hector J. Sussmann, editors
ISBN 3-7643-3091-0, 349 pages, hardcover

PM 28 Tata Lectures on Theta I
David Mumford
ISBN 3-7643-3109-7, 254 pages, hardcover

PM 29 Birational Geometry of Degenerations
Robert Friedman and David R. Morrison, editors
ISBN 3-7643-3111-9, 410 pages, hardcover

PM 30 CR Submanifolds of Kaehlerian and Sasakian Manifolds
Kentaro Yano, Masahiro Kon
ISBN 3-7643-3119-4, 223 pages, hardcover

PM 31 Approximations Diophantiennes et Nombres Transcendants
D. Bertrand and M. Waldschmidt, editors
ISBN 3-7643-3120-8, 349 pages, hardcover

PM 32 Differential Geometry
Robert Brooks, Alfred Gray, Bruce L. Reinhart, editors
ISBN 3-7643-3134-8, 267 pages, hardcover

PM 33 Uniqueness and Non-Uniqueness in the Cauchy Problem
Claude Zuily
ISBN 3-7643-3121-6, 185 pages, hardcover

PM 34 Systems of Microdifferential Equations
Masaki Kashiwara
ISBN 0-8176-3138-0
ISBN 3-7643-3138-0, 182 pages, hardcover

PM 35 Arithmetic and Geometry Papers Dedicated to I. R. Shafarevich on the Occasion of His Sixtieth Birthday Volume I Arithmetic
Michael Artin, John Tate, editors
ISBN 3-7643-3132-1, 373 pages, hardcover

PM 36 Arithmetic and Geometry Papers Dedicated to I. R. Shafarevich on the Occasion of His Sixtieth Birthday Volume II Geometry
Michael Artin, John Tate, editors
ISBN 3-7643-3133-X, 495 pages, hardcover

PM 37 Mathématique et Physique
Louis Boutet de Monvel, Adrien Douady, Jean-Louis Verdier, editors
ISBN 0-8176-3154-2
ISBN 3-7643-3154-2, 454 pages, hardcover

PM 38 Séminaire de Théorie des Nombres, Paris 1981-82
Marie-José Bertin, editor
ISBN 0-8176-3155-0
ISBN 3-7643-3155-0, 359 pages, hardcover

PM 39 Classical Algebraic and Analytic Manifolds
Kenji Ueno, editor
ISBN 0-8176-3137-2
ISBN 3-7643-3137-2, 644 pages, hardcover

PM 40 Representation Theory of Reductive Groups
P. C. Trombi, editor
ISBN 0-8176-3135-6
ISBN 3-7643-3135-6, 308 pages, hardcover

PM 41 Combinatorics and Commutative Algebra
Richard P. Stanley
ISBN 0-8176-3112-7
ISBN 3-7643-3112-7, 102 pages, hardcover

PM 42 Théorèmes de Bertini et Applications
Jean-Pierre Jouanolou
ISBN 0-8176-3164-X
ISBN 3-7643-3164-X, 140 pages, hardcover

PROGRESS IN PHYSICS

Already published

PPh 1 Iterated Maps on the Interval as Dynamical Systems
Pierre Collet and Jean-Pierre Eckmann
ISBN 3-7643-3026-0, 256 pages, hardcover

PPh 2 Vortices and Monopoles, Structure of Static Gauge Theories
Arthur Jaffe and Clifford Taubes
ISBN 3-7643-3025-2, 294 pages, hardcover

PPh 3 Mathematics and Physics
Yu. I. Manin
ISBN 3-7643-3027-9, 112 pages, hardcover

PPh 4 Lectures on Lepton Nucleon Scattering and Quantum Chromodynamics
W.B. Atwood, J.D. Bjorken, S.J. Brodsky, and R. Stroynowski
ISBN 3-7643-3079-1, 574 pages, hardcover

PPh 5 Gauge Theories: Fundamental Interactions and Rigorous Results
P. Dita, V. Georgescu, R. Purice, editors
ISBN 3-7643-3095-3, 406 pages, hardcover

PPh 6 Third Workshop on Grand Unification, 1982
P.H. Frampton, S.L. Glashow, H. van Dam, editors
ISBN 3-7643-3105-4, 388 pages, hardcover

PPh 7 Scaling and Self-Similarity in Physics
(Renormalization in Statistical Mechanics and Dynamics)
J. Fröhlich, editor
ISBN 3-7643-3168-2
ISBN 0-8176-3168-2, 440 pages, hardcover

PPh 8 Workshop on Non-Perturbative Quantum Chromodynamics
K.A. Milton, M.A. Samuel, editors
ISBN 3-7643-3127-5
ISBN 0-8176-3127-5, 284 pages, hardcover

GPSR Compliance

The European Union's (EU) General Product Safety Regulation (GPSR) is a set of rules that requires consumer products to be safe and our obligations to ensure this.

If you have any concerns about our products, you can contact us on ProductSafety@springernature.com

In case Publisher is established outside the EU, the EU authorized representative is:

Springer Nature Customer Service Center GmbH
Europaplatz 3
69115 Heidelberg, Germany

Batch number: 10577896

Printed by Printforce, the Netherlands